ADVANCES IN SUSTAINABLE MICROGRID TECHNOLOGIES

FREQUENCY CONTROL IN MICROGRIDS

Edited by

YOGENDRA ARYA

Department of Electrical Engineering, J.C. Bose University of Science and Technology, YMCA, Faridabad, India

SANDEEP DHUNDHARA

College of Engineering, Faculty of Computing, Engineering and the Built Environment, Birmingham City University, Birmingham, United Kingdom; Department of Basic Engineering, CCS Haryana Agricultural University, Hisar, Haryana, India

HANY M. HASANIEN

Department of Electrical Engineering, College of Engineering, University of Sharjah, Sharjah, United Arab Emirates

ELSEVIER

Elsevier
Radarweg 29, PO Box 211, 1000 AE Amsterdam, Netherlands
125 London Wall, London EC2Y 5AS, United Kingdom
50 Hampshire Street, 5th Floor, Cambridge, MA 02139, United States

ISBN: 978-0-443-36476-1

For information on all Elsevier publications visit our website at https://www.elsevier.com/books-and-journals

Publisher: Priscilla Braglia
Acquisitions Editor: Dan Egan
Editorial Project Manager: Juwairiya Ali
Production Project Manager: Sujatha Thirugnana Sambandam
Cover Designer: Raman Kumar

Typeset by TNQ Tech

Contents

7. Design and implementation of optimized cascaded controllers for frequency stabilization in modern microgrid systems

K. Peddakapu, Dalila Mat Said, P. Srinivasarao and D. J Krishna Kishore

8. Frequency control in multi-microgrid systems

Amil Daraz, Muhammad Majid Gulzar and Suheel Abdullah Malik

9. Hardware and real-time validation of microgrids

Preeti Gupta, Vivek Pahwa and Yajvender Pal Verma

10. Microgrid from perspective of issues of cybersecurity, ethics and regulation

Rahul Saini and Yajvender Pal Verma

Contributors

Sumant Anand Department of Electronics and Telecommunication Engineering, G H Raisoni College of Engineering and Management, Pune, Maharashtra, India

Naladi Ram Babu Department of Electrical and Electronics Engineering, Aditya University, Surampalem, Andhra Pradesh, India; Department of Artificial Intelligence & Machine Learning, Aditya University, Surampalem, Andhra Pradesh, India

Sanjeev Kumar Bhagat Electrical Engineering Department, Sandip University, Madhubani, Bihar, India

Amil Daraz Center for Sustainable Energy Systems, King Fahd University of Petroleum & Minerals, Dhahran, Saudi Arabia

Ark Dev Department of Electronics and Communication Engineering, SR University, Warangal, Telangana, India

Samuel Raafat Fahim Independent Postdoc Researcher, Ain Shams University, Cairo, Egypt

Muhammad Majid Gulzar Center for Sustainable Energy Systems, King Fahd University of Petroleum & Minerals, Dhahran, Saudi Arabia; Department of Control & Instrumentation Engineering, King Fahd University of Petroleum & Minerals, Dhahran, Saudi Arabia; Center for Smart Mobility and Logistics, King Fahd University of Petroleum & Minerals, Dhahran, Saudi Arabia

Preeti Gupta Department of Electrical Engineering, PEC Chandigarh, Chandigarh, India; Department of EEE, UIET, Panjab University, Chandigarh, India

Sachchidanand Jha Electrical Engineering Department, Sandip University, Madhubani, Bihar, India

Muhammad Khalid Center for Sustainable Energy Systems, King Fahd University of Petroleum & Minerals, Dhahran, Saudi Arabia; Electrical Engineering Department, King Fahd University of Petroleum & Minerals, Dhahran, Saudi Arabia

Bhuvnesh Khokhar Department of Electrical Engineering, Galgotias College of Engineering and Technology, Greater Noida, Uttar Pradesh, India

D. J Krishna Kishore Department of Electrical & Electronics Engineering, Sri Vasavi Engineering College, Pedatadepalli, Andhra Pradesh, India

Ashiwani Kumar Department of Electrical Engineering, National Institute of Technology Patna, Patna, Bihar, India

Pradeep Kumar Department of Electrical Engineering, Deenbandhu Chhotu Ram University of Science and Technology, Murthal, Haryana, India

Vineet Kumar Department of Electrical Engineering, Netaji Subhas University of Technology (NSUT), Delhi, New Delhi, India

Suheel Abdullah Malik Department of Electrical and Computer Engineering, Faculty of Engineering and Technology, International Islamic University, Islamabad, Pakistan

Vivek Pahwa Department of Electrical & Electronics Engineering, UIET, Panjab University, Chandigarh, India

Rajneesh Pawar Department of Electrical Engineering, Deenbandhu Chhotu Ram University of Science and Technology, Murthal, Haryana, India

K. Peddakapu Centre of Electrical Energy Systems (CEES), Institute of Future Energy (IFE), Universiti Teknologi Malaysia (UTM), Johor Bahru, Johor, Malaysia; Faculty of Electrical Engineering, Universiti Teknologi Malaysia (UTM), Johor Bahru, Johor, Malaysia

Dalila Mat Said Centre of Electrical Energy Systems (CEES), Institute of Future Energy (IFE), Universiti Teknologi Malaysia (UTM), Johor Bahru, Johor, Malaysia; Faculty of Electrical Engineering, Universiti Teknologi Malaysia (UTM), Johor Bahru, Johor, Malaysia; Department of Engineering, Faculty of Advanced Technology and Multidiscipline, Universitas Airlangga, Surabaya, East Java, Indonesia

Rahul Saini Department of Electrical & Electronics Engineering, UIET, Panjab University, Chandigarh, India

Ravi Shankar Department of Electrical Engineering, National Institute of Technology Patna, Patna, Bihar, India

Pranav Prakash Singh Department of Electrical Engineering, National Institute of Technology Patna, Patna, Bihar, India

K.P. Singh Parmar National Power Training Institute (under Ministry of Power, Government of India), Faridabad, Haryana, India

P. Srinivasarao Faculty of Electrical & Electronics Engineering Tech., Universiti Malaysia Pahang Al-Sultan Abdullah, Pekan, Malaysia

Abhishek Thakur Department of Electrical Engineering, Sandip University, Madhubani, Bihar, India

Muhammad S. Tolba Department of Control & Instrumentation Engineering, King Fahd University of Petroleum & Minerals, Dhahran, Saudi Arabia

Vikash Kumar Verma Department of Electrical Engineering, Parul Institute of Engineering and Technology, Parul University, Vadodara, Gujarat, India

Yajvender Pal Verma Department of Electrical & Electronics Engineering, UIET, Panjab University, Chandigarh, India

Prabhat Kumar Vidyarthi Department of Electrical Engineering, National Institute of Technology Patna, Patna, Bihar, India

Ravindra Kumar Yadav Department of Electrical and Electronics Engineering, Galgotias College of Engineering and Technology Technology, Greater Noida, Uttar Pradesh, India

Preface

The rapid transition toward sustainable and decentralized energy systems has significantly accelerated the development and deployment of microgrids worldwide. Microgrids, integrating renewable energy resources, distributed generation, energy storage systems, and advanced control technologies have emerged as a key solution for enhancing energy resilience, reliability, and sustainability. However, the increasing penetration of renewable energy sources, characterized by intermittency and reduced system inertia, has introduced substantial challenges in maintaining frequency stability. Consequently, load frequency control (LFC) has become one of the most critical technical issues in modern microgrid operation and management.

This edited volume, *Frequency Control in Microgrids*, aims to provide a comprehensive and systematic understanding of frequency regulation challenges, modeling approaches, and advanced control strategies applicable to modern microgrid environments. The motivation behind this book stems from the growing need for consolidated knowledge that bridges theoretical foundations, analytical techniques, and emerging research directions in microgrid frequency control.

This volume begins with an overview of microgrid concepts and fundamental frequency control principles, followed by discussions on key challenges, such as reduced inertia, renewable intermittency, and dynamic load variations, along with their mitigation strategies. It presents the detailed modeling of microgrid components and explores classical, modern, and intelligent control approaches supported by analytical methods for stability and performance evaluation.

The role of energy storage systems in enhancing frequency stability and advanced controller design techniques, including optimized cascaded control structures, are also examined. The book further addresses frequency control in interconnected and multimicrogrid systems, supported by hardware implementation and real-time validation studies.

In addition, cybersecurity, ethical, and regulatory aspects of microgrid operation are discussed, and the book concludes by highlighting the future trends and research directions in microgrid load frequency control.

This book is intended to serve as a reference for graduate students, researchers, and industry professionals working in power systems, renewable energy integration, smart grids, and control engineering. By combining contributions from diverse research perspectives, this book aims to foster interdisciplinary understanding and stimulate further innovation in the development of resilient, secure, and sustainable microgrid systems.

The editors sincerely thank all the contributing authors for their scholarly efforts and commitment to advancing this field, as well as the reviewers and publishing team for their valuable support throughout the preparation of this volume. It is our hope that this book will contribute meaningfully to the ongoing research, technological development, and practical implementation of advanced load frequency control solutions in microgrids worldwide.

Editors

CHAPTER

1

Introduction to microgrids and frequency control

Pradeep Kumar and Rajneesh Pawar

Department of Electrical Engineering, Deenbandhu Chhotu Ram University of Science and Technology, Murthal, Haryana, India

1.1 Introduction

An MG is an integral part of the evolving smart grid, utilizing distributed generation and advanced information technologies to enable a decentralized and automated power delivery system. Global electricity distribution networks are rapidly transforming due to the integration of new DERs, such as MGs. These localized subsystems connect various DERs, including renewable and conventional sources, and represent a pivotal advancement in modernizing electric infrastructure. The growing suite of distributed power strategies is helping to pave the way for widespread MG adoption. MGs play a pivotal role in advancing the transition toward 100% clean and renewable energy while supporting global carbon reduction initiatives (Jinchi et al., 2019). Engineered to seamlessly integrate diverse RESs, including solar, wind, hydro, and biomass, MGs significantly reduce dependence on fossil fuels, which are primary contributors to greenhouse gas emissions. MGs are capable of functioning as self-sufficient systems that are well-regulated. They can be connected to the main grid or operate independently. By doing so, MGs enhance system efficiency, increase energy security for large loads, and improve power quality (Shahzad et al., 2023). Because of the growing awareness of the environmental impacts of traditional power generation methods, the development of more efficient small-scale power generation technologies has spurred the rapid expansion of MGs. Maintaining stable frequency operation in an MG requires continuous and instantaneous balancing of power generation and consumption. Any power imbalance

Frequency Control in Microgrids
https://doi.org/10.1016/B978-0-443-36476-1.00010-9

caused by disturbances can result in undesirable frequency deviations, compromising overall power quality. In autonomous MGs, disturbance rejection is a critical control objective to ensure frequency stability. Unlike grid-connected operations, islanded MGs lack support from the main power grid to compensate for imbalances, making them more susceptible to severe frequency instability and power quality issues. Additionally, the inherent characteristics of RESs in MGs, such as short time constants and low inertia, further complicate the design of robust frequency control mechanisms (Hu et al., 2023).

1.1.1 Overview of MGs

The power system community is increasingly in charge of managing the vast and complex network that is the power system, with an increasing emphasis on the integration of RESs. The need for sustainable electricity is the main factor driving the global switch to renewable energy generation, with the incorporation of renewables into distribution networks being accelerated by environmental concerns and technological advancements (Uddin et al., 2023). The need for sustainable electricity generation is driving the global shift toward renewable energy. Environmental concerns and technological advancements are accelerating the integration of renewable energy into distribution networks. One notable development in this area is the rise of MGs localized power systems of varying sizes that operate within distribution networks. These MGs are becoming increasingly popular due to their numerous advantages.

RERs such as solar, wind, and hydro power offer affordable solutions for meeting energy demand. They provide several key benefits: 1) they enhance energy stability and efficiency, 2) they reduce capital expenditures associated with building power plants and purchasing equipment, 3) they enable ride-by capability with energy storage, and 4) they mitigate the impact of unplanned power disruptions (Hirsch et al., 2018). When connected to the main grid, renewable-based MGs require no further frequency controls. Therefore, MGs also face challenges, including stability problems, bidirectional power movements, load variations, and uncertainties. Additionally, the use of distributed generators (DGs) may introduce complexities that occasionally outweigh their benefits. Despite these challenges, MGs small-scale energy systems that integrate renewable generation and storage technologies are increasingly capable of meeting regional energy demand efficiently (Kaur et al., 2025). Designed as either DC or AC systems, MGs, utilizing multiconverter devices, hold substantial potential for advancing future energy systems by enhancing reliability, efficiency, and power quality (Abbasi et al., 2023).

1.1.2 Definition and concept

An MG is a network of interconnected loads and DERs inside definite electrical restrictions, operating as a single, manageable entity in connection to the main grid. With its ability to connect to or disconnect from the grid, an MG can operate in both grid-connected and islanded modes, allowing flexibility in power delivery. Essentially a hybrid electric network, an MG consists of DERs, local loads, and energy storage systems (ESSs) designed to supply power to targeted areas or remote locations, with the primary goal of ensuring system stability during various network disturbances (Shahgholian, 2021).

MGs are sections of the electricity distribution system that include DGs, storage devices, and controllable loads, all of which can be managed in a coordinated manner. This coordinated control can be maintained whether the MG is connected to the main power network or operating autonomously in islanded mode. Thus, an MG functions as a small-scale power grid encompassing DERs, loads, and control systems. One of its significant advantages lies in its ability to efficiently produce, distribute, and control power flow to local customers, whether integrated with the main grid or functioning independently (Abbasi et al., 2023; Shrivastwa et al., no date). A grid-connected MG model incorporates various power-generating units alongside multiple ESSs is illustrated in Fig. 1.1. The grid-connected MG integrates distributed generation sources, including a biogas genset and a photovoltaic (PV) system. Additionally, it features ESSs such as battery banks to efficiently meet load demands and a converter for power conversion.

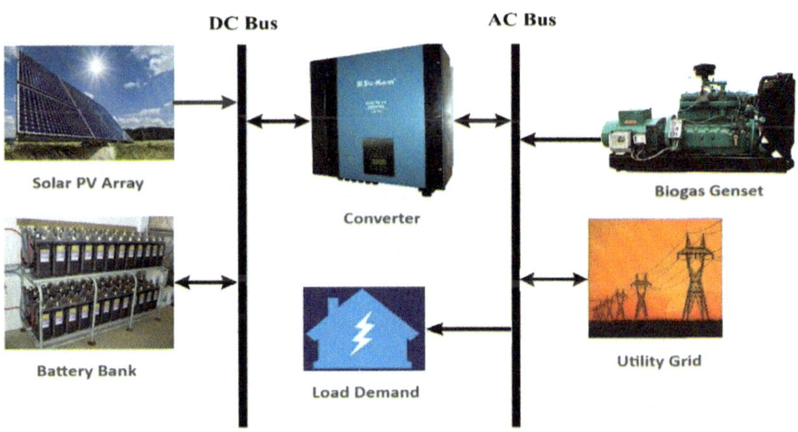

FIGURE 1.1 MG system with RES.

1.1.3 Key components of MGs

An MG system comprises numerous key components that work together to safeguard consistent and effective energy generation, distribution, and management. These components comprise:

- **Generation**: An MG generation system comprises both transferable and nontransferable sources. Transferable sources comprise natural gas generators, biogas generators, and combined heat and power (CHP) systems. Nontransferable sources mainly encompass RESs such as solar, wind, hydro, and biofuels.
- **ESS**: The ESS provides multiple functions within an MG, such as enhancing power quality, reducing peak loads, regulating frequency, stabilizing renewable energy output, and delivering backup power. It also plays a critical role in optimizing the overall cost-efficiency of the MG (Abbasi et al., 2023). A detailed types of ESSs are shown in Fig. 1.2.
- **Energy Management System (EMS)**: The EMS enables intelligent management of the MG through energy meters. It regulates generation and load dispatch within the MG, making decisions based on both economic and reliability criteria.
- **Loads**: MGs manage two main types of loads: (i) critical loads, which must be powered under all circumstances, and (ii) deferrable loads, which can be adjusted to balance the MG load and contribute to cost-effective power generation.
- **Controller**: The MG controller is responsible for supervising the system's real-time operation, ensuring that the MG functions as expected.

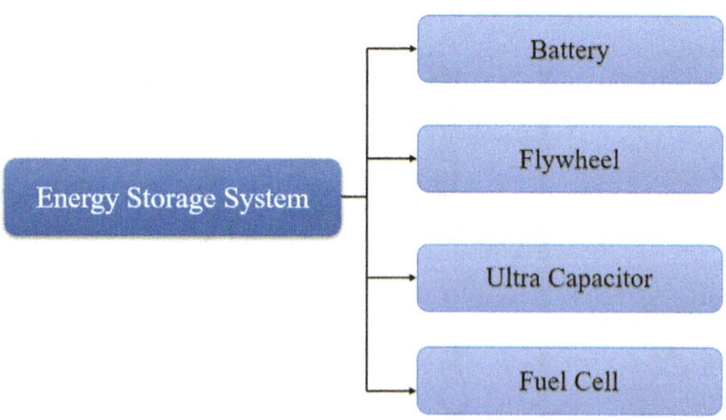

FIGURE 1.2 Various ESSs used in MG systems.

- **Point of Common Coupling (PCC)**: The PCC represents the physical link between the MG and the main grid, enabling power exchange. It includes essential apparatuses like circuit breakers, relays, and synchronization equipment. In standalone MGs, no PCC exists, as these systems function self reliantly of the main grid (Abbasi et al., 2023).

1.2 Importance of frequency control

The primary aim of frequency control in MGs is to align the operating frequency of distributed generation resources (DGRs) and RESs to a unified system frequency. In power systems utilizing synchronous machines, frequency management is inherently achieved by coupling mechanical rotation speed with electrical frequency (Dhundhara et al., 2024). However, RESs and DGRs do not support these traditional regulation mechanisms, creating frequency synchronization challenges in modern power networks (Kumar and Pawar, 2023). Several reasons, including the growing integration of DGRs and RESs, load variations, rising power demand, and voltage instability, can cause frequency deviations in MG integrated power systems. The operation of an independent MG can be challenging due to the intermittent nature of RES and the absence of backup power generation. The output power generation of RES is highly variable, which can cause frequency and voltage deviations and even power outages (Zhang et al., 2018). Additionally, the random load deviations can cause significant disruptions to the system's dynamic response and potentially destabilize the entire system. The generators based on RES in the MG have low inertia, leading to significant frequency variations. Therefore, it is essential to have a robust and automated control mechanism, particularly in autonomous situations, to provide load frequency regulation and to maintain the robust operation of the MG system (Zhang et al., 2018).

1.2.1 Role in power system stability

The capability of an MG to resume normal operation after a disturbance is referred to as stability. This concept has two main components: steady-state stability and dynamic stability. Steady-state stability refers to the MGs ability to consistently maintain voltage and frequency within specified limits, even during both normal and unusual conditions.

Dynamic stability, on the other hand, refers to the system's capacity to recover to steady state circumstances following a fluctuation, such as

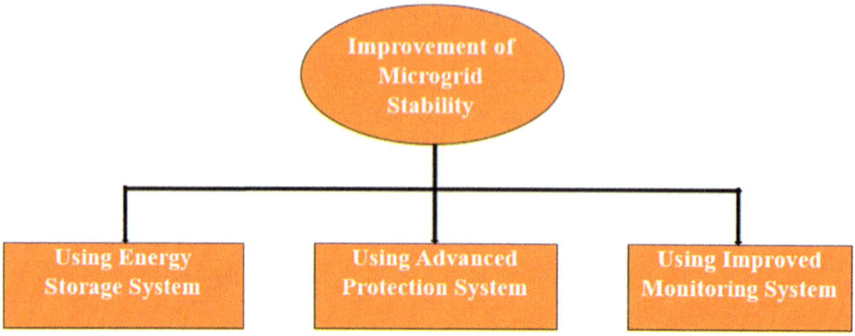

FIGURE 1.3 Strategies to improve MG stability.

variations in supply or load. To enhance MGs stability and reduce harmonics, three key strategies can be implemented: using ESSs, establishing advanced protective mechanisms, and strengthening system monitoring. By incorporating ESS, such as batteries, the MG can store excess energy and provide backup during outages. Additionally, these systems can help reduce frequency and voltage oscillations (Safder et al., 2023). To further improve overall stability, ESS also helps in frequency regulation. Another approach is provided by advanced protection systems, which identify and isolate grid disruptions like faults. By promptly detecting and resolving these problems, sophisticated protection systems stop disruptions from distribution and causing additional grid damage (Hirsch et al., 2018). Fig 1.3 shows some methods for improving MG stability.

1.2.2 Impact on grid reliability

Frequency regulation is crucial for grid reliability as it balances electricity supply and demand, improving grid stability and resilience (Kim and Park, 2021). Ensuring stable MG operation requires the development of advanced control strategies for effective frequency regulation. The dynamic behavior of RESs and load variations significantly impact the operational MGs characteristics. One of the key challenges in integrated MGs is the fluctuation of wind power, which is inherently dependent on wind speed. These variations can lead to substantial frequency deviations and, in extreme cases, grid instability or outages. Therefore, the implementation of a robust control mechanism capable of managing uncertainties, disturbances, and dynamic fluctuations is essential for maintaining reliable and resilient MGs performance (Wang et al., 2023). Here are some of the main effects of effective frequency regulation on grid reliability:

- **Maintains Power Quality and Stability**: Frequency control ensures power quality by maintaining a steady frequency around 50–60 Hz. Deviations can cause swings in voltage and frequency, disrupting equipment cause wear and possibly result in outages or brownouts.
- **Prevents Equipment Damage**: Frequency fluctuations can damage sensitive equipment, particularly in industrial environments where machines and processes rely on a consistent supply. Effective frequency management prevents such damage and reduces maintenance Costs and Downtime.
- **Balances Supply and Demand**: Grid dependability relies on maintaining a balance between generation and demand. Automatic generation control (AGC) allows for real-time adjustment of generating output to match changing conditions in demand, avoiding imbalances that could result in blackouts.
- **Enables Integration of Renewable Energy**: As the use of RESs with variable outputs increases, effective frequency control becomes more critical. Proper control mechanisms ensure that the grid can accommodate these variations while maintaining stability.
- **Supports Grid Resilience and Black Start Capability**: Frequency control is essential for stabilizing generation units after significant outages, thereby enhancing grid resilience. Maintaining frequency stability is critical during black start operations, which restore power without relying on the external grid.
- **Reduces Need for Load Shedding**: Effective frequency regulation allows the grid to handle rapid changes in load or generation, thus eliminating the need for load shedding, which disconnects parts of the grid to prevent larger system failures.

1.3 Configuration of MGs

Technological advancements and declining costs are accelerating the adoption of MGs in the commercial sector. By integrating distributed RERs such as PV systems and wind energy, MGs offer enhanced flexibility and reliability. Their growing popularity is driven by the pursuit of energy independence, economic benefits, and the increasing demand for sustainable energy solutions, especially in regions with well-developed electrical infrastructure.

To achieve technological feasibility and cost-effectiveness, good system design requires selecting the right combination of generation units and storage devices. As nations throughout the world shift to renewable and sustainable energy sources, the idea of an MG with controlled loads, energy storage, and distributed generation is becoming increasingly popular.

MGs can operate independently, off-grid, or connected to the grid, and are also essential for improving the resilience of electricity systems. An MG consists of a number of distributed generation (DG) sources, including solar panels, wind turbines, and biofuels, as well as the loads connected to them, operating as a single, integrated system (Anusha et al., 2022). An islanded MG system is a self sufficient power generation and distribution network planned to function autonomously. It integrates small scale DERs, including wind turbines, PV panels, and biodiesel/diesel generators, alongside ESSs and smart grid technologies for efficient power management within a localized area. When connected to the main grid, the MG can function as a standalone system with bidirectional power flow capability, enabling seamless energy exchange and enhanced operational flexibility (Du et al., 2018).

1.3.1 Islanded MGs

MGs are gradually replacing centralized electricity producing facilities. These systems rely on many energy sources, including internal combustion engine generators, solar PV, and wind power. Limited reactive and current power capacity in islanded MGs might cause system instability due to their lack of inertia compared to a main power grid. MGs can operate independently in islanded mode or connect with the main grid, offering flexibility in operation and management (Prevedello and Werth, 2021). By integrating DERs, MGs provide a more resilient and adaptable power system. A utility MG closely resembles an island MG, with the primary difference being that community MGs are generally not directly connected to the primary power grid as illustrated in Fig. 1.4. However, if an island is near the mainland, a connection to the mainland utility grid via cable may be feasible. For islands with existing power infrastructure, decision making around MG setup can often be faster (Singh et al., 2016). In remote or isolated communities and developing regions, off grid MGs prioritize integrating diverse, distributed energy sources. Many of these MGs are designed with the potential to connect to a larger grid as electrical infrastructure advances in these areas, while others are intended to remain self-sufficient, providing energy security and reliability for their communities (Kamal et al., 2023).

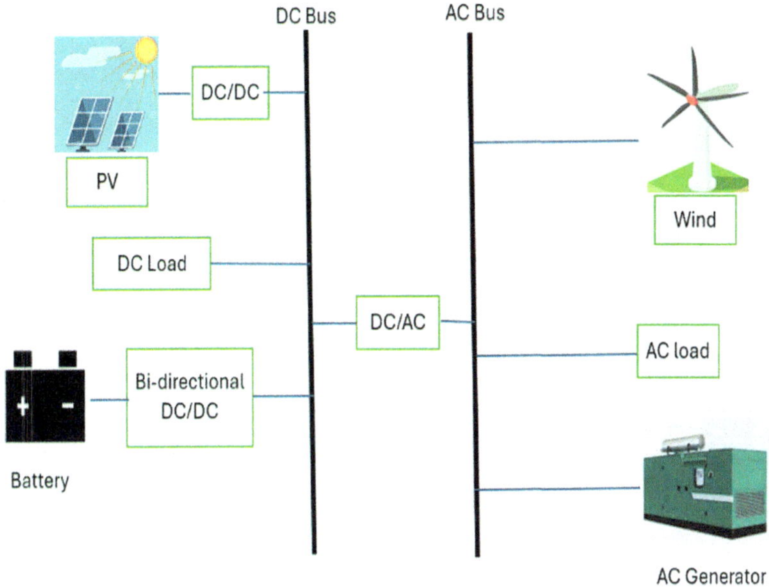

FIGURE 1.4 Islanded MG.

1.3.2 Grid-connected MGs

In grid-connected mode, most of the operational limitations present in islanded mode are eliminated. Over time, the power sector has achieved high standards of power quality and reliability, and consumers now expect this level of performance. While islanded MGs strive to meet these standards, they typically cannot fully achieve them. Connection to the main grid can bridge this gap, as it offers a stable supply with minimal voltage and frequency variations, providing far more power than a small MG can on its own (Mahmud et al., 2022). The ability to switch seamlessly between islanded and grid-connected modes makes MGs highly flexible and reliable. However, to enable this transition, MGs must meet several key requirements, as well as specific operating conditions and specialized functions. Many requirements for grid-connected mode differ significantly from those for islanded operation. Transitioning between operating modes in an MG involves more than simply engaging an interconnection switch. Given that an MG includes multiple distributed generation units operating in parallel through power converters, grid connection requires synchronized and coordinated adjustment of all control mechanisms. In islanded mode, voltage and frequency references are generated internally by the MG, whereas in grid-connected mode, these references are externally set by the main grid as shown in Fig. 1.5 (Bifaretti et al., 2017a).

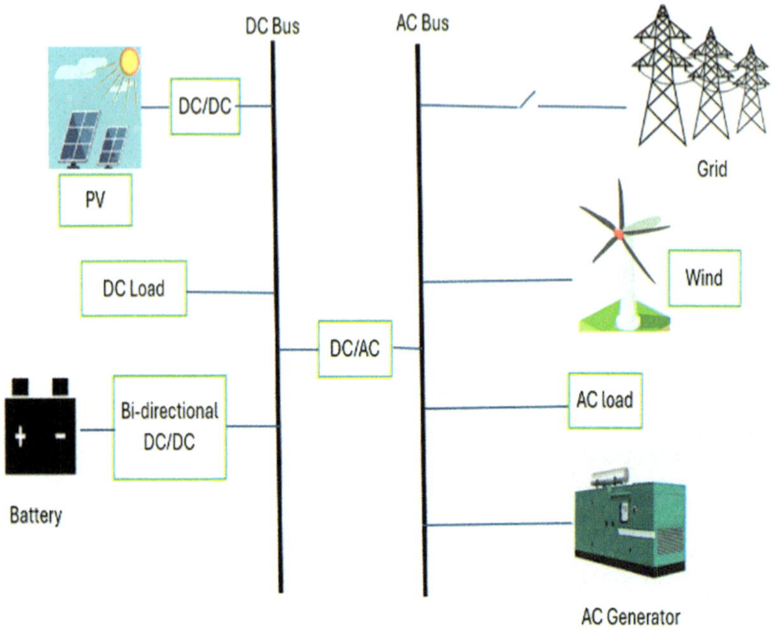

FIGURE 1.5 Grid connected MG.

1.4 Types of MGs

MGs are categorized by power type into AC, DC, or hybrid systems, each offering distinct advantages and challenges depending on the application. In an AC MG, all RERs and loads are connected to a joint AC bus. A major challenge of AC MGs is their complex control and operational requirements. Based on the distribution system design, AC MGs can be classified into three types: single-phase, three-phase without neutral lines, and three-phase with neutral lines. In contrast, DC MGs provide higher reliability, greater efficiency, and simpler integration with various DERs (Dhundhara et al., 2019). Hybrid MGs combine AC and DC components to enhance overall network efficiency by minimizing conversion stages, improving reliability, reducing the need for interfacing devices, and lowering energy costs (Kumar and Pawar, 2024).

1.4.1 AC MGs

An AC MG is an electrical system that uses an alternating current (AC) power source to supply AC loads. It can function independently or connect to the main grid through an authorized point of common coupling (PCC) as shown in Fig. 1.6. This configuration employs an AC bus to link power sources, storage systems, and other components in order to fulfill

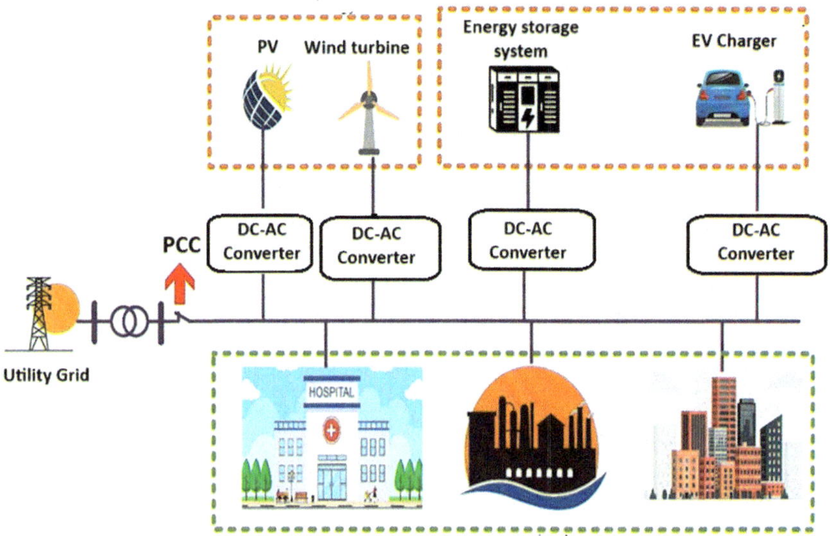

FIGURE 1.6 AC MG model.

the AC load requirements (Jinchi et al.,2019). They are primarily categorized into three types: single-phase, grounded three-phase, and ungrounded three-phase. Additionally, AC MGs are classified by frequency into high-frequency, low-frequency, and standard-frequency systems. Among different MG types, AC MGs are largely used and widely adopted in real-world applications (Planas et al., 2015). However, AC MGs face challenges in synchronizing with the main grid, as they must maintain consistent voltage, phase angle, and frequency. They also tend to have lower efficiency and reliability, often requiring complex architectures and control systems (Gerber et al., 2023).

1.4.2 DC MGs

DC MGs offer significant advantages over AC MGs, as they require no synchronization and rarely face power quality issues, eliminating concerns about power factor correction. However, several power electronics devices and converters are needed to connect DC MGs to existing distribution systems. Compared to AC MGs, DC MGs attain better efficiency and require fewer conversions when supplying DC loads (Lotfi and Khodaei, 2017a). They are widely used in telecommunications, electric vehicles (EVs), and other commercial settings. DC MGs are generally categorized as monopolar, bipolar, or homopolar. The key advantages of DC MGs are their capability to connect DC loads right to the DC bus, diminishing the need for power converters, as shown in Fig. 1.7. However, DC MGs lack standardized voltage, often requiring additional conversion steps to

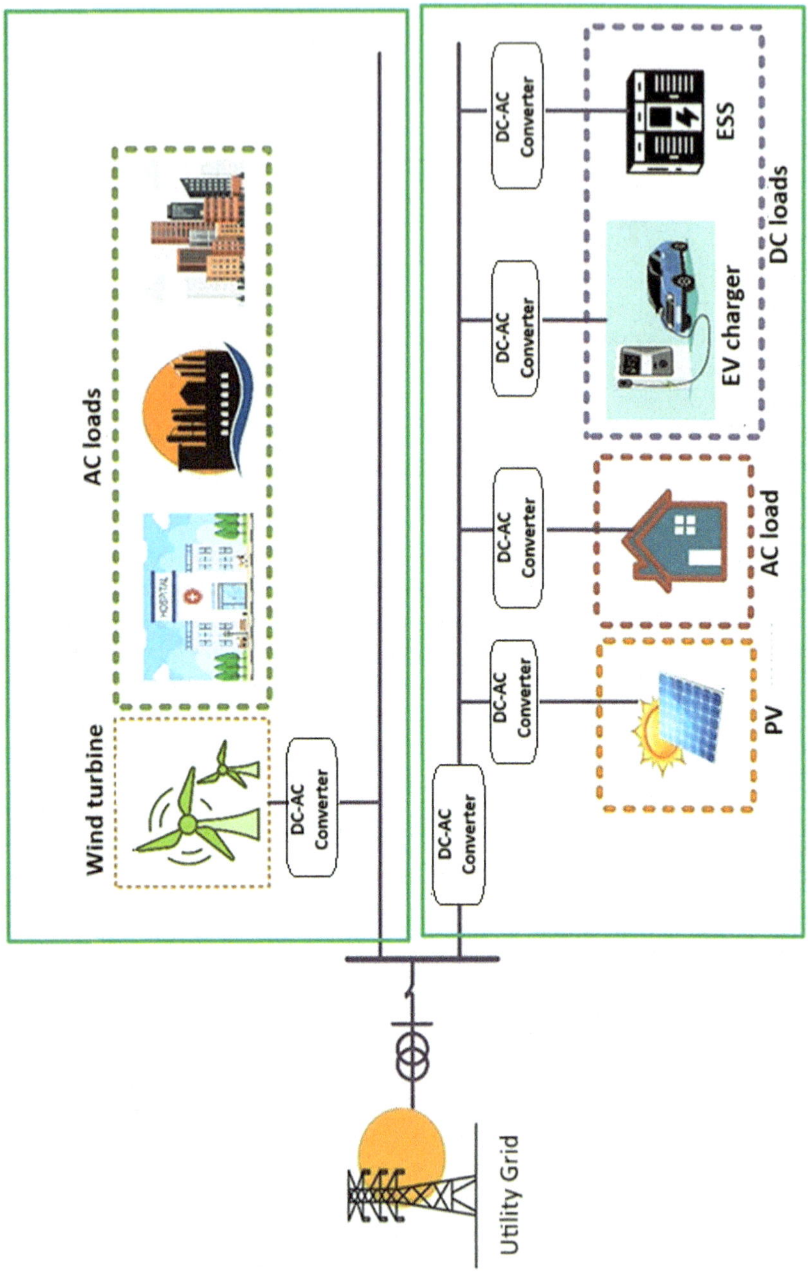

FIGURE 1.7 DC MG model.

produce AC power. They cannot be easily reconfigured from the present grid, and their safety systems are complex (Kabalcı, 2018).

1.4.3 Hybrid MG

Hybrid MGs is a combination of both AC and DC systems to supply connected loads via both types of distribution systems. The primary goal of a hybrid MG is to minimize power conversion stages and interface devices, thereby reducing energy costs while enhancing overall efficiency and reliability. These systems support both AC and DC loads, permitting consumers to modify their power usage according to specific needs, as illustrated in Fig. 1.8. Power electronic converters are used to separate the AC and DC sections of the MG. DG units in AC-DC hybrid MGs can connect directly to either AC or DC networks without requiring synchronization (Lotfi and Khodaei, 2017b). However, this formation does not automatically reduce energy losses, as inefficiencies can result from converters, transmission, and control limitations. Hybrid MGs require advanced control and management systems, particularly when functioning in islanded mode. They may also have lower reliability than AC MGs due to the use of interface converters for DC link generation within the distribution network. Nonetheless, reducing the number of converter stages can enhance the reliability of interconnected devices (Jain and Saxena, 2023).

1.5 Interconnection with the main grid

As an evolving concept, the smart grid primarily aims to manage the behavior of its connected users. A key device in this system is the smart meter, which enables users to independently monitor and control their energy consumption, thus promoting efficient use of resources and electricity. This adjustment is expected to improve power plant load factors, reduce the carbon footprint, and increase the transfer capacity of transmission systems. Consequently, interconnected smart grids are anticipated to enhance power plant load factors, reduce the power system's carbon footprint, improve management of systems with high renewable energy integration, and expand retail electricity markets. Smart grids must also adapt to new types of demand, such as during the charging of battery ESSs (Kumar and Pawar, 2021). When an MG connects to the distribution system, it can meaningfully affect these cumulative effects. For example, when PV generation is closely matched with air conditioning loads, network peak currents may be reduced, yielding positive outcomes. However, there can also be challenges; for instance, maximum PV generation often occurs during low-load periods, which can reduce

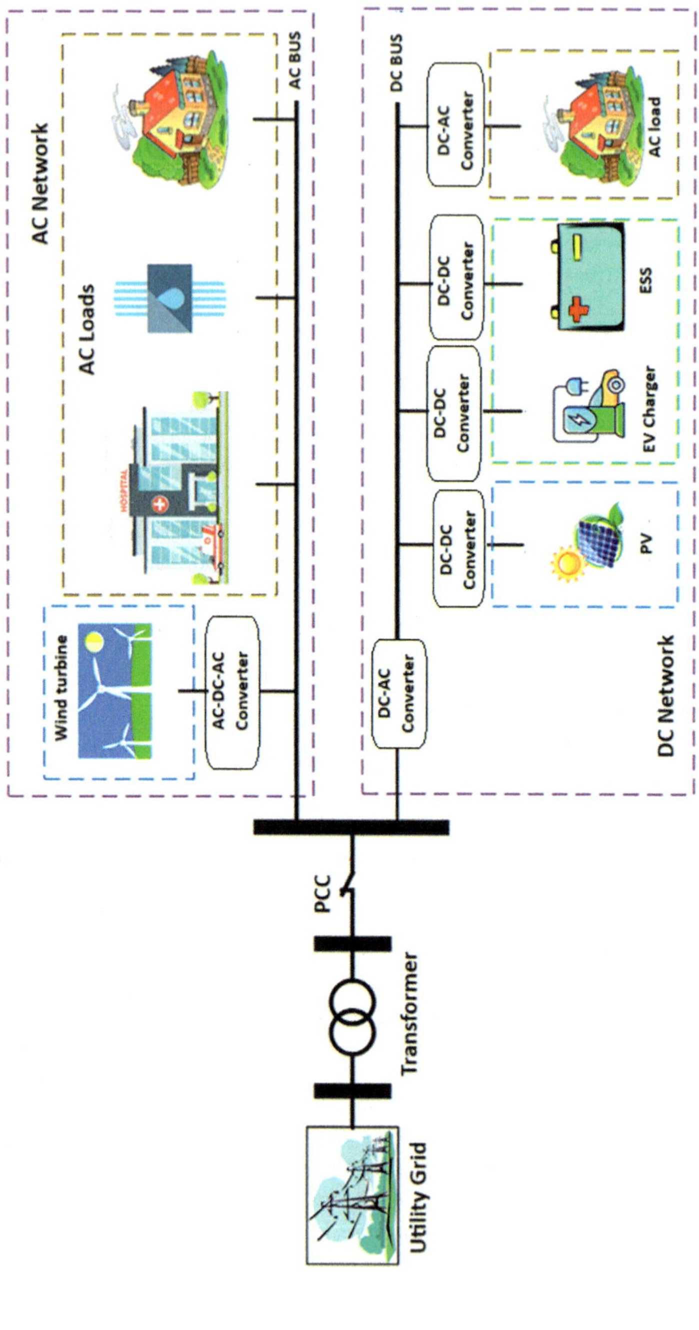

FIGURE 1.8 Hybrid MG model.

current flows and increase network voltage. Additional challenges with distributed generation include unintended islanding, fault currents, grounding issues, and fluctuations in power output, all of which can impact power quality, system safety, and control. These considerations are essential for the successful combination of DG into the grid (Bifaretti et al., 2017b). Considering a network comprising 1 to n interconnected MGs. These MGs are not only connected to the main power grid but are also linked with one another, enabling both power and information exchange through a shared power bus and a communication network. Fig 1.9 illustrates the structure of such a system. It is assumed that each MG can schedule its power supply and flexible loads based on forecasted daily renewable energy generation. Any discrepancies arising from forecast errors are managed through transactions in the real-time market.

1.5.1 Challenges and benefits of MG interconnection

Interconnected power networks offer advantages like cost savings and enhanced energy stability, but they also present challenges (Shahzad et al., 2023).

Some of the difficulties faced by interconnected electrical systems are the following:

- **Transmission Losses**: Transmission losses occur during long-distance power transfers, leading to significant energy losses. This makes it challenging to deliver energy from resource-rich areas to places with high demand.
- **Grid Stability**: Grid stability can be affected by interconnected systems that may experience disruptions due to voltage and frequency fluctuations. Factors such as weather patterns and sudden changes in energy supply or demand can also impact stability. Weather events and fluctuations in energy supply or demand can affect stability.
- **Complex planning and coordination**: Managing an interconnected grid necessitates collaboration among utilities and government organizations, which makes ensuring smooth and efficient operations challenging.
- **Regulatory challenges**: Effective regulation of interconnected power grids requires coordination across different regions.
- **Cyber security**: Effective cybersecurity procedures are crucial to safeguard these systems.

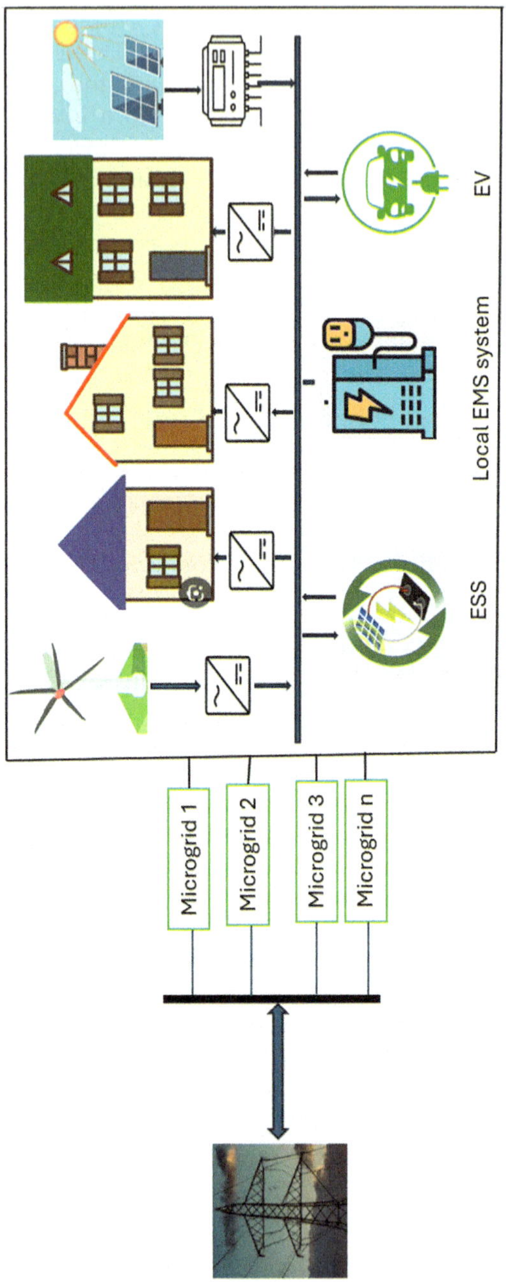

FIGURE 1.9 Interconnection of MGs with the main grid.

Below are some key benefits and issues associated with interconnection.

- **Cost Stability**: Owning an MG provides a safeguard against unpredictable and potentially high costs for backup or emergency power. It also shields your operations from fluctuations in electricity rates, offering more consistent energy expenses.
- **Improved Power Quality**: Certain operations require power quality that the standard grid may not reliably supply. An MG gives you greater control over power quality, which is essential for sensitive environments like laboratories, advanced manufacturing, healthcare facilities, and other critical applications.
- **Improved energy reliability**: Interconnected systems provide more reliable electricity delivery. In case of an outage or shortage, one region can obtain power from another, ensuring stability.
- **Reduced costs**: Interconnected networks enable utilities to acquire electricity from more economical regions, resulting in lower overall prices.

1.6 Frequency control in MGs

MG resources are inherently distributed, but the variable and intermittent nature of RESs adds complexity to effective MG operations, necessitating a robust control strategy to ensure stable and continuous power flow. Typically, an MG Central Controller (MGCC) is employed to manage and regulate the MG, with installation options at either a local control center or a distribution substation (Heidary et al., 2022). The MGCC coordinates the operation of local DG units and ESS to maintain system stability and reliability, communicating with lower level controllers. Additionally, MGs can be managed through decentralized approaches, such as droop control or agent based systems. A number of frequency control methods used in MG systems are discussed in Table 1.1

1.6.1 Significance of frequency regulation

Frequency regulation is a serious concern of MG action, providing reliability, stability, and an effective power supply. Meanwhile, MGs integrate various energy sources and work in both autonomous and grid connected means, sustaining a steady frequency is crucial to avoid power imbalances and failure of the system. The main significance of frequency regulation is shown in Fig. 1.10.

- **Power Balance**: Smooth transition between islanded and grid-tied modes enables the MG to function autonomously or remain

TABLE 1.1 Frequency control methods/techniques used in different MG systems.

Authors name	Source of power generation	Controller type	Optimization method	References
B. Singh et al.,	BDG (bio diesel generator), wind turbine generator (WTG), PV	PIDF-PDF	Modified African Buffalo Optimization	Singh et al. (2023)
K. Peddakapu et al.,	PV, WTG, DG (diesel generator)	CFOID-FOPIDN	Barnacle mating optimizer	Peddakapu et al. (2022)
S. Patel et al.,	WTG, PV	Hybrid fuzzy integrated fractional order cascaded PDF (1 + PI) controller	Modified sin cosine algorithm-pattern search (m-SCA-PS)	Patel et al. (2023)
R.K. Khadanga et al.,	WTG, PV, TPP (thermal power plant)	Multi-stage PID controller (MS-PID)	Sine augmented scaled arithmetic optimization algorithm (SCA-AOA)	Khadanga et al. (2023)
Nirojakanta Swain et al.	DG, WTG, PV system	Two-degree of freedom PID (2DOFPID) controller	African vultures optimization algorithm (AVOA) technique	Swain et al. (2023)
H. Shayeghi et al.	DG, WTG, dish-Stirling solar power generation (DSPG)	Tilt–FO derivative with a filter cascaded to the FOPI controller (TOFDn-FOPI)	Prairie dog optimization (PDO) algorithm	Barik and Das (2018)
A. Elsawy Khalil et al.	DG, WTG, PV system	Proportional-integral (PI) and proportional integral derivative (PID) controllers	Artificial rabbits optimization algorithm (ARO)	Khalil et al. (2023)
M. Joshi et al.	Hydro power plant	Fuzzy-PSO-PID	Particle swarm optimization (PSO)	Joshi et al. (2023)
A. Elsawy Khalil et al.	DG, WTG, PV system	Cascade PID controller, cascade loop controller	Artificial rabbits optimization algorithm (ARO), canonical dandelion optimizer (DO)	Khalil et al. (2024); Elsawy Khalil et al. (2024)

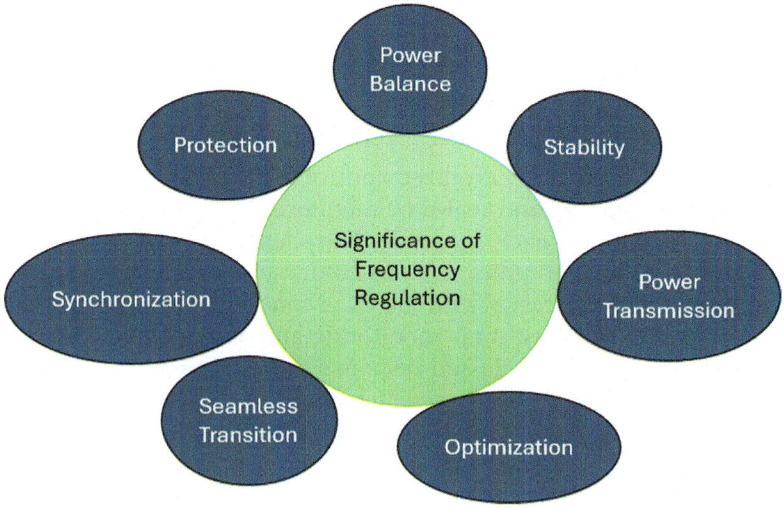

FIGURE 1.10 Significant of frequency regulation.

connected to the main grid as needed. This adaptability allows for continued operation without service disruptions. Even during external grid interruptions or scheduled transitions.

- **Protection**: Monitoring energy flow and key system components is essential for detecting flaws and protecting the MG from damage. This safety layer allows for a swift response to concerns and avoids cascading failures in both the MG and the broader grid.
- **Power Transmission**: Controls power exchange between the MG and the main grid, allowing the MG to draw energy from or supply excess electricity to it. This bidirectional flow promotes energy balance, maximizes resource utilization, and improves total system stability.
- **Optimization**: Enhances the MG's dispatch strategy to maximize economic benefits by considering energy costs, demand, and available resources. By improving dispatch, the system reduces costs and minimizes energy waste, leading to greater energy efficiency under varying conditions.
- **Synchronization**: Synchronisation aligns the voltage, phase, and frequency of the MG with the main power grid, ensuring efficient power transfer without causing interruptions during connection or disconnection.
- **Stability**: Stability regulates voltage and frequency to maintain the stability of MGs across different operating modes. It supports both AC and DC components, ensuring a steady power supply for sensitive equipment (Safder et al., 2023).

1.6.2 Control techniques

Control techniques in MG systems are important for confirming reliable, stable, and efficient operation of MG, particularly when integrated with RERs. These control techniques can be classified as shown in Fig. 1.11

- **Centralized Control**: Centralized control administration allows for efficient positioning and real-time monitoring of the entire MG system. In this approach, a central controller (CC) oversees all dispersed generation units. Each DG unit includes a local controller (LC) can connect directly with the CC. Modern computing technology enables real-time monitoring and analysis of data from LCs, resulting in a simple and responsive deployment for management. Centralized control has limits, especially in large hybrid systems where a failure might affect the entire system. Furthermore, this strategy lacks flexibility and scalability, which may limit adapting to the increasing MG structure (Abbasi et al., 2023).
- **Decentralized Control**: In recent years, decentralized control has improved the autonomy of micro-sources and loads in MGs. This strategy prioritizes stability, cost-effectiveness, and dependability by allowing control decisions to be made locally using on-site measurements with few linkages required.

This design reduces the need for advanced computer equipment and significant networking. Decentralized control can provide stability and autonomy, but it may not necessarily result in a globally optimal solution for micro grid (Abbasi et al., 2023).

- **Distributed Control**: In distributed control, controllers share some information with each other, allowing them to understand the actions

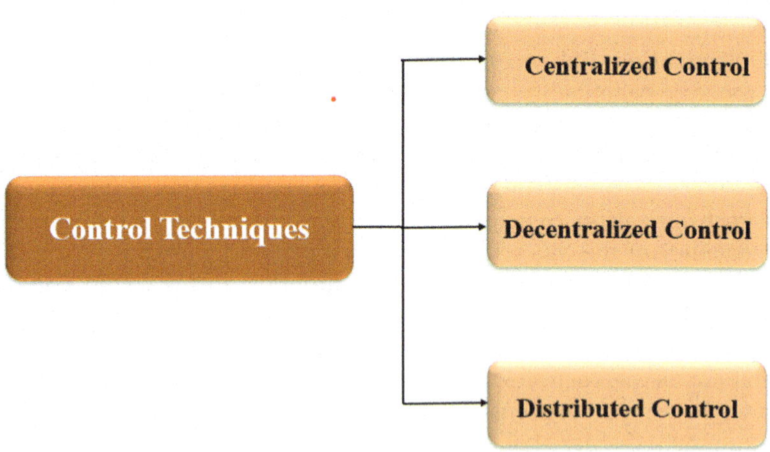

FIGURE 1.11 Control techniques in MG system.

and behavior of other controllers. This cooperative approach enhances the overall performance of the MG, as each controller operates with a level of awareness of the others, promoting more efficient and balanced system management (Abbasi et al., 2023).

1.7 Future trends in MG technology

MGs represent a dynamic and promising area of research within power engineering. While significant advances have been made in recent MG studies, several critical challenges remain. Both AC and DC MGs present open research areas that need to be addressed for future progress (Wu et at., 2023).

- **Modeling**: As future MGs aim to operate with power sources, ESS, and loads as plug and play components, their complexity will increase. To manage this, future MGs will require a fundamental redesign.
- **Mode of Operation**: An important area for future research is the development of systems that enable MGs to transition seamlessly between grid connected and autonomous operation.
- **Protection**: Traditional protection methods in distribution networks rely on fixed relay settings, which may not be sufficient for MGs. Research is needed to develop adaptive protection mechanisms for safe operation in all modes and transitions.
- **Control**: Efficient control techniques are crucial for maintaining system parameters in standalone mode, as MG components have different operating characteristics.
- **Energy Management**: Energy management involves careful coordination of renewable sources, storage systems, and load. Effective solutions are necessary to balance power distribution among renewable sources, generators, and ESSs, enabling efficient resource utilization.
- **Generation Load Stability**: In standalone mode, MGs need detailed control strategies to effectively balance energy generation and load.
- **Storage Units**: Future study should focus on utilizing various ESS technologies to improve MG stability and inertia.
- **Integration of EVs**: Further research on merging EVs with MGs could provide useful insights for power system improvements as EVs become more prevalent.

These areas represent critical challenges and opportunities for the advancement of MG technology, and addressing them could significantly impact the evolution of modern power systems (Uzair et al., 2023).

1.7.1 Developments in control systems

Recent developments in MG control and management systems are primarily driven by the maximizing integration of RESs, the requirement for improved grid flexibility, and the rising complexity of MG operations. These innovations focus on improving the reliability, and efficiency of MGs, facilitating unified combination of DERs and operative management of grid actions (Mahmoud et al., 2014). Key developments in MG control and management systems include

- **Intelligent EMS**: EMS now utilize artificial intelligence, and innovative optimization algorithms to manage energy generation, storage, and consumption within MGs more effectively. These systems overlook and predict energy demand and generation in real-time, dynamically improve power dispatch, and support decision-making processes to enhance overall operational performance.
- **Advanced ESS Management**: Advanced control strategies are enhancing the utilization and effectiveness of energy ESS in MGs. These strategies implement intelligent scheduling based on real-time capacity demands, renewable energy availability, and grid conditions, allowing for efficient energy balancing, peak load reduction, load shifting, and overall grid support.
- **Demand Response (DR) Integration**: MG control systems are increasingly integrating demand response (DR) mechanisms, enabling consumers to actively engage in load management. Advanced DR algorithms and communication protocols allow real-time interaction between MG operators and end users, supporting load shedding or load shifting while demand is high and optimizing overall energy consumption.
- **Cyber Physical Security**: With MG control systems progressively interconnected and dependent on digital communication, cybersecurity and resilience are critical. Advanced security measures, including encryption, validation protocols, irregularity detection, and intrusion safety systems, are being incorporated into control and supervisory systems to guard against cyber threats and strengthen system resilience.

1.7.2 Integration of RESs

Renewable MGs are those powered by distributed RERs. Renewable MGs combine RESs and battery storage devices to generate low-carbon electricity. This makes them increasingly popular and widely used around the world. However, the variable and intermittent nature of RES output complicates the efficient operation of MGs. Meeting fluctuating demand in isolated MGs poses a substantial difficulty. ESS are a promising technology for maximizing renewable energy usage and are extensively used to balance supply and demand in MGs (Aziz et al., 2020). Nonetheless, a major challenge remains in coordinating storage systems, distributed RES, and fluctuating power demand. Therefore, an is essential for renewable EMS MGs.

- **Solar MGs:** Solar MGs are highly adaptable and scalable, making them a viable option for rural electrification globally. They perform best in regions with ample sunlight. To address cloudy weather, most solar MGs include storage to ensure operation during low-sun periods. Solar MGs are environmentally friendly but face the challenge of managing the fluctuating output of solar PV (Seo et al., 2022). Effective control systems are necessary to address this variability. Common applications of solar MGs include powering schools, street lighting, homes, and irrigation systems.
- **Wind MGs:** Wind MGs use a network of wind turbines connected to loads, functioning as a unified power source within limited electrical boundaries. Wind-based MGs generally include ESS to smooth power supply and store surplus power for later use, enhancing reliability.
- **Biomass MGs:** Biomass MGs generate electricity by using biomass gasifiers to produce syngas, which is burned in an engine to power a generator. Although less common than solar or wind-based MGs, biomass MGs are gaining traction, especially in areas with ample biomass resources. Biomass systems are generally less capital-intensive than solar PV systems but require considerable feedstock, storage facilities, and labor. Operational challenges, such as tar buildup, wet biomass, and frequent maintenance needs, can affect performance (Singh et al., 2021).
- **Micro hydro MGs:** Micro hydro MGs are typically run-of-the-river systems that use flowing water to generate electricity. They are cost-effective in areas with reliable water flow, offering low-cost energy per kWh. However, they are location-dependent, requiring proximity to suitable water sources.
- **Hybrid MGs:** Hybrid MGs are designed to supply power to remote areas using a combination of renewable sources, like solar, wind, and micro-hydro. These systems provide more stable and reliable power by diversifying energy sources.

A brief integration of RERs with different MG systems is described in Table 1.2

Each type of renewable MG has unique benefits and challenges, but with the appropriate technology and management systems, they can significantly contribute to sustainable energy goals worldwide.

TABLE 1.2 Integration of RESs with different MG systems comprises of various ESS.

References	Source of power generation	MG type	ESS	Authors name
Pahasa and Ngamroo (2016)	Thermal generator, WTG	Autonomous MG	EVs	J. Pahasa et al.
Annamraju and Nandiraju (2018)	DG, WTG, PV system	Autonomous MG	Battery energy storage system (BESS), fuel cell (FC)	A. Annamraju et al.
Srivastava et al. (2022)	BDG, WTG, archimedes wave power generation	Autonomous MG	FC	A. K. Srivastava et al.
Khokhar et al. (2021)	MT, DG, WTG, PV system	Autonomous MG	Flywheel energy storage system (FESS), BESS, FC, ultra capacitor (UC)	B. Khokhar et al.
(Khadanga et al., 2023)	TPP, WTG, PV system	Autonomous MG	BESS	R.K. Khadanga et al.
Prusty et al. (2022)	MT, DG, PV system	Autonomous MG	BESS, FESS	U.C. Prusty et al.
Yildirim et al. (2023)	Micro-hydro turbine generator (MHTG), PV	Autonomous MG	BESS, FESS,	Burak Yildirim et al.
Yang et al. (2023)	DG, WTG	Multi-MG	FC	Fan Yang et al.
Kumar and Pawar (2023)	BDG, WTG, PV system	Autonomous MG	EV, FC, UC	P. Kumar et al.

1.8 Conclusion

This chapter provides an overview of MGs, highlighting their potential significance in modern energy systems. It surveys the key technologies in MGs and categorizes them into several groups based on application, infrastructure, and customer requirements. Additionally, this review examines various control strategies for MGs to give a deeper understanding of these approaches. The development of MGs promises a more ecologically friendly and enhanced power supply facilities, yet remain several critical areas for improvement. This chapter identifies major research challenges and issues facing MGs and outlines future research directions to address these gaps. By presenting the latest advancements in MG technology, this chapter aims to gain the interest of experts, and researchers in this area. However, it does not cover crucial communication systems needed for MG operation, which is a vital area for future research to assess the current progress in MG communication technologies.

References

Abbasi, M., Abbasi, E., Li, L., Aguilera, R.P., Lu, D., Wang, F., 2023. Review on the microgrid concept, structures, components, communication systems, and control methods. Energies 16 (1). https://doi.org/10.3390/en16010484.

Annamraju, A., Nandiraju, S., 2018. Robust frequency control in an autonomous microgrid: a two-stage adaptive fuzzy approach. Electr. Power Compon. Syst. https://doi.org/10.1080/15325008.2018.1432723.

Anusha, U., Jagadeesh, S., Rao, K.V., Umavani, M., 2022. Energy management of an off-grid and grid connected hybrid renewable energy source micro-grid system for commercial load. In: MysuruCon 2022 - 2022 IEEE 2nd Mysore Sub Section International Conference. Institute of Electrical and Electronics Engineers Inc., India https://doi.org/10.1109/MysuruCon55714.2022.9972673.

Aziz, M.S., Khan, M.A., Khan, A., Nawaz, F., Imran, M., Siddique, A., 2020. Rural electrification through an optimized off-grid microgrid based on biogas, solar, and hydro power. In: International Conference on Engineering and Emerging Technologies, ICEET 2020. Institute of Electrical and Electronics Engineers Inc., Pakistan https://doi.org/10.1109/ICEET48479.2020.9048222.

Barik, A.K., Das, D.C., 2018. Expeditious frequency control of solar photovoltaic/biogas/biodiesel generator based isolated renewable microgrid using grasshopper optimisation algorithm. IET Renew. Power Gener. 12 (14), 1659–1667. https://doi.org/10.1049/iet-rpg.2018.5196.

Bifaretti, S., Cordiner, S., Mulone, V., Rocco, V., Rossi, J.L., Spagnolo, F., 2017. Grid-connected microgrids to support renewable energy sources penetration. Energy Proc. 105, 2910–2915. https://doi.org/10.1016/j.egypro.2017.03.658.

Dhundhara, S., Sharma, M., Guéniat, F., Arya, Y., 2024. Overview of the renewable-dominated power systems and their frequency regulation issues. In: Dhundhara, S., Arya, Y., Bansal, R.C. (Eds.), Advanced Frequency Regulation Strategies in Renewable-Dominated Power Systems. Academic Press, pp. 1–19.

Dhundhara, S., Verma, Y.P., Williams, A., 2019. Techno-economic evaluation of AC and DC microgrid systems. In: Lecture Notes in Electrical Engineering. Springer Verlag, pp. 265–280. https://doi.org/10.1007/978-981-13-6772-4_24.

Du, Y., Wang, Z., Liu, G., Chen, X., Yuan, H., Wei, Y., Li, F., 2018. A cooperative game approach for coordinating multi-microgrid operation within distribution systems. Appl. Energy 222, 383–395. https://doi.org/10.1016/j.apenergy.2018.03.086.

Elsawy Khalil, A., Boghdady, T.A., Alham, M.H., Ibrahim, D.K., 2024. A novel cascade-loop controller for load frequency control of isolated microgrid via dandelion optimizer. Ain Shams Eng. J. 15 (3). https://doi.org/10.1016/j.asej.2023.102526.

Gerber, D.L., Nordman, B., Brown, R., Poon, J., 2023. Cost analysis of distributed storage in AC and DC microgrids. Appl. Energy 344. https://doi.org/10.1016/j.apenergy.2023.121218.

Heidary, J., Gheisarnejad, M., Rastegar, H., Khooban, M.H., 2022. Survey on microgrids frequency regulation: modeling and control systems. Elec. Power Syst. Res. 213. https://doi.org/10.1016/j.epsr.2022.108719.

Hirsch, A., Parag, Y., Guerrero, J., 2018. Microgrids: a review of technologies, key drivers, and outstanding issues. Renew. Sustain. Energy Rev. 90, 402–411. https://doi.org/10.1016/j.rser.2018.03.040.

Hu, S., Ge, X., Chen, X., Yue, D., 2023. Resilient load frequency control of islanded AC microgrids under concurrent false data injection and denial-of-service attacks. IEEE Trans. Smart Grid 14 (1), 690–700. https://doi.org/10.1109/TSG.2022.3190680.

Jain, D., Saxena, D., 2023. Comprehensive review on control schemes and stability investigation of hybrid AC-DC microgrid. Elec. Power Syst. Res. 218. https://doi.org/10.1016/j.epsr.2023.109182.

Jinchi, H., Chengyun, L., Chao, M., 2019. Energy Efficiency Comparison of AC and DC Microgrids Pplied Energy Symposium 2019: Low Carbon Cities and Urban Energy Systems, pp. 1–6.

Joshi, M., Sharma, G., Çelik, E., 2023. Load frequency control of hydro-hydro power system using fuzzy-PSO-PID with application of UC and RFB. Elec. Power Compon. Syst. 51 (12), 1156–1170. https://doi.org/10.1080/15325008.2023.2196663.

Kabalcı, E., 2018. An islanded hybrid microgrid design with decentralized DC and AC subgrid controllers. Energy 153, 185–199. https://doi.org/10.1016/j.energy.2018.04.060.

Kamal, M.M., Ashraf, I., Fernandez, E., 2023. Optimal sizing of standalone rural microgrid for sustainable electrification with renewable energy resources. Sustain. Cities Soc. 88. https://doi.org/10.1016/j.scs.2022.104298.

Kaur, M., Dhundhara, S., Verma, Y.P., Chauhan, S., Arya, Y., 2025. Impact of anaerobically digested pretreated crop residue on performance and economics of solar photovoltaic-biomass microgrid system. Sustain. Energy Technol. Assess. 76.

Khadanga, R.K., Das, D., Kumar, A., Panda, S., 2023. Sine augmented scaled arithmetic optimization algorithm for frequency regulation of a virtual inertia control based microgrid. ISA (Instrum. Soc. Am.) Trans. 138, 534–545. https://doi.org/10.1016/j.isatra.2023.02.025.

Khalil, A.E., Boghdady, T.A., Alham, M.H., Ibrahim, D.K., 2023. Enhancing the conventional controllers for load frequency control of isolated microgrids using proposed multi-objective formulation via artificial rabbits optimization algorithm. IEEE Access 11, 3472–3493. https://doi.org/10.1109/ACCESS.2023.3234043.

Khalil, A.E., Boghdady, T.A., Alham, M.H., Ibrahim, D.K., 2024. A novel multi-objective tuning formula for load frequency controllers in an isolated low-inertia microgrid incorporating PV/wind/FC/BESS. J. Energy Storage 82. https://doi.org/10.1016/j.est.2024.110606.

Khokhar, B., Dahiya, S., Singh Parmar, K.P., 2021. Load frequency control of a microgrid employing a 2D Sine Logistic map based chaotic sine cosine algorithm. Appl. Soft Comput. https://doi.org/10.1016/j.asoc.2021.107564.

Kim, J.J., Park, J.H., 2021. A novel structure of a power system stabilizer for microgrids. Energies 14 (4). https://doi.org/10.3390/en14040905.

Kumar, P., Pawar, R.D.S., 2021. Biomass availability and biomass-based generating units in north india: a review. In: IEEE 2nd International Conference on Electrical Power and Energy Systems, ICEPES 2021. Institute of Electrical and Electronics Engineers Inc., India https://doi.org/10.1109/ICEPES52894.2021.9699499. http://ieeexplore.ieee.org/xpl/mostRecentIssue.jsp?punumber=9698816.

Kumar, P., Pawar, R., 2023. Robust control strategy for power/frequency regulation in autonomous microgrid system. In: Energy Sources, Part A Recovery, Utilization and Environmental Effects, pp. 6834–6855. https://doi.org/10.1080/15567036.2023.2215184.

Kumar, P., Pawar, R., 2024. A comparative study of AC and DC microgrid system with distributed energy resources. Int. J. Techn. Res. Sci. 9 (Spl), 27–35. https://doi.org/10.30780/specialissue-iset-2024/020.

Lotfi, H., Khodaei, A., 2017a. AC versus DC microgrid planning. IEEE Trans. Smart Grid 8 (1), 296–304. https://doi.org/10.1109/TSG.2015.2457910.

Lotfi, H., Khodaei, A., 2017b. Hybrid AC/DC microgrid planning. Energy 118, 37–46. https://doi.org/10.1016/j.energy.2016.12.015.

Mahmoud, M.S., Azher Hussain, S., Abido, M.A., 2014. Modeling and control of microgrid: an overview. J. Franklin Inst. 351 (5), 2822–2859. https://doi.org/10.1016/j.jfranklin.2014.01.016.

Mahmud, D.M., Ahmed, S.M.M., Hasan, S., Zeyad, M., 2022. Grid-connected microgrid: design and feasibility analysis for a local community in bangladesh. Clean Energy 6 (3), 447–459. https://doi.org/10.1093/ce/zkac022.

Pahasa, J., Ngamroo, I., 2016. Coordinated control of wind turbine blade pitch angle and PHEVs using MPCs for load frequency control of microgrid. IEEE Syst. J. https://doi.org/10.1109/JSYST.2014.2313810.

Patel, S., Ghosh, A., Ray, P.K., 2023. Improved power flow management with proposed fuzzy integrated hybrid optimized fractional order cascaded proportional derivative filter (1+ proportional integral) controller in hybrid microgrid systems. ISA (Instrum. Soc. Am.) Trans. 136, 742–754. https://doi.org/10.1016/j.isatra.2022.11.005.

Peddakapu, K., Srinivasarao, P., Mohamed, M.R., Arya, Y., Krishna Kishore, D.J., 2022. Stabilization of frequency in multi-microgrid system using barnacle mating optimizer-based cascade controllers. Sustain. Energy Technol. Assessments 54. https://doi.org/10.1016/j.seta.2022.102823.

Planas, E., Andreu, J., Gárate, J.I., Martínez De Alegría, I., Ibarra, E., 2015. AC and DC technology in microgrids: a review. Renew. Sustain. Energy Rev. 43, 726–749. https://doi.org/10.1016/j.rser.2014.11.067.

Prevedello, G., Werth, A., 2021. The benefits of sharing in off-grid microgrids: a case study in the philippines. Appl. Energy 303. https://doi.org/10.1016/j.apenergy.2021.117605.

Prusty, U.C., Nayak, P.C., Prusty, R.C., et al., 2022. An improved moth swarm algorithm based fractional order type-2 fuzzy PID controller for frequency regulation of microgrid system. Energ. Source. Part A. https://doi.org/10.1080/15567036.2022.2038735.

Safder, M.U., Sanjari, M.J., Hamza, A., Garmabdari, R., Hossain, M.A., Lu, J., 2023. Enhancing microgrid stability and energy management: techniques, challenges, and future directions. Energies 16 (18). https://doi.org/10.3390/en16186417.

Seo, S.B., Go, E.S., Ling, J.L.J., Lee, S.H., 2022. Techno-economic assessment of a solar-assisted biomass gasification process. Renew. Energy 193, 23–31. https://doi.org/10.1016/j.renene.2022.05.033.

Shahgholian, G., 2021. A brief review on microgrids: operation, applications, modeling, and control. Int. Trans. Electr. Energy Syst. 31 (6). https://doi.org/10.1002/2050-7038.12885.

Shahzad, S., Abbasi, M.A., Ali, H., Iqbal, M., Munir, R., Kilic, H., 2023. Possibilities, challenges, and future opportunities of microgrids: a review. Sustainability 15 (8). https://doi.org/10.3390/su15086366.

Shrivastwa, R.R., Hably, A., Melizi, K., Bacha, S., Understanding Microgrids and Their Future Trends.

Singh, S., Singh, M., Kaushik, S.C., 2016. Feasibility study of an islanded microgrid in rural area consisting of PV, wind, biomass and battery energy storage system. Energy Convers. Manag. 128, 178–190. https://doi.org/10.1016/j.enconman.2016.09.046.

Singh, M., Singh Brar, Y., Singh, H., 2021. Critical assessment of biomass material for power generation in Punjab India. Mater. Today Proc. 48, 927–931. https://doi.org/10.1016/j.matpr.2021.04.414.

Singh, B., Bishnoi, S.K., Sharma, M., Singh, P., Dhundhara, S., 2023. An application of nature inspried algorithm based dual-stage frequency control strategy for multi micro-grid system. Ain Shams Eng. J. 14 (9). https://doi.org/10.1016/j.asej.2023.102125.

Srivastava, A.K., Latif, A., Shaoo, S.C., et al., 2022. Analysis of GOA optimized two-stage controller for frequency regulation of grid integrated virtual power plant. Energy Rep. https://doi.org/10.1016/j.egyr.2021.11.117.

Swain, N., Sinha, N., Behera, S., 2023. Stabilized frequency response of a microgrid using a two-degree-of-freedom controller with African vultures optimization algorithm. ISA (Instrum. Soc. Am.) Trans. 140, 412–425. https://doi.org/10.1016/j.isatra.2023.05.009.

Uddin, M., Mo, H., Dong, D., Elsawah, S., Zhu, J., Guerrero, J.M., 2023. Microgrids: a review, outstanding issues and future trends. Energy Strategy Rev. 49. https://doi.org/10.1016/j.esr.2023.101127.

Uzair, M., Li, L., Eskandari, M., Hossain, J., Zhu, J.G., 2023. Challenges, advances and future trends in AC microgrid protection: with a focus on intelligent learning methods. Renew. Sustain. Energy Rev. 178. https://doi.org/10.1016/j.rser.2023.113228.

Wang, G., Chen, Z., Wang, Z., Xu, Z., Zhang, Z., Liu, Y., Zhang, X., 2023. Research and implementation of frequency control strategy of islanded microgrids rich in grid-connected small hydropower. Energy Rep. 9, 5053–5063. https://doi.org/10.1016/j.egyr.2023.04.011.

Wu, Y., Cui, J.Z., Liu, C., 2023. State-of-the-art review on energy management systems, challenges and top trends of renewable energy based microgrids. EAI Endorsed Trans. Energy Web 10, 1–17. https://doi.org/10.4108/EW.4124.

Yang, F., Yu, S., Zhao, J., et al., 2023. Distributed reconfigurable model predictive control based load frequency control of microgrid clusters with dynamic topology. Electr. Power Syst. Res. https://doi.org/10.1016/j.epsr.2023.109445.

Yildirim, B., Razmi, P., Fathollahi, A., et al., 2023. Neuromorphic deep learning frequency regulation in stand-alone microgrids. Appl. Soft Comput. https://doi.org/10.1016/j.asoc.2023.110418.

Zhang, B., Li, Q., Wang, L., Feng, W., 2018. Robust optimization for energy transactions in multi-microgrids under uncertainty. Appl. Energy 217, 346–360. https://doi.org/10.1016/j.apenergy.2018.02.121.

CHAPTER

2

Frequency control challenges and mitigation strategies in microgrids

Samuel Raafat Fahim[1], *and Preeti Gupta*[2]

[1] Independent Postdoc Researcher, Ain Shams University, Cairo, Egypt;
[2] Department of EEE, UIET, Panjab University, Chandigarh, India

2.1 Overview of frequency control in microgrids

The fundamental principle for load frequency control (LFC) is based on coordinating electricity generation and consumption within the electrical grid to maintain frequency within its nominal range. For frequency stability, supply-demand balance is essentially required to be maintained in the real-time frame. This LFC principle, which is applied to large grids, is equally applicable to microgrids as well. Similar to large conventional power grids, frequency control in microgrids is implemented through three hierarchical mechanisms, namely primary, secondary, and tertiary control. This section briefly presents the three control types (Khalghani et al., 2021):

- *Primary control:* It is also known as droop control and is the initial response for small frequency deviations. In this control, the power output from generators is adjusted to compensate for load variations. It is done using governor systems. It is usually automatic and characterized by its fast-acting nature, in the first few seconds, to stabilize the frequency disturbance. For microgrids, this principle is expanded to inverter-based RES, which is programmed to emulate the droop characteristics of rotating machines (Sharma et al., 2025).
- *Secondary control:* It is also known as automatic generation control (AGC) and is characterized by its slow response (tens of seconds to minutes) that aims to restore system frequency to its nominal value.

Frequency Control in Microgrids
https://doi.org/10.1016/B978-0-443-36476-1.00004-3

It also manages the power flow through tie-in points present in interconnected systems. This control can coordinate the power output between several generators by sharing the loading more evenly, thus achieving the required operating point by the electrical grid, which applies to microgrids but also among the distributed energy resources (DER) (Singh et al., 2023).

- *Tertiary control:* It is a long-term response that may range from several minutes to hours. The main objectives of tertiary control include managing system unbalances and ensuring security. It also covers optimal power flow and economic dispatch. This is implemented via either generation schedule adjustments or deploying demand side management programs. In microgrids, based upon generation resources economic optimization, this control selects the operating modes that are either grid-connected or islanded (Abubakr et al., 2022).

Microgrid dynamics are more complex compared to traditional electrical grids; microgrids mainly contain DER and energy storage. Microgrid as a localized energy system, consists of several entities that are connected to each other either in a radial or loop manner. The main system components are the generators (DER in specific), energy storage systems, loads, and control systems. These components shall work together in harmony to ensure a reliable, stable, and optimized power supply. This section demonstrates each component briefly, which ultimately helps in building effective LFC (Abazari et al., 2019):

- *Generators:* They are the main power source of the electrical energy in the microgrid. They encompass all the generation technologies from the traditional fossil fuel types to the renewable resource's ones. The most famous renewable types are the photovoltaic panel and wind turbines, which are characterized by their inherently variable output. Depending on their geographical location, microgrids may also include small-scale hydroelectric and biomass-based generators. In microgrids, the LFC strategy is strongly influenced by the presence of inertia, which is inherently present in rotating generators. Therefore, systems dominated by inverter-based generators require special control strategies to ensure stable frequency regulation (Shahgholian, 2021).
- *Energy Storage Systems:* Microgrid systems typically have high penetration levels of renewable energy resources (solar and wind) that are associated with intermittency and cannot be fully dispatchable. Therefore, the presence of energy storage capacity is very important for the microgrid operation as it provides necessary buffering and flexibility to achieve the supply and demand balancing. A variety of energy storage options is now being adopted,

such as flywheels, super capacitors, and battery energy storage systems etc. Flywheels are more suitable for power-intensive and short-duration applications. Whereas, Super capacitors and BESSs are characterized with high energy storage density and fast dynamic response. BESS and SCSS possess the capability to substantially compensate and respond to the transient requirement of the system and regulate the frequency under load and generation variability. BESS can be used as a source of virtual inertia in the inverter-based generation and supports the system capability for fast frequency response (Sharma et al. 2022a).

- *Loads:* There are several types of loads connected to the power networks. These can be broadly classified into critical and non-critical categories. Critical loads, like hospitals, telecom networks and data centers, have strict frequency, voltage regulation, and power reliability. Critical loads are prioritized as per their importance and must be kept running. On the other hand, non-critical loads are the potential demand response resource for the load managing entities and can be used to balance the local system during islanded operation or otherwise. The integration of smart meters and the Internet of Things (IoT) enables LFC to deploy sophisticated load control strategies and demand response programs, allowing frequency control in real time. Further, electric vehicles (EVs) can be utilized in dual mode; mainly, they are a load that needs a grid connection for charging. They can also be used as a source of stored electrical energy via vehicle-to-grid (V2G) technology. Giving LFC more versatile control solutions (Keivanimehr et al., 2021).
- *Control Systems:* In the microgrid operation, the control unit plays the pivotal role in enabling LFC to achieve its design objectives. Accordingly, several design architectures, such as centralized, decentralized or local are employed (Keivanimehr et al., 2021; Sharma et al., 2022b). The control unit can operate in different modes, such as master-slave, peer to peer, and hybrid configurations. It can use diverse control methodologies without compromising system reliability, which may or may not employ a communication framework. The control structure is typically organized into hierarchical levels, namely primary, secondary, and tertiary control (Dhundhara and Verma, 2020). A more comprehensive overview of the control system from various points of view is depicted in Fig. 2.1 Fig. 2.1.

In a decentralized architecture, individual controllers (Local controllers) are assigned to each DERs and major load centers. Dedicated separate local controllers are responsible for maintaining the performance of the local set points within pre-set limits. Also, responding to the other control signals

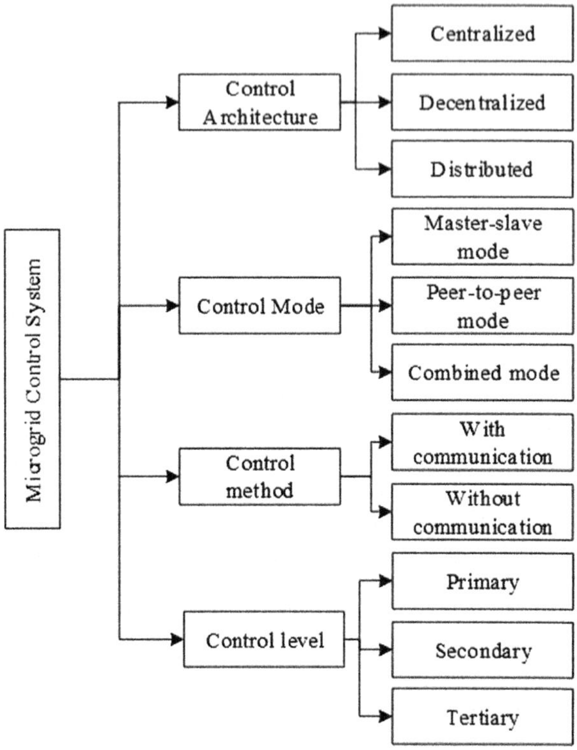

FIGURE 2.1 Microgrid control systems overviews.

from the system. The local controllers have no intercommunication link. Distributed controllers are interconnected local controllers (for subsystems) via communication protocols. In this control architecture, each subsystem/microgrid works independently, but can coordinate its own limits based on the data received from other subsystem/microgrids. Centralized controllers are also called microgrid central controller (MGCC). This overarching control type coordinates all the local controllers of the microgrid components. The MGCC, as a part of the LFC strategy, uses secondary and tertiary control to adjust the power generation and manage generation among RES, thus achieving frequency stability:

- *Energy management system:* This control system works in conjunction with MGCC to optimize the overall operation of the microgrid. It incorporates economic dispatch and load forecasting both, thereby indirectly influencing the performance of LFC.
- *Communication infrastructure* is essential to the control systems to have a communication network that is characterized by robustness and low latency.

Microgrid's DER such as PV systems, wind turbines, and battery energy storage systems, are classified s inverter-based resources. These resources rely on power electronic interfaces to connect to the electrical grid, in contrast to conventional synchronous generators that possess inherent rotational inertia. High inertia of large synchronous generators is very important to resist is frequency deviations in the system caused by unbalances between generation and consumption. RES lack the natural inertia effect of traditional synchronous generators. Renewable resources like wind and solar are integrated to the system through power electronic converters. These devices are free from mechanical inertia. Consequently, the adoption of virtual inertia concepts and fast-control mechanisms becomes essential to maintain overall power system stability and reliability (Santra and De, 2023).

The increasing penetration of renewable energy generation further introduces significant variability due to its decentralized deployment and intermittency issues. The integration of energy storage systems provides an effective means to mitigate these challenges; however, their benefits can be fully realized through fast and coordinated control strategies. Moreover, microgrids have different modes of operation, namely grid-connected or island mode, and may dynamically transit between them. Therefore, LFC schemes must be adaptive and flexible, accounting for both modes of operation and enabling an easy and smooth transition between them while maintaining frequency stability (Sepehrzad et al., 2022).

In Table 2.1, a quick summary of the functions of the three control hierarchies: primary, secondary, and tertiary is presented. In a microgrid dominated by inverter-based DERs, the LFC must compensate for the physical lack of inherent inertia. Some techniques could be accounted for in this regard. For emulating the synchronous generator inertia, both Virtual Synchronous Machine (VSM) and Droop-Based Virtual Inertia (DBVI) could be of great help. While techniques like Model Predictive Control (MPC) and Adaptive PI/PID controllers can be used to robustly face uncertainty, especially when dynamic response is addressed. The aforementioned techniques mitigate frequency deviation in a short-term window. Unfortunately, their performance is affected by communication delays, along with scalability issues and modeling inaccuracies. Therefore, moving forward toward an intelligence-based and data-driven control paradigm/framework became imperative.

2.2 Key issues in microgrid frequency control

This section discusses the key issues that impact the overall operation of a microgrid and play a major role in aggravating the load frequency

TABLE 2.1 – Function summary for hierarchical control.

#	Feed-back signals	Feed-in signals	Control Hierarchy	Function
	From microgrid	To microgrid	Primary	• Give provision for voltage stability, • Preserving frequency stability, • Enable plug & play capability of DERs, • Protection of devices, • Avoiding the circulation of currents among DERs.
	From microgrid	To primary control	Secondary	• Voltage deviation compensation caused by primary control, • Frequency deviation compensation caused by primary control, • Grid synchronizing.
	From both microgrid & main power grid	To secondary control	Tertiary	• Power flow control in grid-tied mode, • Optimal operation in both operating modes (on-grid/off-grid).

regulation problem. In general, it has been considered that the microgrid possesses renewable-based generation capability (wind, solar) only.

2.2.1 Low inertia and reduced damping

Microgrids dominated by renewable generators, especially wind and solar provide little or no rotational inertia. Although non-synchronous RES offer many advantages such as decarbonization, reduced greenhouse effect, and electrification of remote areas, their large-scale replacement of conventional generators introduces critical operational challenges. The reduction in system inertia complicates frequency regulation, leading to an increased rate of change of frequency (RoCoF). Without mechanical inertia, any imbalance between generation and load leads to rapid frequency excursions, reducing system stability margins. Low system inertia increases the vulnerability of the power system to disturbances such as faults, short circuits or component outages. The resulting transient processes take place rapidly and cause large deviations from nominal conditions. It complicates the function of automated protection and control system to respond effectively, thereby increasing the risk of instability and component damage. The relationship between system inertia and frequency dynamics is described by the swing equation (Kundur et al., 1994), given as:

$$\frac{d\omega}{dt} = \Delta P \frac{\omega s}{2H} \tag{2.1}$$

Here ΔP is the difference between input mechanical power and electrical power output (per unit). ω_s is angular synchronous speed in radians/seconds, and H represents system inertia in seconds.

Eq. (2.1) reveals that a decrease in inertia results in an increased rate of change of frequency. It is indicated that for same magnitude of power imbalances larger frequency drops will take place within shorter time periods, leading to several operational challenges. It reduces the power system stability margin and makes the system prone to instability. An increase in RoCoF may lead to triggering of protective relays and a cascaded effect can result in multiple trippings and system blackout. In addition, the conventional frequency control techniques lose their efficacy, as these techniques rely mainly on the electromechanical response of synchronous machines. This necessitates that RES provide support the power system under disturbances and play an active role in frequency regulation. This mandates the shift toward synthetic/virtual approaches. For example: virtual synchronous machines, droop-based virtual inertia, and fast-acting inverter controls (Magdy et al. 2021) (Mohan et al., 2023).

Another adverse effect of reduced inertia is the enhanced complexity of interaction among various DERs and microgrids in multi-microgrid systems. It often leads to suboptimal power sharing among generators and other multiple incidents, such as unintended resonance phenomena and an increase in the magnitude of oscillations. To effectively manage these challenges and maintain frequency stability in the face of sudden disturbances; integration of fast local control actions with hierarchical coordination is essentially required. This would further enhance system resilience (Habibullah et al., 2021).

2.2.2 Intermittency of renewable generation

Solar and wind power are inherently variable and unpredictable. Their stochastic nature leads to frequent frequency oscillations, making it difficult to maintain steady-state stability. The variability demands fast-acting control and energy storage solutions. The nature of the generated power is influenced by the stochastic nature of the weather conditions. In return, the output as well has fast transients. Thus, the deterministic classical control is insufficient in capturing the dynamics simultaneously, especially if long-term stability is targeted (Rameshrao et al., 2022). To buffer/absorb these fast transients, energy storage systems (ESS) are introduced to the network. The super-capacitors and/or batteries are examples of ESS. They can handle short-term imbalances as well. On the other hand, their limited capacity and degrading characteristics limit continuous reliance on them.

Consequently, in addition to energy storage capabilities, fast-acting and adaptive control strategies are often required to achieve operational objectives. Advanced control schemes such as model predictive control and adaptive inverter-based frequency support provide enhanced dynamic performance (Patwary et al., 2024). Further, connecting renewable microgrids at different nodes of a power network exposes the system to simultaneous generation fluctuation. These fluctuations are generally diversified in both spatial and temporal dimensions, and they interact in a very complex and nonlinear manner. This leads to potential amplification in frequency deviations. As a result, probabilistic forecasting and scenario-based optimization have gained considerable attention from both the network planner and system operators. These tools provide an anticipation of the likelihood and impact of critical events. Moreover, the presence of advanced sensor networks aided with high-speed communication facilities, enables coordinated real-time corrective actions to effectively manage these variations and enhance system stability (Dhundhara and Verma, 2018).

2.2.3 Lack of coordinated control

Microgrids as a concept, are based on connecting the generation side, which could be DER (distributed energy resources), and the demand side, which is the grid consumers. This is done through controllers that are required to coordinate the operation of the two sides (Magdy et al., 2018). Both the generation as well as demand sides have an inherently variable nature. On the generation side, it could be of renewable origin, which is inherently intermittent. While the demand side behavior is affected by the consumer's time-dependent demands.

Here, controller action is to balance the electricity demand with the available generation. Another key factor is the controller configuration; that may be centralized or decentralized. In the decentralized topology of microgrids there are be multiple entities. This further adds to the operational complexity. Nevertheless, any mismatch in the generation and demand that could directly affect the overall electrical performance of the microgrid. The imbalance could be deepened by the lack of coordination between the different controllers. Microgrid systems already have the difficulty of low inertia introduced by the inverter-based renewable generation. Therefore, the role of load frequency control becomes more challenging, which must harmonize the performance of the three parts of the microgrid. As uncontrolled interaction among microgrid components will eventually lead to bottleneck instances and compromises the grid reliability and voltage-frequency stability.

There are factors that affect the coordinated control, as highlighted previously; the presence of multiple control entities within the same microgrid introduces the need for a robust communication structure. Also, this can introduce communication latency and result in uncoordinated behavior of multiple controllers. Further the ill-designed communication structure will lead to congestion in the data transfer throughout the microgrid. There should be an overarching coordination control, via several local controllers, to address system disturbances and prevent conflicting actions. Thus, pushing away, the poor power sharing, voltage/frequency excursions, circulating currents, oscillatory behavior, and/or load imbalance. In the same regard, grid storage capacities can be prone to premature depletion or underutilization. By keeping most of these under control, the microgrid stability can be ensured (Abbasi et al., 2023).

The intermittency of the renewable energy is unavoidable, especially in the short-term window. However, the generating units spread spreading at different locations in the microgrid may be different weather conditions. Also, the generation from solar and wind units can also be affected differently. Therefore, coordination among different generation units geographically distributed along the microgrid, together with the ESS, is crucial for mitigating the impact of generation uncertainty and load variability. The role of demand-side resources for providing required flexibility is also very important. Demand side flexibility reduces the energy storage capacity requirements by compensating for generation variability and reducing overall power mismatch (Acharya et al., 2018).

2.2.4 Communication latency and cybersecurity

Frequency control in modern microgrids depends heavily on communication networks for distributed coordination. Latency, packet loss, or cyber attacks can degrade control performance or jeopardize system reliability. As discussed in earlier sections, the need for a sophisticated control structure is extremely important.

An added set of challenges is introduced by cyber-attacks that can cause data spoofing, denial-of-service, or unauthorized control interventions (Kondoro et al., 2021). To manage these risks, a multifaceted approach is essential, like robust control algorithms that are capable of tolerating communication delays. Along with this, a resilient network structure that is complete with redundant routes/controllers and anomaly detection systems is also very important. Adoption of secure authentication protocols for command and telemetry signals is also essentially required to reduce the cyber-threats. Both the control strategies, distributed or decentralized, can enhance the resilience, in comparison with the centralized scheme (Singh et al., 2025). Nevertheless, achieving an effective balance between cyber-security, communication delays and control design requires a holistic approach. The frequency regulation in the microgrid is directly influenced by the underlying communication infrastructure, making operational performance and secure data exchange inseparable (Hoummadi et al., 2024).

2.2.5 Economic and policy constraints

One of the major constraints in the microgrids is the heavy cost of components, which includes ESS, DERs, power electronic interfaces, phasor measurement units, and cyber-security devices. Usually, their prices exceed the budget of small/rural microgrids. This puts the decision makers in a harsh trade-off between basic reliability performance and meeting the sophisticated requirements of frequency regulations. These constraints inhibit the deployment of robust microgrids in many areas (Billanes et al., 2025).

Another key issue is the maintenance cost of the installed hardware and software license renewal of the LFC in microgrids. Also, this includes the prices of replacing the controllers and sometimes the hardware itself (ex, degraded batteries), coupled with the fuel cost of diesel generators and monitoring systems. The microgrid operators must consider revenue streams either from tariff-based cost recovery or from formal ancillary service payments. If the cost recovery mechanisms are unclear, the microgrid operators may face financial uncertainty. This inhibits the adoption of required advanced LFC control schemes (Khooban et al., 2017). Unfortunately, only microgrids with large synchronous machines have been included in frequency regulation markets. While the inverter-based and/or small-scale microgrids are excluded. This gap means that not all types of microgrids can be monetized. This, in return, weakens the frequency support capabilities, in case fast response devices were needed. This makes the LFC consider only low-cost controllers that are based on PID schemes, rather than intelligence-based control schemes (Abbood, 2025). A microgrid lies in a gray regulatory zone, as it is not a utility-scale grid. It is rather an independent grid that can either be combined with the main grid or work independently. This limits the decision makers when reviewing its compliance to the utility grid connection specifications, in regard to frequency limits, inertia contribution, protection coordination, and finally it's interconnection. This makes the microgrid operators stick more to conservative strategies to align more with interoperability (Amir et al., 2024).

Environmental and emissions regulations add up to the economic constraints. All these have pushed the microgrids to adopt inverter-based generation (i.e., renewables). And replace the fossil fuel units, which are rich in inertia. In return, this mandates poor inertia resources to produce synthetic inertia, to cope with the fast dynamics of the microgrid and keep grid frequency within safe limits. Further, to provide sufficient flexibility to the system this shift requires widespread ESS capacities to accommodate variability and uncertainty of both load and generation. However, implementation of all the above remedies has economic implications as these require huge investment. In the absence of modern and sophisticated control equipment and infrastructure for LFC, microgrids with large renewable penetration are prone to disturbances and frequency excursions beyond specified range every now and then. Therefore, frequency stability, reliability and quality of power supply are at a great risk in such cases (Rasolomampionona et al., 2024). Nevertheless, the development of a standardized framework requires coordinated efforts among various parties, including technology developers, system operators, and regulatory bodies. Leading to establish a market with clear structures, develop cost recovery mechanisms, and performance standards. Developing innovative financing models, sharing infrastructure investments, and the presence of regulatory support will eventually lead to bridging the gap between technical capability and practical implementation (Hoummadi et al., 2024).

2.3 Identified gaps and emerging needs/remedies/techniques to handle

2.3.1 Gap#1: Low Inertia System Challenges

Inertia reflects the capacity of the system to sustain disturbances and is vital for stable system operation. To avoid frequent frequency oscillations caused by unbalances between electricity generated and load, it is very much required that system inertia remains high.

Microgrid concept has provided a thrust to the utilization of renewable electricity, and their penetration in the power system is increasing. The shifting of the microgrid topology toward inverter-based DERs introduced a new challenge to the designers/operators of microgrids. This is mainly due to the lack/loss of rotational inertia, which is a key feature in traditional fossil fuel synchronous generators. This inertial loss impacts the frequency stability of the microgrid. This leads to a new set of challenges in regard of frequency regulation, including increasing the Rate of Change of Frequency (RoFoC). The problem further aggravates when the microgrid experiences a sudden disturbance due to load/generation changes. In these situations restoring frequency within specified operational band becomes more challenging and slower. This may result in large deviations from the preset frequency margins. If the system is not properly damped or damping is insufficient, the frequency fluctuations may persist for longer duration. This can lead to unwanted load shedding and affects overall system reliability. Even in using synthetic inertia capabilities from the inverter-based resources, the frequency levels must be precise to avoid the occurrence of any unwanted events. In addition, even if the frequency measurements are precise, they are still vulnerable to signal noise distortion (Agupugo et al., 2024).

In traditional microgrids, the damping of frequency fluctuations relies on the fast responding gas turbines or hydroelectric facilities. But with the shift toward HRES, the battery energy storage systems along with flexible and smart demand side management, are expected to play that role. However, although BESS with bidirectional power flow capability provides flexibility quite effectively but they also have limitations. The batteries have limited capacities, and they are prone to degradation due to continuous cycles of charging/discharging. In addition, the demand side management needs robust communication infrastructure to harness expected benefits (Nema et al., 2022). In addition to the above, the communications and cyber-security arise as an additional challenge. As the large reliance on communication, to counter the low inertia system, make latency, especially in losing the small packets of data increases the response time and eventually jeopardizes overall performance. Along with this is the cyber-security that can cause unauthorized interference and deepen the imbalance/mismatch (Yusuf et al., 2025). The operating modes also bring another challenge to the microgrid performance. As in islanded mode, the microgrid with large penetration of renewable DERs, and limited ESS capacities will face severe frequency instability, especially when the transition is sudden. This must be backed up with sophisticated LFC, based on AI, to enable the microgrid to restore its balanced operation without overloading resources or depleting the ESS reserve (Alizadeh et al., 2019).

Addressing all the above-mentioned concerns properly results in complex system modeling, and sophisticated modern techniques and algorithms are required to provide accurate and efficient results in real-time operation (Abubakr et al., 2022; Rajamand, 2021).

Emerging needs

- Virtual or synthetic inertia using inverter control algorithms.
- Fast-acting storage systems such as batteries or flywheels.

- Adaptive droop control to dynamically tune frequency response.

2.3.2 Gap#2: Intermittency and Uncertainty of RES.

Electricity generated and consumed must be balanced in the power systems at all time periods to maintain stability. Any generation-load mismatch leads to frequency deviations from its nominal value. Therefore, for frequency stability, it is mandatory that supply supply-demand balance is maintained continually. In addition, large frequency variations from the set value can cause the operation of frequency relays, resulting in unintended load shedding. Thus continuous power balance is very important for power system frequency stability and reliability.

Conventionally, frequency stability, which is directly associated with supply-demand balancing, is achieved by adjusting the generation in response to load variations. In the renewable-dominated systems, the implementation of this approach alone may not be sufficient under varying operating conditions. With the increased penetration of RES share of synchronous generators declines, resulting in limited power reserve capacity for effective primary and secondary frequency control response.

Also, the conventional fossil fuel generation units and renewable generators are fundamentally different in the sense that the electricity generation from the former is predictable and dispatchable; while the renewable generation is associated with a high degree of stochasticity (Gupta et al., 2022). The power outputs from solar and wind units are directly influenced by the weather conditions and their accurate predictability is challenging. Although advanced weather forecasting systems are now available but still the uncertainty and the intermittency cannot be completely eliminated. Therefore, conventional control strategies such as proportional control, are insufficient in maintaining the frequency variations within desired range. These concerns necessitate the use of advanced preventive and predictive control strategies for a dependable and stable system operation. Although, during the past few decades, this has attracted a lot of attention from both the research community and industry, and large advancements in technology is also occurring in this direction, still there remains a scope in this field (Hu et al., 2024). Hybrid control strategies like droop-based control along with model predictive control (MPC), can be utilized to forecast the generation shortfalls/surpluses. Based upon that, surplus energy can either be dispatched to storage or the inverter output can be adjusted accordingly. These strategies balance generation variability and positively influence real-time frequency dynamics (Shobug et al., 2024).The stochastic optimization framework can also be used to forecast uncertainty (Khokhar and Singh Parmar, 2024).

Emerging needs

- Predictive control leveraging weather and load forecasts.
- Integrated storage for proactive balancing.
- Hybrid control blending droop and MPC techniques.

2.3.3 Gap#3: Limitations in Coordinated Control

The existing centralized control architecture faces limitations regarding scalability, reliability, and robustness. They also require high-speed communications and are prone to heavy computational burdens, which are affected exponentially with the number of DERs. This eventually creates bottlenecks and single points of failure (Khosravi et al., 2025). Considering the growth in size and complexity of microgrids, the limitations lead to delayed corrective actions. One of the solutions is the emerging framework named multi-agent system (MAS), which enables the coordination of individual DERs, as agents, following consensus protocols toward achieving global objectives. MAS framework has many advantages such as: reduced burden on centralized computation, improved fault tolerance limits, and permitting local decision-making based on information from the neighborhood. This helps in maintaining frequency stability even if partial communication failures occur. The event triggers communication strategies, enabling the data transition when there is a significant deviation occurs. This adheres to the available bandwidth capacity, prevents data congestion, and minimizes latency-induced errors. On the other hand, decentralized control architecture prevents the spreading of misbehaving DERs throughout the entire system. Thus, maintaining the microgrid's overall performance (Saxena et al., 2024).

Emergng needs

- Multi-Agent Systems (MAS) for distributed consensus-based coordination.
- Event-triggered communication to reduce data overhead.
- Fault-tolerant decentralized architectures.

2.3.4 Gap#4: Integration of Storage and Inverter-Based Systems

Solar photovoltaic generation and wind generators predominate electricity generation in the microgrids. However, solar and wind are intermittent sources of energy, resulting in fluctuations in electricity generation. Therefore, their integration into a reliable, stable power grid is a major challenge.

Introducing inverter-based DERs to microgrids mandates the deployment of energy storage systems. Moreover, the synergetic operation between the two is pivotal in regard to frequency stabilization. But this integration comes with real technical and operational challenges. On the other hand, inverter-based DERs need precise synchronization and robust control. It is essential to avoid operational conflicts and ensure effective power sharing. Although ESS provides rapid active power support to the system, it has capacity constraints. Also, they are prone to degradation and have a limited intervention scale. Addressing these challenges requires intelligent storage management algorithms. This optimizes the charging/discharging cycles, which prolongs the device lifespan (Zuo and Wu, 2022). Integrating a VSM enables the inverter to emulate the actual synchronous machine inertia and damping characteristics. This enhances the ESS performance in regard to dynamic frequency support (Izmirlioglu et al., 2024). Integrating multiple inverter-based DERs with ESS creates a hybrid resource that both offers resilience and support to the microgrid. This requires predictive control and high-fidelity state estimation. IT provides resilience to the microgrid in both normal conditions and in contingencies. Eventually, it will mitigate the voltage/frequency oscillations. Finally, this transforms the performance of inverter-based DERs and ESS from reactive to proactive stabilizing agents (Hasheminasab et al., 2024).

Emerging needs

- Intelligent storage management algorithms.
- Virtual synchronous machine (VSM) integration.
- Hybrid resource optimization frameworks.

2.3.5 Gap#5: Cyber-security and Communication Constraints

For their efficient and reliable working, microgrids depend greatly on the communication networks. Heavy utilization of remote sensing and control; and information and communication facilities is inevitable in the overall microgrid operational management. This introduces additional challenges such as data congestion and latency in the communication infrastructure. In addition, this makes the whole system vulnerable to cyber-security attacks. The communication system design shall balance between operation reliability, system security, and real-time responsiveness (Shobug et al., 2024).

Emerging needs

- Secure communication protocols and redundancy.
- Cyber-resilient control algorithms.
- Real-time monitoring and anomaly detection.

2.3.6 Gap#6: Lack of Standardization and Policy Support

The development and wide-spread implementation of advanced frequency control technologies depends on developing regulatory frameworks, market incentives, and technical standards. Also, the presence of compensation mechanisms enhances the investment and adoption of frequency ancillary services (Hossain et al., 2024). Defining clear control standards between regulators, utilities, and technology developers promotes best practices and establishes economic frameworks. To ensure the interoperability of various DERs, there is a need to develop standardized regulations and technical frameworks. Thus, to enable coordinated control and introduce incentivized participation for microgrids' ancillary service provision (Mutluri and Saxena, 2024).

Emerging needs

- Regulatory standards for DER integration.
- Market incentives for frequency ancillary services.
- Interoperable communication and control standards.

2.4 Emerging technological solutions

This section presents some of the emerging technological solutions, in a summarized form, that enable microgrid designers and operators to have a better microgrid with coordinated LFC performance compared to the traditional solutions. Among the technologies are the following:

- **Adaptive/non-linear strategies:** These strategies influence real-time learning and system feedback to maintain stable operation under uncertain and dynamic conditions. Sliding-mode control (SMC), model-predictive control (MPC), adaptive neuro-fuzzy control (ANF), and nonlinear Lyapunov-based schemes are some of the commonly implemented methods. The controller parameters are continuously updated to respond to load variations, renewable intermittency, and changes in network topology. MPC can provide control actions that are based on predictive future disturbances, while SMC provides robust control actions that can adapt the continuously varying parameters of microgrid. By accounting for nonlinear dynamics of inverter-based resources and distributed generations, these approaches are more accurate than linear methods, enhancing robustness against disturbances and improving transient response. Overall, adaptive nonlinear controllers provide a flexible, resilient, and high-performance solution for maintaining frequency stability in modern microgrids with high renewable penetration (Trivedi et al., 2022).

- **AI/ML-based Controllers:** AI/ML techniques play a crucial role in enhancing microgrid frequency stability. Using AI/ML, more accurate and faster algorithms can be designed, which give better solutions than the conventional approaches under variable, real-time operating conditions. For example, traditional droop control methods often face accuracy and performance degradation issues in real-time dynamic operating conditions. In contrast, machine learning techniques-such as artificial neural networks (ANNs) and reinforcement learning (RL) are more efficient and work effectively under real-time operation. AI techniques can be utilized to adjust droop coefficients in real time and provide faster frequency regulation. There are several AI-based heuristic optimization techniques available in the literature that can be effectively used to finely adjust controller parameters, resulting in better performance. Various studies have proved that the controllers designed using AI/ML techniques have considerably reduced settling time, smaller overshoots, and better frequency regulation.

In general, the application of AI, especially machine learning (ML), is increasing in multiple areas of the power system operation and control. ML provides tools for decision making and hence control actions, based upon real-time operational data. This makes them more suitable for microgrids with having higher penetration level of intermittent renewables. The ML-based algorithms give more accurate predictions of load variability and generation forecasts of renewable resource availability. This provides an opportunity to better estimate the overall variability and hence, manage it before it escalates or propagates in the microgrid (Suryani et al., 2024; Joshal and Gupta, 2023):

- **Grid Forming Inverters:** These are voltage source inverters and play very important role in islanded operation of microgrids. These types of inverters are suitable for microgrids that have a large penetration of renewable energy generation. Systems incorporating a significant proportion of renewable resources possess reduced inertia and thus, exhibit poor frequency regulation. However, power electronic converters used in grid forming control (GFM) mode exhibit voltage source characteristics. Along with ESS installed within the

microgrids, such as flywheel, lithium-ion batteries, etc., these inverters can be controlled in such a manner so as to compensate for mechanical inertia. They are equipped with specialized algorithms that emulate the virtual inertial effects of synchronous machines and help in stabilizing frequency disturbances. They modify the active power output in response to load variations and hence reduce frequency deviations. They can also be utilized for droop control by adjusting output active/reactive power to regulate frequency/voltage like synchronous generators (Kumar et al., 2022).
- **Hybrid Control Architectures:** The traditional load frequency control approach is to adjust generator output to balance load variations, ensuring system stability. Any variation in load leads to speed variation or change in frequency, which is detected by the speed governor. Generator speed governor mechanism detects the load variations and responds quickly by adjusting turbine valve opening to stabilize frequency deviations. Followed by secondary control action for the sustained fluctuations and errors through AGC/LFC mechanisms. These mechanisms work upon simple deterministic mathematical models and provide clear decisions to adjust generator output to neutralize the generation-load imbalance. While in microgrid systems, in addition to load variations, there can be fluctuations and unpredictability in generator output also. In such situations, to ensure generation-load balance, intelligent control methods are more efficient and accurate in comparison to conventional approaches. They use operational historical values and use ML to learn from past data and devise patterns. However, these are more complex and require technological and infrastructural sophistication.

Nevertheless, a judicious combination of traditional and intelligent control strategies allows the system to take the benefit of the strengths of both approaches. This approach is termed as hybrid control, where both methods are employed in a coordinated manner. Traditional controls provide fast and deterministic responses, while intelligent controls contribute long-term adaptive optimization Ahmad et al., (2025):

- **IoT Integration:** Microgrids have a high share of renewables, which are intermittent and are characterized by low inertia. This results in increased frequency deviations than in conventional grids. Microgrids with higher levels of RES share often possess little flexibility on the generation side. Under such situation, coordinated control of distributed resources and loads becomes essential for reliable and stable system operation. Advancements in IoT technologies provide the required observability and helps in implementing control actions in real time. Extensive deployment of

IoT technologies and sensor networks in microgrids facilitates the generation and continuous transmission of important operational parameters. Thus, provides critical data for improving the microgrid overall performance.

IoT technologiesmake it possible to continuously monitor the operating conditions of the system; thus enabling real-time data-driven decision-making. In other words, it improves situational awareness, remote sensing, and forecasting. IoT also facilitates the implementation of demand side management and market participation in microgrids by utilizing smart metering, allowing bidirectional interaction, and distributed control. IoT helps in monitoring and transmitting important operational data throughout the microgrid, such as voltage levels, frequency, power quality, and the state of charge (SoC) of ESS. The offered high-speed data exchange helps in achieving more coordinated LFC decisions Babu et al., (2024):

- **Digital Twins (DTs):** Digital twin refers to the technique of creating a virtual replica of the actual system comprising complete functional details, and provides operational information about the system as if it were actual system or product. IoT acts as the enabling technology for DTs. Principally, a digital twin is a cloud-based application that uses IoT framework. In other words, IOT infrastructure forms the primary layer of the DT architecture. Recently, this technology is also gaining importance in various applications of power system, mainly for fault diagnosis, online analysis and smart grids. DT technology, when integrated with the smart grid carries out simulation and analysis in addition to the primary IoT ability of monitoring and control. As DTs replicate the actual system through physics-based model, they are capable of mimicking the exact output given the same input. Therefore, these can be effectively deployed to identify any events of frequency instability when applied in conjunction with the frequency regulation mechanism. In addition, optimization program can be designed using DT to test a number of droop coefficient values to give the optimal value. Such optimization can be used to provide the necessary control signals for adjusting the generator droop coefficients (Bazmohammadi et al., 2022) (Dev et al., 2025).

All these techniques help in shifting the frequency control in microgrids from just responding to disturbances to actively predicting, optimizing, and mitigating them. The points mentioned in section 2.4 must be deployed effectively to have a resilient, scalable, and intelligent microgrid, and can sustain frequency stability under various operating conditions and with high renewable penetration.

2.5 Conclusions

Microgrid frequency instabilities challenge the reliability of microgrids because of the increasing number of renewable energy penetrations, the decrease in system inertia, and the limited controllability of distributed energy resources. These characteristics lead to rapid frequency transients during disturbances and load changes. The development of microgrid frequency controls can be linked to advances made in all three arenas: the integration of renewables into electric grids, advances in technologies, and advances in control theory. The gaps currently identified (i.e., low inertia and cybersecurity risk) indicate that there exists an urgent requirement for intelligent adaptive-coordinated control systems. Continued research should emphasize the development of co-designed control-communication systems, integration of artificial intelligence-based optimisation, and development of suitable regulatory and market structures. Collectively, these innovations represent critical steps toward creating microgrids that operate in a reliable, sustainable, and secure manner, as well as toward developing future electricity systems that are capable of tolerating significant levels of renewable energy generation.

References

Abazari, A., Monsef, H., Wu, B., 2019. Coordination strategies of distributed energy resources including FESS, DEG, FC and WTG in load frequency control (LFC) scheme of hybrid isolated micro-grid. Int. J. Electr. Power Energy Syst. 109, 535–547. https://doi.org/10.1016/j.ijepes.2019.02.029.

Abbasi, M., Abbasi, E., Li, L., Aguilera, R.P., Lu, D., Wang, F., 2023. Review on the microgrid concept, structures, components, communication systems, and control methods. Energies 16 (1). https://doi.org/10.3390/en16010484.

Abbood, Z.A., 2025. Enhanced PV network controller for optimizing performance of LFC power systems. Babyl. J. Netw. 2025, 1–13. https://doi.org/10.58496/bjn/2025/001.

Abubakr, H., Guerrero, J.M., Vasquez, J.C., Mohamed, T.H., Mahmoud, K., Darwish, M.M.F., Dahab, Y.A., 2022. Adaptive LFC incorporating modified virtual rotor to regulate frequency and tie-line power flow in multi-area microgrids. IEEE Access 10, 33248–33268. https://doi.org/10.1109/ACCESS.2022.3161505.

Acharya, S., El Moursi, M.S., Al-Hinai, A., 2018. Coordinated frequency control strategy for an islanded microgrid with demand side management capability. IEEE Trans. Energy Convers. 33 (2), 639–651. https://doi.org/10.1109/TEC.2017.2763186.

Agupugo, C.P., Obatoyinbo Ajayi, A., Nwanevu, C., Oladipo, S.S., 2024. Policy and regulatory framework supporting renewable energy microgrids and energy storage systems. Eng. Sci. Technol. J. 5 (8), 2589–2615. https://doi.org/10.51594/estj.v5i8.1460.

Ahmad, G., Hassan, A., Islam, A., Shafiullah, M., Abido, M.A., Al-Dhaifallah, M., 2025. Distributed control strategies for microgrids: a critical review of technologies and challenges. IEEE Access 13, 60702–60719. https://doi.org/10.1109/ACCESS.2025.3552940.

Alizadeh, G.A., Rahimi, T., Nozadian, M.H.B., Padmanaban, S., Leonowicz, Z., 2019. Improving microgrid frequency regulation based on the virtual inertia concept while considering communication system delay. Energies 12 (10). https://doi.org/10.3390/en12102016.

Amir, M., Zaery, M., Singh, K., Suhail Hussain, S.M., Abido, M.A., 2024. Enhancement of frequency regulation by TFOID controller in hybrid renewable energy with battery storage system-based multi-area microgrids. IEEE Access 12, 110813–110828. https://doi.org/10.1109/ACCESS.2024.3439738.

Babu, V.V., Preetha Roselyn, J., Nithya, C., Sundaravadivel, P., 2024. Development of grid-forming and grid-following inverter control in microgrid network ensuring grid stability and frequency response. Electronics 13 (10), 1958. https://doi.org/10.3390/electronics13101958.

Bazmohammadi, N., Madary, A., Vasquez, J.C., Mohammadi, H.B., Khan, B., Wu, Y., Guerrero, J.M., 2022. Microgrid digital twins: concepts, applications, and future trends. IEEE Access 10, 2284–2302. https://doi.org/10.1109/ACCESS.2021.3138990.

Billanes, J.D., Jørgensen, B.N., Ma, Z., 2025. A framework for resilient community microgrids: review of operational strategies and performance metrics. Energies 18 (2). https://doi.org/10.3390/en18020405.

Dev, A., Kumar, V., Khare, G., Giri, J., Amir, M., Ahmad, F., Jain, P., Anand, S., 2025. Advancements and challenges in microgrid technology: a comprehensive review of control strategies, emerging technologies, and future directions. Energy Sci. Eng. 13 (4), 2112–2134. https://doi.org/10.1002/ese3.2095.

Dhundhara, S., Verma, Y.P., 2018. Capacitive energy storage with optimized controller for frequency regulation in realistic multisource deregulated power system. Energy 147, 1108–1128. https://doi.org/10.1016/j.energy.2018.01.076.

Dhundhara, S., Verma, Y.P., 2020. Application of micro pump hydro energy storage for reliable operation of microgrid system. IET Renew. Power Gener. 14 (8), 1368–1378. https://doi.org/10.1049/iet-rpg.2019.0822.

Gupta, A.K., Singh, R.K., Mitra, U., 2022. A Study of real-time Human machine interface for scheduling smart microgrids. Int. Conf. Emerg. Trends Eng. Med. Sci. 135–140. https://doi.org/10.1109/ICETEMS56252.2022.

Habibullah, M., Bhumkittipich, K.N., Mithulananthan, N., Sharma, R., Zare, F., 2021. Damping oscillation and removing resonance in a RE based DC microgrids. IEEE Access 9, 163516–163525. https://doi.org/10.1109/ACCESS.2021.3135033.

Hasheminasab, S., Alzayed, M., Chaoui, H., 2024. A review of control techniques for inverter-based distributed energy resources applications. Energies 17 (12). https://doi.org/10.3390/en17122940.

Hossain, M.T., Hossen, M.Z., Badal, F.R., Islam, M.R., Hasan, M.M., Ali, M.F., Ahamed, M.H., Abhi, S.H., Islam, M.M., Sarker, S.K., Das, S.K., Das, P., Tasneem, Z., 2024. Next generation power inverter for grid resilience: technology review. Heliyon 10 (21). https://doi.org/10.1016/j.heliyon.2024.e39596.

Hoummadi, M.A., Aroussi, H.A., Bossoufi, B., Karim, M., Mobayen, S., Zhilenkov, A., Alghamdi, T.A.H., 2024. A review of constraints and adjustable parameters in microgrids for cost and carbon dioxide emission reduction. Heliyon 10 (6). https://doi.org/10.1016/j.heliyon.2024.e27489.

Hu, Z., Li, Q., Zhang, P., Wang, R., Zhang, K., 2024. A novel handling method to intermittent feedback in load frequency regulation for renewable energy-dominated microgrids. IEEE Trans. Instrum. Meas. 73. https://doi.org/10.1109/TIM.2024.3457958.

Izmirlioglu, Y., Pham, L., Son, T.C., Pontelli, E., 2024. A survey of multi-agent systems for Smartgrids. Energies 17 (15). https://doi.org/10.3390/en17153620.

Joshal, K.S., Gupta, N., 2023. Microgrids with model predictive control: a critical review. Energies 16 (13). https://doi.org/10.3390/en16134851.

Keivanimehr, M., Chamorro, H.R., Zareian-Jahromi, M., Segundo-Sevilla, F.R., Guerrero, J. M., Konstantinou, C., 2021. Load shedding frequency management of microgrids using hierarchical fuzzy control. Proc. World Autom. Congr. 216–221. https://doi.org/10.23919/WAC50355.2021.9559468.

Kundur, P., and B. Gao. "Long term stability of power systems." (1994).
Khalghani, M.R., Solanki, J., Khushalani Solanki, S., Sargolzaei, A., 2021. Stochastic secondary frequency control of islanded microgrid under uncertainties. IEEE Syst. J. 15 (1), 1056–1065. https://doi.org/10.1109/JSYST.2020.3014642.
Khokhar, B., Singh Parmar, K.P., 2024. Frequency regulation in a microgrid integrating redox flow battery by utilizing an optimal predictive control approach. Electr. Eng. 106 (4), 4257–4275. https://doi.org/10.1007/s00202-023-02222-6.
Khooban, M.H., Niknam, T., Shasadeghi, M., Dragicevic, T., Blaabjerg, F., 2017. Load frequency control in microgrids based on a stochastic noninteger controller. IEEE Trans. Sustain. Energ. 9 (2), 583–861.
Khosravi, N., Çelik, D., Bevrani, H., Echalih, S., 2025. Microgrid stability: a comprehensive review of challenges, trends, and emerging solutions. Electr. Power Energy Syst. 170.
Kondoro, A., Dhaou, I.B., Tenhunen, H., Mvungi, N., 2021. A low latency secure communication architecture for microgrid control. Energies 14 (19). https://doi.org/10.3390/en14196262.
Kumar, N., Prabhansu, Krishnaveni, S., Aruna, R., 2022. Artificial Intelligent Approaches for Load Frequency Control in Isolated Microgrid with Renewable Energy Sources Renewable Energy Technologies: Advances and Emerging Trends for Sustainability. Wiley, pp. 511–537. https://doi.org/10.1002/9781119827634.ch15.
Magdy, G., Shabib, G., Elbaset, A.A., Mitani, Y., 2018. Optimized coordinated control of LFC and SMES to enhance frequency stability of a real multi-source power system considering high renewable energy penetration. Springer, Japan Prot. Control Mod. Power Syst. 3 (1). https://doi.org/10.1186/s41601-018-0112-2.
Magdy, G., Hossam, A., Xu, D., 2021. Effective control of smart hybrid power systems: Cooperation of robust LFC and virtual inertia control systems. CSEE J. Power Energy Syst. 8 (6), 1583–1593.
Mohan, A.M., Meskin, N., Mehrjerdi, H., 2023. LQG-based virtual inertial control of islanded microgrid load frequency control and DoS attack vulnerability analysis. IEEE Access 11, 42160–42179. https://doi.org/10.1109/ACCESS.2023.3271012.
Mutluri, R.B., Saxena, D., 2024. A comprehensive overview and future prospectives of networked microgrids for emerging power systems. India Smart Grid Sustain. Energy 9 (2). https://doi.org/10.1007/s40866-024-00218-0.
Nema, S., Prakash, V., Pandzic, H., 2022. Adaptive synthetic inertia control framework for distributed energy resources in low-inertia microgrid. IEEE Access 10, 54969–54979. https://doi.org/10.1109/ACCESS.2022.3177661.
Patwary, A.K., Sayem, M.A., Hossain, M.A., Halim, M.A., 2024. A review of energy storage systems (ESS) for integrating renewable energies in microgrids. Control Systems and Optimization Letters 2 (1), 102–112.
Rajamand, S., 2021. Load frequency control and dynamic response improvement using energy storage and modeling of uncertainty in renewable distributed generators. J. Energy Storage 37. https://doi.org/10.1016/j.est.2021.102467.
Rameshrao, A.G., Koley, E., Ghosh, S., 2022. A LSTM-based approach for detection of high impedance faults in hybrid microgrid with immunity against weather intermittency and N-1 contingency. Renew. Energy 198, 75–90. https://doi.org/10.1016/j.renene.2022.08.028.
Rasolomampionona, D.D., Połecki, M., Zagrajek, K., Wróblewski, W., Januszewski, M., 2024. A comprehensive review of load frequency control technologies. Energies 17 (12). https://doi.org/10.3390/en17122915.
Santra, S., De, M., 2023. Grey wolf optimization approach for enhancing the transient stability of microgrid using fractional-order PID-based inertia injection controller. Electr. Eng. 105 (6), 4361–4376. https://doi.org/10.1007/s00202-023-01946-9.

Saxena, A., Shankar, R., Zaabi, O.A., Hosani, K.A., Muduli, U.R., 2024. Rule-based adaptive frequency regulation with real stochastic model intermittency in a restructured power system. IEEE Trans. Ind. Inf. 20 (2), 2907–2919. https://doi.org/10.1109/TII.2023.3299422.

Sepehrzad, R., Nakhaeisharif, S., Al-Durra, A., Allahbakhshi, M., Moridi, A., 2022. Islanded micro-grid frequency control based on the optimal-intelligent lyapunov algorithm considering power dynamic and communication uncertainties. Elec. Power Syst. Res. 208. https://doi.org/10.1016/j.epsr.2022.107917.

Shahgholian, G., 2021. A brief review on microgrids: operation, applications, modeling, and control. Int. Trans. Electr. Energy Syst. 31 (6). https://doi.org/10.1002/2050-7038.12885.

Sharma, M., Dhundhara, S., Singh, R., 2022a. Impact of hybrid electrical energy storage system on realistic deregulated power system having large-scale renewable generation. Sustain. Energy Technol. Assessments 56.

Sharma, M., Dhundhara, S., Arya, Y., Prakash, S., 2025. Harris hawks optimization assisted fuzzy- PI λ controller for frequency stabilization in renewable integrated power systems. Optim. Control Appl. Methods 46 (6), 2925–2941. https://doi.org/10.1002/oca.70037.

Sharma, M., Saxena, S., Prakash, S., Dhundhara, S., Arya, Y., 2022b. Frequency stabilization in sustainable energy sources integrated power systems using novel cascade noninteger fuzzy controller. Energy Sources, Part A Recovery, Util. Environ. Eff. 44 (3), 6213–6235. https://doi.org/10.1080/15567036.2022.2091693.

Shobug, M.A., Chowdhury, N.A., Hossain, M.A., Sanjari, M.J., Lu, J., Yang, F., 2024. Virtual inertia control for power electronics-integrated power systems: challenges and prospects. Energies 17 (11). https://doi.org/10.3390/en17112737.

Singh, B., Bishnoi, S.K., Sharma, M., Singh, P., Dhundhara, S., 2023. An application of nature inspried algorithm based dual-stage frequency control strategy for multi micro-grid system. Ain Shams Eng. J. 14 (9). https://doi.org/10.1016/j.asej.2023.102125.

Singh, R., Ramesh, K., Raj, U., Shankar, R., 2025. Robust load frequency control in cyber-vulnerable smart grids with renewable integration. Energies 18 (11), 2899. https://doi.org/10.3390/en18112899.

Suryani, A., Sulaeman, I., Rosyid, O.A., Moonen, N., Popovic, J., 2024. Interoperability in microgrids to improve energy access: a systematic review. IEEE Access 12, 64267–64284. https://doi.org/10.1109/ACCESS.2024.3396275.

Trivedi, R., Patra, S., Sidqi, Y., Bowler, B., Zimmermann, F., Deconinck, G., Papaemmanouil, A., Khadem, S., 2022. Community-based microgrids: literature review and pathways to decarbonise the local electricity network. Energies 15 (3), 918. https://doi.org/10.3390/en15030918.

Yusuf, S.S., Kunya, A.B., Abubakar, A.S., Salisu, S., 2025. Review of load frequency control in modern power systems: a state-of-the-art review and future trends. Electr. Eng. 107 (5), 5823–5848. https://doi.org/10.1007/s00202-024-02828-4.

Zuo, K., Wu, L., 2022. A review of decentralized and distributed control approaches for islanded microgrids: novel designs, current trends, and emerging challenges. Electr. J. 35 (5). https://doi.org/10.1016/j.tej.2022.107138.

CHAPTER

3

Modeling microgrid components for LFC studies

Ravindra Kumar Yadav[1], *K.P. Singh Parmar*[2], *and Bhuvnesh Khokhar*[3]

[1] Department of Electrical and Electronics Engineering, Galgotias College of Engineering and Technology Technology, Greater Noida, Uttar Pradesh, India; [2] National Power Training Institute (under Ministry of Power, Government of India), Faridabad, Haryana, India; [3] Department of Electrical Engineering, Galgotias College of Engineering and Technology, Greater Noida, Uttar Pradesh, India

Nomenclature and Abbreviations

BESS battery energy storage system
BMS battery management system
DER distributed energy resource
DRL deep reinforcement learning
DSM demand side management
ECM equivalent circuit model
EMS energy management system
ESS energy storage system
FESS flywheel energy storage system
LFC load frequency control
MG microgrid
MPC model predictive control
MPPT maximum power point tracking
PMSM permanent magnet synchronous machine
PV photovoltaic
RES renewable energy resource
SoC state-of-charge
SoH state-of-health
VSC voltage source inverter
WT wind turbine

3.1 Introduction to microgrid component modeling

MGs are smaller energy systems that can operate with the utility grid or independently (Khokhar et al., 2020, 2021). For effective design, operation, and control, precise component modeling is a must. Component modeling

Frequency Control in Microgrids
https://doi.org/10.1016/B978-0-443-36476-1.00003-1

is the process of creating mathematical representations of the various parts that comprise an MG, including loads, DERs, ESSs, and PEDs. These models are essential for analyzing the system's behavior, predicting its performance, and guaranteeing its stability under a range of operating conditions (Khokhar and Singh Parmar, 2024; Khokhar et al., 2024).

3.1.1 Importance of accurate component models

Accurate component models are vital for several reasons as stated below:

1. *System stability and control:* Accurate models facilitate the development of efficient control systems to uphold voltage and frequency stability, particularly when incorporating the RESs with fluctuating outputs (AlMuhaini et al., 2023).
2. *Optimal scheduling and demand side management:* Reliable load models are crucial for effective scheduling and successful demand side management (DSM) in MGs. Precise load modeling facilitates comprehension of load profiles, which may fluctuate considerably, thereby contributing to efficient energy management (Boennec et al., 2023).
3. *Design and planning:* High-fidelity models facilitate the meticulous design of MGs by establishing the requisite level of model precision, particularly in battery models, to guarantee efficient and dependable operation (Boennec et al., 2023).
4. *Integration of energy management systems:* To prevent discrepancies between modeled and actual results, it is essential for MG designers to comprehend and accurately model the variables within the energy management systems (EMSs) they want to utilize in the operational MG (Fontenot and Dong, 2019).

3.1.2 Role of modeling in LFC studies

LFC is crucial for sustaining equilibrium between power supply and demand, hence assuring stable system frequency. In MGs, precise component modeling is essential for LFC analyses:

1. *Controller design:* Precise models are required to depict the complex dynamics of modern power systems, including the interactions between DERs, loads, and ESSs, in order to create controllers that efficiently control frequency fluctuations. For these models to ensure reliable performance in a variety of operating scenarios, nonlinearities, time delays, and uncertainties must be taken into consideration. Advanced LFC techniques have been developed to improve system performance by leveraging artificial intelligence

techniques like deep reinforcement learning and fuzzy logic control, as well as cutting-edge algorithms like adaptive control, robust control, and model predictive control (MPC). By effectively coordinating multiple energy resources, these advanced techniques optimize frequency regulation, enhance system resilience to disturbances, and enable instantaneous modifications (Davoudkhani et al., 2024).

2. *Dynamic analysis:* Understanding system behavior during transient events and creating successful LFC strategies depend on the ability to analyze dynamic interactions among components, which is made possible by precise modeling. Accurate modeling makes it easier to determine key factors that affect frequency stability, such as demand fluctuations, energy storage dynamics, governor response time, and inertia constants. Additionally, it helps evaluate the impact of communication barriers, uncertainty, and time delays related to RESs like solar and wind. Researchers are able to more effectively simulate real-world operational scenarios by incorporating accurate models of MG components, demand response mechanisms, and virtual inertia. This makes it possible to create efficient control strategies that ensure quick frequency recovery, lessen overshoots, and increase system resilience to sudden changes in load or interruptions in generation (Xiong et al., 2020).
3. *Optimization of control strategies:* The optimization of control strategies, like MPC, to enhance frequency regulation in islanded MGs is made easier by accurate models. These models provide a thorough representation of the dynamics of the system, making it easier to predict frequency variations and implement preventative control strategies. Since islanded MGs often have less inertia and are more sensitive to changes in load demand and renewable energy generation, detailed modeling is necessary to create controllers that can effectively handle uncertainties and disturbances. Additionally, by enabling quick and effective power adjustment, the inclusion of various ESSs in the model - such as batteries and supercapacitors - improves frequency regulation. By employing accurate models, advanced control techniques can significantly improve frequency stability, reduce overshoots and settling times, and maximize energy usage in MGs, thereby enhancing their dependability and efficiency (Tudu et al., 2024).

3.2 Generation component modeling

For performance analysis, stability assurance, and the development of suitable control strategies, modeling generation components, or DERs, within an MG is essential. The modeling of conventional generators, the addition of RESs, and the dynamic properties and parameters pertinent to these elements are all covered in this section. The process of creating mathematical representations for the various DERs in an MG is known as generation component modeling. These models are crucial for ensuring reliable operation under a variety of circumstances, optimizing performance, and simulating system behavior.

3.2.1 Modeling conventional generators

The two most commonly used conventional generators in an MG are diesel engine generators and gas turbines. Synchronous machine models are typically used to depict these two. The models include the electrical and mechanical dynamics of the generator, including excitation systems, damping, and rotor inertia. Evaluating transient stability and creating control systems inside the MG require accurate modeling of these components.

3.2.1.1 Diesel engine generator

Fig. 3.1 displays the schematic diagram of a conventional diesel engine generator (Bangash et al., 2019). Many crucial parts work together to ensure efficient power generation and grid stability in the generator that is connected to the power grid. Diesel fuel is burned in the internal combustion engine to produce mechanical energy, which is then converted into electrical energy by the synchronous generator. In order to create the magnetic field necessary for inducing voltage in the stator winding, the excitation system supplies the generator's rotor winding with the necessary DC current. In order to maintain a steady voltage output and adjust to changes in load, the voltage regulator modulates the excitation current. This ensures that the generator's terminal voltage satisfies grid requirements. In order to keep the generator running at a steady speed, which is necessary to maintain the intended frequency of the electrical output, the speed regulator or governor simultaneously controls the engine's fuel input. When connected to the power grid, this system ensures that the electricity

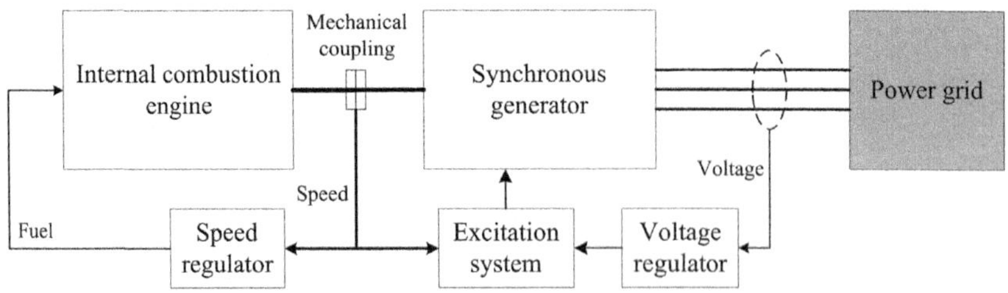

FIGURE 3.1 Schematic diagram of a conventional diesel engine generator.

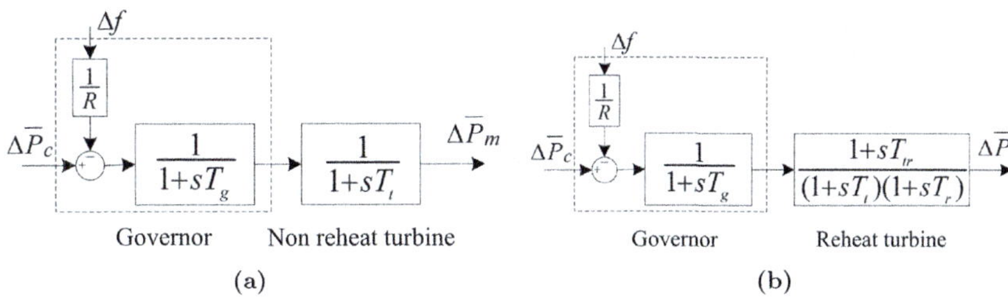

FIGURE 3.2 Transfer function model of a conventional diesel engine generator with (a) non-reheat turbine and (b) reheat turbine.

produced matches the grid's voltage, frequency, and phase, allowing for smooth power integration.

Most commonly employed transfer function models of a conventional diesel engine generator for the LFC analysis with non-reheat turbine and reheat turbine are shown in Fig. 3.2(a) and (b), respectively (Bevrani, 2014). Here, T_g is the governor time constant, T_t is turbine time constant, and R is the speed regulation constant. Both are first-order transfer function models. In LFC studies, first-order transfer function models are favored since they provide a simplified yet sufficiently accurate depiction of the dominant dynamics of the system. Without the complexity of higher-order systems, these models capture the fundamental behavior of components such as governors, turbines, and generators, therefore greatly simplifying control design, stability analysis, and real-time implementation.

3.2.1.2 Gas turbine

Schematic diagram of a gas turbine is displayed in Fig. 3.3(a) (Paul et al., 2012). Through the use of several crucial components, the system converts fuel energy into mechanical and electrical power. The compressor first takes in outside air, compresses it, and raises its temperature and pressure. After entering the combustion chamber, the high-pressure air mixes with fuel and ignites, releasing high-energy exhaust gases. The gas turbine is propelled and rotated by the rapid expansion of the hot gases. To guarantee continuous operation, the turbine is mechanically coupled to a generator and the compressor. The generator converts the kinetic energy of the turbine into electrical power, which is then sent to the motor. Because of its effectiveness and quick response to changes in load, this system is widely used in power plants, aircraft, and industrial settings.

Fig. 3.3(b) shows the commonly used transfer function model of a gas turbine for the LFC analysis with T_f as the time constant of the fuel system (Wang et al., 2017). A fuel system ensures consistent and effective

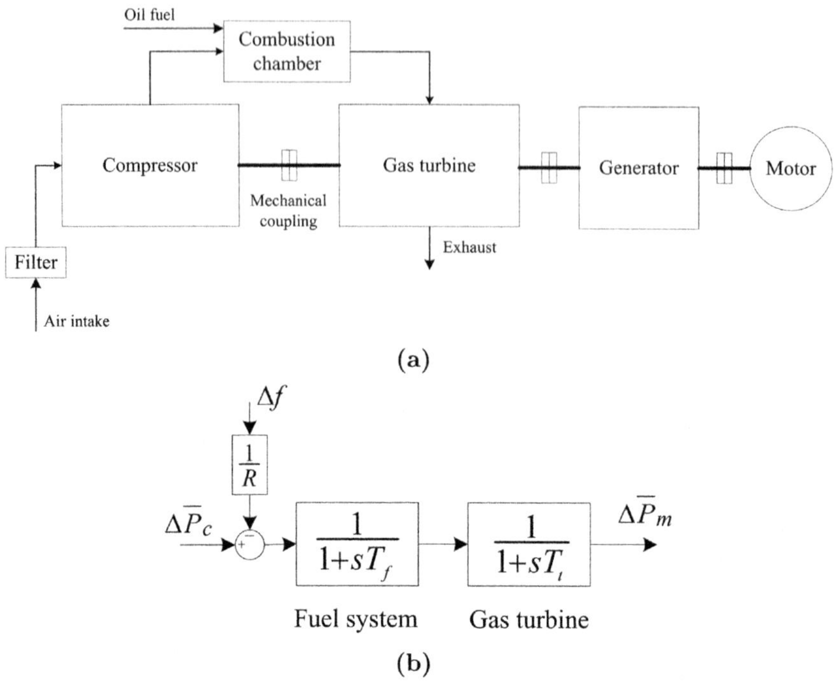

FIGURE 3.3 (a) Schematic diagram and (b) transfer function model of a gas turbine.

operation by delivering the proper amount of fuel to the combustion chamber under different operating conditions.

In MG systems, conventional generators frequently function as primary power sources, particularly during periods of intermittent renewable energy availability. Their models must incorporate governor and excitation control systems to accurately simulate responses to load variations and disturbances.

3.2.2 Integration of RESs

RESs like wind turbines (WTs) and photovoltaic (PV) systems are increasingly integrated into MGs. Modeling these sources requires capturing their intermittent nature and the power electronic interfaces through which they connect to the grid.

3.2.2.1 Wind turbines

WT models include electrical models of the generator, mechanical models of the drive train, and aerodynamic models of the rotor. Stochastic modeling techniques are required to accurately depict power output fluctuations due to wind speed variability. In order to address the unpredictable nature of wind gusts, Bierbooms and Cheng created a stochastic gust model to enhance design calculations for WTs (Bierbooms and Cheng, 2002). Using stochastic approaches to improve performance predictions under variable wind conditions, fluck proposed methodologies for the unsteady aerodynamic study of WT blades (Fluck and Crawford, 2016). In order to accurately model wind conditions for turbine analysis, Song et al. developed a method for simulating wind speed fields by stochastic harmonic function representation (Song et al., 2019). The significance of such detailed models in creating efficient control strategies is further highlighted by a study on dynamic modeling and operation strategies for MGs with wind and PV resources (Bae and Kwasinski, 2012). A schematic diagram of a WT that transforms mechanical power into electrical power is shown in Fig. 3.4(a) (Blaabjerg and Ionel, 2017). While multipole generator systems are made especially for multi-megawatt applications, gearboxes are frequently used in small WTs. To increase flexibility, a power electronic converter could be positioned between the generator and the grid. Fig. 3.4(b) shows the commonly used transfer function model of a WT for LFC analysis. The model comprises a band-limited white noise block for achieving variations in the WT power output ($\Delta\overline{P}_w$) and a low pass filter (LPF) (Khokhar et al., 2020).

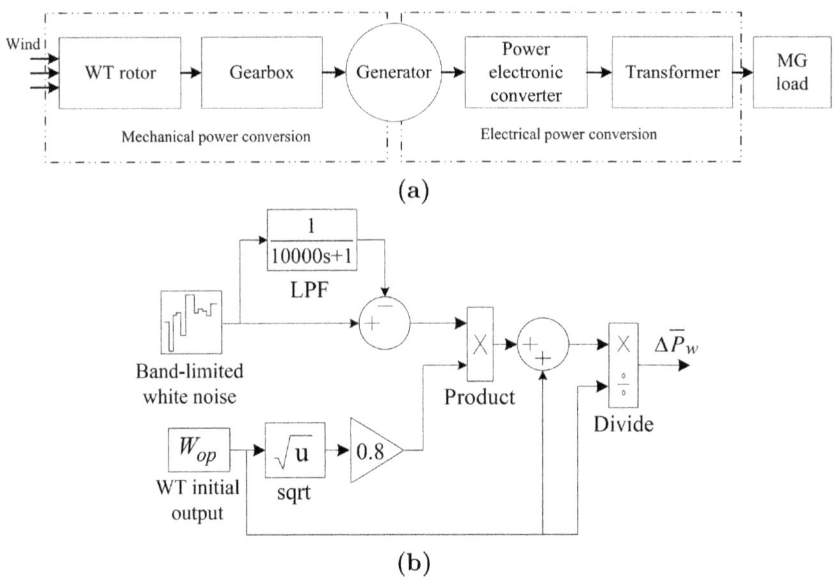

FIGURE 3.4 (a) Schematic diagram of a WT integration with MG and (b) transfer function model of a WT.

3.2.2.2 Photovoltaic systems

PV system models include the dynamics of maximum power point tracking (MPPT) algorithms as well as the electrical properties of solar panels, such as the current-voltage relationship. The impact of environmental variables on power output, such as temperature and irradiance, must also be taken into account in these models. Fig. 3.5(a) displays a general schematic diagram of a solar PV panel integration with MG via a power electronic interface and transformer (Blaabjerg and Ionel, 2017). Other controls include DC voltage and current control in addition to MPPT. Chakraborty et al. used MPPT and voltage-frequency (V-f) control techniques to study a microgrid system with a two-stage integration of PV arrays in a standalone configuration (Chakraborty et al., 2019). With an emphasis on maintaining stability and dependability, Vijay et al. presented a novel control strategy for a microgrid configuration that comprised solar PV arrays, wind energy generation systems, and battery

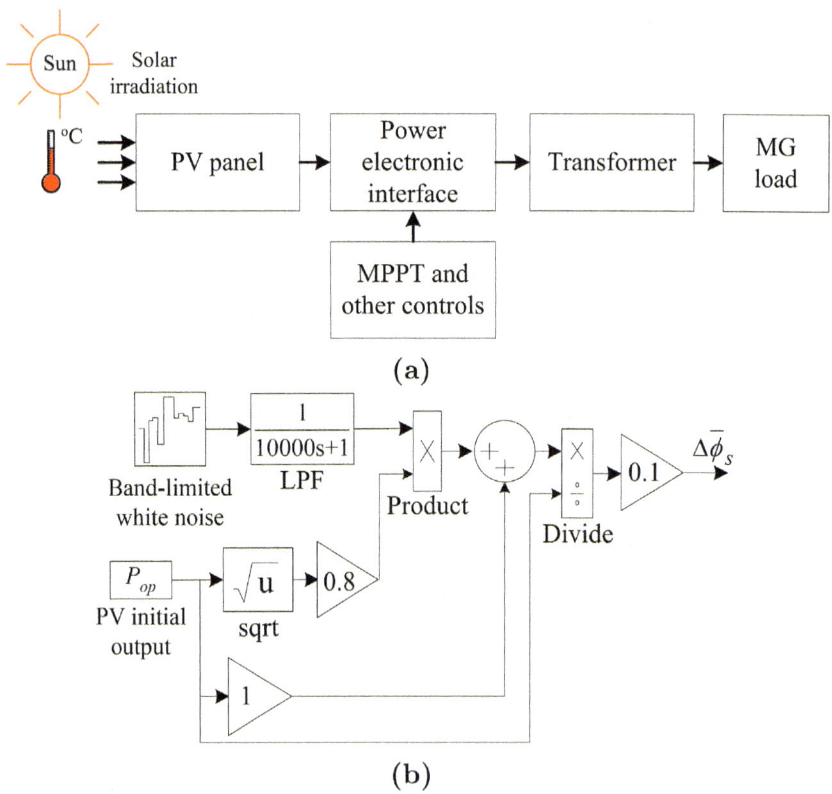

FIGURE 3.5 (a) Schematic diagram of a solar PV panel integration with MG and (b) transfer function model of a solar PV panel.

storage (Bhuvaneswari Singh et al., 2020). Fuzzy-PI controllers were used to effectively regulate voltage and frequency in a DC microgrid that integrated fuel cells and photovoltaic arrays (Nempu et al., 2019). Fig. 3.5(b) shows a generally employed transfer function model of a PV panel for the LFC analysis (Khokhar et al., 2020).

There are several obstacles to overcome when integrating WTs and PV arrays into MGs (Khokhar and Parmar, 2022, 2023). Because wind and solar energy are intermittent and unpredictable, power generation fluctuates, which causes frequency instability in the MG. Furthermore, during islanded operations, low inertia property of RESs could cause noticeable frequency fluctuations. Furthermore, complex EMSs are needed to ensure a reliable power supply due to the inconsistency of different ESSs. Therefore, designing EMSs that can balance supply and demand, guarantee power quality, and preserve system stability requires precise modeling.

MGs rely on a mix of fast-acting ESSs (to be discussed in the subsequent sections), smart control systems, and flexible generation sources (such as conventional diesel generators) to respond to sudden changes in solar or wind power generation. The control system of the MG rapidly identifies the imbalance and delivers stored energy or modifies dispatchable DERs to preserve frequency stability when solar or wind power output falls suddenly because of weather variability. On the other hand, excess power can be stored, curtailed, or used to run extra loads during a rapid rise in renewable generation. This coordinated approach guarantees consistent operation even if the RESs are sporadic.

3.2.3 Dynamic characteristics and parameters

Understanding the dynamic characteristics of DERs is crucial for analyzing MG behavior during transient events and for designing appropriate control strategies. Key characteristics include:

1. *Inertia*: Conventional generators provide system inertia, influencing the MG capacity to withstand frequency fluctuations. Conversely, RESs connected via PEDs generally lack inertia, which may result in accelerated frequency dynamics (Venayagamoorthy et al., 2016).
2. *Response time*: It is the rate at which DERs can modify their output in reaction to control signals or disruptions. Conventional generators exhibit slower response times due to mechanical limitations, however RESs with PEDs can respond more swiftly (Kale et al., 2011).
3. *Control strategies*: These are utilized to regulate voltage, frequency, and power output. Conventional generators employ governor and excitation controls, while RESs rely on inverter or converter control

systems. These control techniques guarantee system stability, improve power quality, and facilitate the seamless integration of renewable energy sources with the grid (Nayak and Gaonkar, 2012).

Precise parameter identification and dynamic modeling are crucial for simulating MG behavior, establishing control systems, and maintaining stable operation across many scenarios. A comparative summary of dynamic characteristics and parameters of various DERs is shown in Table 3.1 (Venayagamoorthy et al., 2016).

TABLE 3.1 Dynamic characteristics and parameters of various DERs.

DER	Dynamic characteristics	Parameters
Conventional diesel	High inertia,	Governor time constant,
Engine generator	Slow response time, and	Electrical time constant, and
	Suitable for primary as well as	Droop characteristics
	Backup energy source	
Gas turbine	Lower inertia compared	Turbine time constant,
	To diesel generator,	Delay in fuel system, and
	Faster response time, and	Droop and ramp rate
	Appropriate for peak load handling	
WT generator	High variability,	Pitch angle,
	Low inertia, and	Tip speed ratio, and
	Restricted direct involvement	Power coefficient
	In frequency regulation	
PV array	Variable output	MPPT time constant,
	No physical inertia, and	Inverter response time, and
	Faster response time	Temperature coefficient
	With inverter	

3.3 Energy storage system modeling

ESSs are essential to contemporary power systems since they improve efficiency, dependability, and RES integration. For efficient design, control, and optimization, these systems must be accurately modeled. In order to model ESSs, mathematical representations of their electrical, thermal, and dynamic behaviors must be made. These models are essential for creating control strategies, simulating performance, and guaranteeing system stability in a range of operating scenarios. The following subsections provide a brief overview of modeling aspects of several widely used ESSs.

3.3.1 Battery energy storage system (BESS)

Modeling BESSs is crucial for optimizing their design, operation, and integration into power systems. Various modeling approaches have been developed to capture the complex behaviors of batteries, including their electrical characteristics, thermal dynamics, and degradation mechanisms.

3.3.1.1 *Equivalent circuit models*

Equivalent circuit models (ECMs) of batteries are widely used due to their simplicity and computational efficiency. In order to simulate dynamic behaviors, they use electrical components such as resistors and capacitors to represent the battery. For real-time applications like state-of-charge (SoC) and state-of-health (SoH) estimation in BMSs, these models are especially helpful. For instance, the *Rint* model, the most basic ECM of a battery, is depicted in Fig. 3.6(a). Here, a voltage source (OCV) and a resistor (Ro) are connected in series. Vop_batt is the battery's output voltage, and I_batt is its current. Similarly, Fig. 3.6(b) illustrates the Thevenin model, another widely used ECM, in which a resistor (Rp) and capacitor (Cp) are connected in parallel. Polarization resistance is referred to as Rp, and capacitance as Cp. The Thevenin model provides a more accurate description of

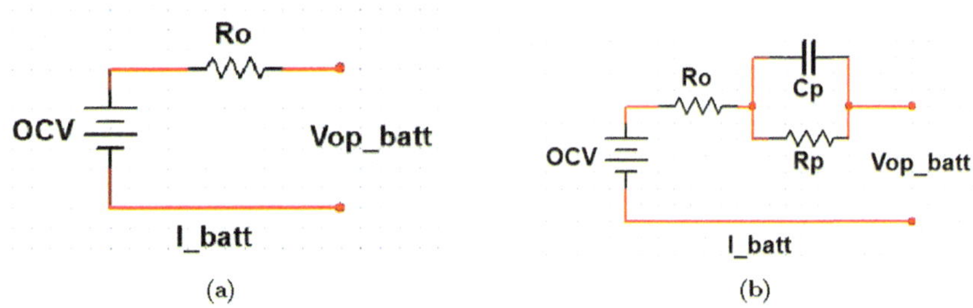

FIGURE 3.6 ECMs of a battery: (a) *Rint* model and (b) Thevenin model.

the dynamic behavior of the ECM than the *Rint* model (Shen and Gao, 2019; Hu et al., 2012). Refer Shen and Gao, (2019) for more information on these two ECMs as well as other advanced ECMs.

3.3.1.2 *Electro-thermal models*

To evaluate energy efficiency and thermal performance, electro-thermal models are employed. These models consider both electrical and thermal aspects, providing insights into temperature effects on battery performance and lifespan. For instance, Schimpe et al. conducted an energy efficiency evaluation of a stationary lithium-ion battery system via electro-thermal modeling (Schimpe et al., 2018). A study by Yonghuang Ye et al. introduced an electro-thermal model that concentrated on spiral-wound lithium-ion batteries, highlighting the significance of porous current collectors in thermal dynamics (Ye et al., 2014).

3.3.1.3 *Degradation and life prediction models*

Understanding battery degradation is essential for reliable operation. Models focusing on degradation mechanisms help in predicting battery lifespan under various operating conditions. Smith et al. developed a life prediction model for grid-connected lithium-ion battery, aiding in maintenance and replacement planning (Smith et al., 2017). Wang et al. presented a model that encapsulated the two-phase degradation behavior of lithium-ion batteries, integrating a dynamic change point to improve predictive accuracy (Wang et al., 2023).

3.3.1.4 *Techno-economic models*

These models integrate technical performance with economic factors to assess the viability of BESS projects. DiOrio et al. suggested a detailed model within the system advisor model (SAM) framework, evaluating the economic feasibility of BESSs by analyzing elements such as system costs, performance, and financial indicators (DiOrio et al., 2015). Mathews et al. formulated a model that assesses the utilization of second-life lithium-ion batteries in extensive solar power applications, integrating battery aging dynamics to forecast performance and economic results (Mathews et al., 2020). Most commonly employed transfer function model of a BESS for LFC analysis is shown in Fig. 3.7(a) (Khokhar and Parmar, 2025).

3.3.2 Flywheel and supercapacitor models

Flywheels and supercapacitors are ESSs known for high power density and rapid charge-discharge capabilities. Their modeling is discussed briefly below.

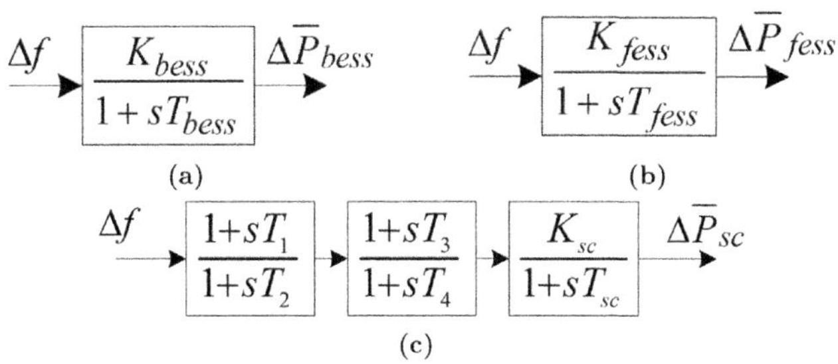

FIGURE 3.7 Transfer function model of (a) BESS, (b) FESS, and (c) supercapacitor.

3.3.2.1 *Flywheel energy storage system (FESS)*

Modeling FESSs encompasses the mechanical dynamics of a rotating mass and the electrical interface for energy transfer. Essential factors include rotational inertia, frictional losses, and the interconnection between mechanical and electrical domains. The mechanical model of an FESS delineates the dynamics of the rotating mass and its interaction with the system. The energy stored within the flywheel can be calculated as (Arani et al., 2017):

$$\overline{E} = \frac{1}{2}\overline{J}\omega^2 \tag{3.1}$$

where $\overline{J}$ is the flywheel inertia (kg-m^2), $\overline{E}$ is the stored kinetic energy (joules), and ω is the angular velocity of the flywheel (rad/s). The power provided by the flywheel is dictated by the rate of change in ω, expressed as:

$$\overline{P} = \frac{d\overline{E}}{dt} \tag{3.2}$$

Efficacy of an FESS is contingent upon frictional losses (air resistance and bearing losses) and power conversion losses. The flywheel, in general, is a permanent magnet synchronous machine (PMSM) and is frequently paired with an electrical machine (motor/generator). The relationship between voltage and current in a PMSM can be expressed through its dq model as follows Saleh et al., (2019):

$$\begin{cases} \overline{v}_q = R_{stat}\overline{i}_q + \frac{d}{dt}\overline{\tau}_q - \omega_e\overline{\tau}_d \\ \overline{v}_d = R_{stat}\overline{i}_d + \frac{d}{dt}\overline{\tau}_d - \omega_e\overline{\tau}_q \end{cases} \tag{3.3}$$

where $\overline{v}_d$ and $\overline{v}_q$ represent the dq stator voltage components, whereas $\overline{i}_d$ and $\overline{i}_q$ denote the dq stator current components. ω_e represents the angular velocity of the electromagnetic field within the rotor. R_{stat} denotes the resistance of the stator. The flux linkages $\overline{\tau}_d$ and $\overline{\tau}_d$ are defined as:

$$\begin{cases} \overline{\tau}_d = L_d\overline{i}_d + \overline{\tau}_f \\ \overline{\tau}_q = L_q\overline{i}_q \end{cases} \tag{3.4}$$

In this context, $\overline{\tau}_f$ represents the flux linkage of the permanent magnet and L_d and L_q denote the dq-axis inductances. For further modeling details, interested readers can refer (Saleh et al., 2019). Most commonly employed transfer function model of a FESS for LFC analysis is shown in Fig. 3.7(b) (Khokhar and Parmar, 2025).

3.3.2.2 *Supercapacitors*

Based on their electrochemical characteristics, supercapacitors are modeled with an emphasis on capacitance, leakage currents, and equivalent series resistance. They are appropriate for applications needing short energy bursts because of their quick charging and discharging capabilities. When incorporating supercapacitors into hybrid ESSs, like electric vehicles that combine batteries and supercapacitors, accurate models are crucial.

Similar to the plate separation distance, the capacitance of a typical capacitor is inversely proportional to the distance between the charges and directly proportional to the effective area of the electrodes. The capacitance per unit volume of supercapacitors is higher than that of regular capacitors. In comparison to traditional capacitors, this is mainly due to the electrodes' porous structure, which offers a substantially greater effective surface area, and, secondarily, the shorter distance between the ionic and electric charges. A supercapacitor's capacitance is not constant; it varies greatly with voltage. Its voltage either rises or falls when current is applied or removed, which changes the charge concentration on the Helmholtz layer (San Martín et al., 2013). According to the equation below, this causes the capacitance to change as the distance of charge separation either increases or decreases Rafik et al., (2007)

$$D_c = \sqrt{\frac{\varepsilon_0\varepsilon_r\overline{K}_bT}{2\overline{n}^2\overline{C}_e}} \tag{3.5}$$

where D_c is the Debye length contingent upon the electrolyte concentration $\overline{C}_e$, and ε_0 and ε_r are the dielectric constants of vacuum and the separation medium, respectively, $\overline{K}_b$ is the Boltzmann constant, and T is absolute temperature. The actual capacitance of the supercapacitor (C_{act}) can be derived in function of its voltage ($\overline{v}_c$) via a suitably accurate linear equation as (Rafik et al., 2007):

$$C_{act} = C_0 + \overline{k}\overline{v}_c \tag{3.6}$$

where C_0 represents the capacitance of the supercapacitor when $\overline{v}_c$ equals 0 V, and $\overline{k}$ is a constant. For further modeling details, interested readers may refer (Rafik et al., 2007). Most commonly employed transfer function model of a supercapacitor for LFC analysis is shown in Fig. 3.7(c) (Khokhar and Parmar, 2025).

3.4 Load modeling for LFC

Load modeling is a critical aspect of LFC in MGs, as it directly influences system stability, control strategies, and the integration of DERs. The different kinds of loads in MGs, the development of dynamic load models, and the effects of distributed generation on load modeling are all covered in this section. Making mathematical representations of an MG's demand side is known as load modeling in the context of LFC. Predicting system behavior, creating efficient control systems, and guaranteeing dependable operation in a variety of scenarios all depend on accurate load models.

3.4.1 Types of loads in microgrids

Types of loads in MGs can be broadly classified into three types based on their electricity demand as shown in Fig. 3.8 (Roosa, 2020).

Loads are frequently classified as sheddable, adjustable, or sensitive. Tier one (sensitive) loads, such as emergency lighting systems, elevators, and refrigeration units, are necessary and must run continuously without interruption. Any disruption to these loads brought on by power outages could result in significant costs, potential equipment damage, and safety hazards. Non-essential Tier two (adjustable) loads can be temporarily reduced or changed to match the available generation capacity. Air conditioners, certain fans, and household water heaters are a few examples.

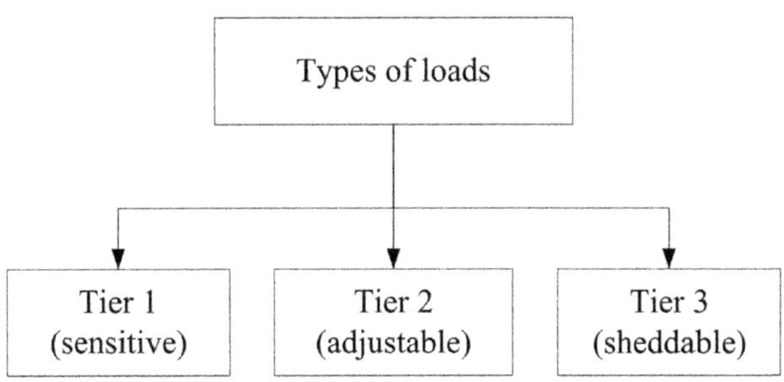

FIGURE 3.8 Classification of loads in an MG.

During emergencies brought on by unanticipated or partial generation failure, Tier three (sheddable) loads may be decreased. Depending on particular needs, these loads could include backup power equipment, interior lighting, and kitchen appliances.

3.4.2 Dynamic load models

For MGs to accurately simulate and control the transient behaviors of various loads and thereby increase system stability and efficiency, dynamic load modeling is essential. The time-varying characteristics of loads are often not well represented by traditional static load models, especially in dynamic operating conditions. By incorporating the time-sensitive responses of loads to changes in voltage, frequency, and other system parameters, dynamic load models help to alleviate this limitation.

3.4.2.1 *Dynamic load models classification*

1. *Induction motor models:* Induction motors, prevalent in industrial and commercial settings, demonstrate dynamic characteristics like slip variations and transient responses to voltage swings. Modeling these dynamics is essential for evaluating the influence of motor-driven loads on MG stability. Mahdavian et al. performed a small-signal stability analysis of MGs, incorporating dynamic load models such as induction motors, to assess their impact on system stability (Mahdavian et al., 2021).
2. *Thermal load models:* Thermal loads, including heating, ventilation, and air conditioning systems, exhibit thermal inertia that influences their reaction to fluctuations in power supply. The dynamic modeling of these loads is crucial for precise load forecasting and demand response techniques. Raju and Jain formulated and verified a generalized modeling methodology for islanded inverter-based MGs with both static and dynamic loads, encompassing thermal loads, to improve the precision of load modeling in MG simulations (Raju and Jain, 2019).
3. *Lighting load models:* Lighting systems, particularly those utilizing fluorescent or LED technology, may demonstrate dynamic behaviors influenced by ballast characteristics and dimming controls. Modeling these dynamics is crucial for comprehending their influence on MG performance. Roos et al. consolidated component-based grid-feeding DERs and load models, encompassing lighting loads, to simulate MG islanding transients, thereby elucidating the dynamic interactions between DERs and loads (Roos et al., 2020).

3.4.3 Dynamic demand response strategies

Usually, conventional load modeling in MGs depends on static or aggregated depictions that fail to show the control of contemporary electrical loads. But dynamic demand response (DR) strategies have to be included into load models as RESs' rising variability drives more of them. DR-enabled loads can actively change their consumption patterns in reaction to MG conditions including frequency variations, voltage levels, or real-time price signals. Dynamically modeling these flexible loads, including smart HVAC systems, electric vehicles, and thermostatically controlled appliances, using time-varying functions, differential equations, or price-elastic demand curves (Choi et al., 2006).

Including DR into load modeling increases MG flexibility, stability, and resilience. It lets demand-side resources engage in frequency and voltage control, lowers the load on supply-side reserves, and helps to integrate more consistent renewable sources such wind and solar. Many studies have shown that demand response can greatly lower frequency variations and enhance transient stability in isolated MG settings (Rahimi and Ipakchi, 2010).

3.4.4 Impacts of DERs on load modeling

Load modeling is significantly impacted by the integration of DERs into MGs, necessitating the use of complex techniques to accurately represent the dynamic interactions between loads and generating sources. DERs affect load profiles and system stability by adding unpredictability and variability to the power supply.

Unpredictable and irregular generation patterns are exhibited by distributed generation sources, particularly renewable energy sources like wind and solar. Because of this variability, traditional load profiles are altered because the net load, the difference between the total load demand and the output of distributed generation, becomes more dynamic. Majzoobi and Khodaei investigated how MGs could reduce net-load ramping in distribution networks caused by distributed solar output variability, highlighting the need for dynamic load modeling to control these variations (Majzoobi and Khodaei, 2016). Furthermore, power can flow in multiple directions thanks to DG sources, which makes load modeling more difficult and calls for more sophisticated analytical techniques. Additionally, these sources might interact with pre-existing control mechanisms, affecting frequency management strategies and requiring load model adjustments to account for these interactions.

In general, dynamic load models must be developed in order to accurately represent the complex relationships between generating sources and loads when integrating DERs into MGs. Accurate load forecasting, demand response strategies, and guaranteeing the general stability and dependability of MG systems all depend on these models.

3.5 Power electronic devices modeling

PEDs are essential in modern MGs, enabling efficient energy conversion, renewable energy integration, and system stability. Accurate modeling of these devices helps in analyzing their behavior under various operational scenarios and designing effective control strategies.

3.5.1 Inverters and converters modeling

Inverters and converters are the primary PEDs in MGs. They serve as interfaces for energy conversion and play a critical role in maintaining power quality and system stability.

3.5.1.1 Inverter modeling

1. Role of inverters in LFC of MG

For MG control, especially for LFC, inverter modeling is crucial. Connecting different DERs, ESSs, and the AC grid is made easier by inverters. By adjusting their active power output in response to frequency variations, they play a crucial role in maintaining frequency stability. For grid compatibility, inverters transform DC power into AC power. They model: i) switching mechanisms using semiconductor devices like metal oxide semiconductor field effect transistors (MOSFETs) and insulated gate bipolar transistors (IGBTs); ii) equivalent circuit models with harmonic suppression filters; and iii) loops for voltage and current control (Guerrero et al., 2010; Rocabert et al., 2012).

2. General control structure of an inverter

The control architecture of an inverter generally has two primary loops (Bevrani and Shokoohi, 2013):

1. External control loop: It produces the reference current for the internal loop based on active power and reactive power references i.e., $\overline{P}_{ref}$ and $\overline{Q}_{ref}$, respectively. The active power control loop is essential for LFC.
2. Internal control loop: This loop modulates the output current to align with the reference current. It also facilitates rapid and precise monitoring of the reference signals.

3. *Droop control and inverter dynamic model:* Droop control is a decentralized approach employed in MGs to distribute power among inverters and uphold frequency stability. The droop equation for regulating active power is (Majumder et al., 2009; Kundur, 1994):

$$\overline{P}_{ref} = \overline{P}_{set} - K_p.\Delta f \tag{3.7}$$

where, $\overline{P}_{set}$ is set point power, K_p is droop coefficient, and Δf is frequency deviation. The droop coefficient K_p quantifies the variation in active power production in relation to frequency variations. An increased K_p leads in a more pronounced droop characteristic.

The dynamics of the inverter can be characterized as a first-order system as (Majumder et al., 2009; Kundur, 1994):

$$\frac{d\overline{P}}{dt} = \frac{1}{T_i}\left(\overline{P}_{ref} - \overline{P}\right) \tag{3.8}$$

where, $\overline{P}$ is actual active output power and T_i is inverter time constant. Taking Laplace transform of the above equation, we get:

$$\frac{\overline{P}(s)}{\overline{P}_{ref}(s)} = \frac{1}{1 + sT_i} \tag{3.9}$$

3.5.1.2 Converter modeling

To maintain frequency stability, a converter in LFC of an MG controls power exchange between DERs, ESSs, and the grid. It functions as an interface that uses control mechanisms like droop control, secondary PI-based frequency restoration, and tertiary economic dispatch to modulate the injection of active and reactive power in response to frequency variations. To reduce fluctuations brought on by changes in load and the intermittent nature of RESs, the converter dynamically controls power flow. Additionally, it ensures the seamless integration of ESSs, promoting quick frequency response and improving MG stability in general. A voltage source converter (VSC) is the most common converter used in MGs (Xu, 2013).

Fig. 3.9 shows a schematic diagram of a converter. The AC side dynamics of a converter connected to an MG are given as (Ndeke et al., 2024):

$$\overline{V}_{ac} = L\frac{d\overline{I}_{conv}}{dt} + R\overline{I}_{conv} + \overline{V}_{conv} \tag{3.10}$$

where, $\overline{V}_{ac}$ is the generator side voltage, $\overline{V}_{conv}$ is the converter output voltage, $\overline{I}_{conv}$ is the converter current, and L and R are the line inductance

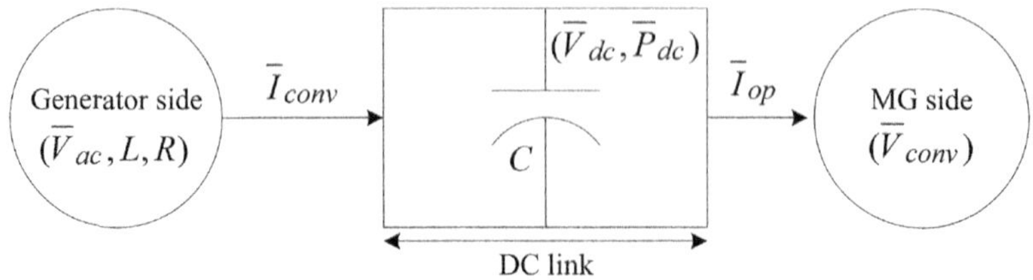

FIGURE 3.9 Schematic diagram of a converter.

and resistance, respectively. Further, the output voltage (DC side voltage) of the converter is given by Ndeke et al. (2024):

$$C\frac{d\overline{V}_{dc}}{dt} = \frac{\overline{P}_{dc}}{\overline{V}_{dc}} - \overline{I}_{op} \tag{3.11}$$

where, $\overline{V}_{dc}$ is the DC-link voltage, C is the DC-link capacitor, $\overline{P}_{dc}$ is the power from the DC source, and $\overline{I}_{op}$ is the output DC of the converter.

3.5.2 Control system dynamics

LFC schemes now include new dynamics brought about by the integration of PEDs, such as converters and inverters, into power systems. These devices, which are renowned for their controllability and quick reaction times, offer both possibilities and difficulties for frequency management.

PEDs can quickly adjust their output in response to changes in frequency. This feature is particularly helpful in low-inertia MGs, where the use of RESs in place of synchronous generators reduces conventional mechanical inertia. The goal of suggested advanced control strategies, such as reinforcement learning-based techniques, is to quickly regulate frequency by leveraging the grid-forming VSCs' quick reaction times. By preventing frequency threshold breaches and enhancing system stability, these data-driven methods enable VSCs to modulate their power output in real-time (Chen et al., 2024; Yan and Xu, 2018).

However, PEDs' quick switching can produce electromagnetic interference and high-frequency harmonics, which lowers overall power quality. Complex control techniques, such as MPC, have been used for LFC problems in order to overcome these constraints. MPC effectively manages the complex dynamics offered by power electronic devices by using a dynamic model of the power system to predict future behavior and optimize control actions (Stanojev et al., 2020; Aguirre et al., 2020).

Furthermore, because PEDs have nonlinear characteristics, robust control strategies that can handle system uncertainties and disruptions must be developed. For model-free LFC in nonlinear power systems, deep reinforcement learning (DRL) has emerged as a promising technique. By interacting with the power system environment, DRL-based controllers can learn the best control policies and adapt to the complex and nonlinear dynamics that PEDs present (Khooban and Gheisarnejad, 2020; Muduli et al., 2024).

The dynamics of LFC are significantly impacted by the addition of PEDs to MGs. These devices offer enhanced controllability and quick response times, but they also come with drawbacks that call for complex control schemes to preserve system stability and maximize performance.

3.5.3 Impact on frequency control

PEDs are essential for LFC in MGs. Their quick response capabilities and exact regulation of power flows provide substantial benefits in preserving frequency stability, particularly in systems with a high integration of RESs. Simultaneously, these devices also pose certain challenges in LFC. Key benefits and challenges are shown in Fig. 3.10 and discussed below in brief.

3.5.3.1 *Benefits*

Inverter-based resources provide effective primary frequency control by quickly adjusting their output to reduce frequency fluctuations. In

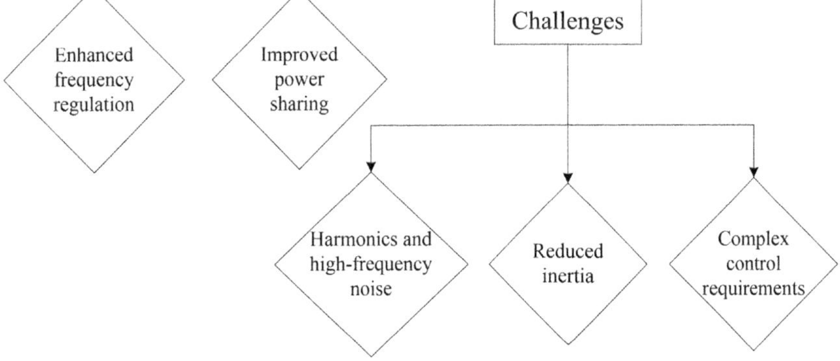

FIGURE 3.10 Impact of PEDs on frequency control.

MGs with low inertia, where traditional mechanical systems might not have the speed required to maintain stability, this quick response is crucial (Ojo et al., 2023). The proportionate distribution of active and reactive power among DERs is made possible by sophisticated control algorithms. This increases overall system reliability by preventing individual units from being overloaded and ensuring fair load distribution (Chang and Zhang, 2016).

3.5.3.2 *Challenges*

The switching actions characteristic of PEDs can generate harmonics and high-frequency noise within the system. These disruptions may affect sensitive equipment performance and power quality (Ghosh and Zare, 2022). MGs that rely heavily on inverter-based DERs often exhibit lower system inertia, making them more susceptible to fluctuations in frequency. To guarantee frequency stability, sophisticated control systems must be established (Ojo et al., 2023). Sophisticated control algorithms that can handle the nonlinear and time-varying characteristics of PEDs are required for effective LFC in MGs with a variety of DERs. Such controls can be complex and resource-intensive to set up and implement (She et al., 2023).

3.6 Integration of component models for system-level simulation

System-level simulation of MGs necessitates the incorporation of diverse component models, including DERs, ESSs, PEDs, and loads. This complete method offers insights into the relationships between components, facilitating thorough performance analysis, control design, and optimization.

3.6.1 Co-simulation approaches

Co-simulation is a methodology that combines different models or simulation tools, each designed for different parts of an MG, to enable thorough system-level analysis. This approach preserves the distinct complexities and particularities of various models while facilitating their interaction. Co-simulation involves combining various simulators or models, each of which solves its own domain-specific equations and exchanges data at predetermined intervals. Co-simulation techniques have been crucial for analyzing and enhancing LFC in MGs. By combining multiple simulation platforms, it enables a thorough examination of the intricate relationships between the electrical and communication systems in an MG.

The functional mock-up interface (FMI) standard was used in a noteworthy study by Nguyen et al. to assess the impact of communication networks on the secondary control of islanded MGs (Nguyen et al., 2017). This approach made it easier to thoroughly evaluate how communication delays and disruptions affect frequency regulation, providing important information for creating more reliable control systems.

By developing a distributed control framework for frequency and voltage regulation in AC MGs Kölsch et al., (2020), Kolsch et al. made a significant contribution. Their co-simulation approach made it possible to assess the controller's effectiveness and resilience to changes in the demand for active and reactive power, demonstrating the viability of their control strategy in maintaining system stability.

Furthermore, Singhal et al. presented a synchronized control method for grid-forming and grid-following inverters in inverter-based MGs (Singhal et al., 2020). They demonstrated the importance of coordinated management in achieving accurate power distribution and frequency regulation by co-simulating their approach across a variety of disturbances and communication degradation scenarios.

These results demonstrate how crucial co-simulation is to improving LFC methods in MGs. The development of robust control mechanisms that are essential for maintaining frequency stability in modern power systems is facilitated by co-simulation, which provides a platform for modeling and assessing the complex dynamics of electrical and communication systems.

Additional important uses of co-simulation in MGs include: i) evaluating how RESs and load demand interact, ii) investigating how communication delays affect system stability, and iii) developing synchronized control strategies for ESSs and DERs.

3.6.2 Challenges in integrating diverse models

The integration of models from many domains and technologies into co-simulation frameworks poses numerous technical and practical issues that must be resolved for effective system-level simulation.

1. *Diversity of models:* Models for DERs, ESSs, loads, and PEDs frequently utilize distinct mathematical formulations, solvers, and temporal scales. For example, a generator model may function on millisecond dynamics, whereas a load model may encompass steady-state activity spanning several minutes. This discrepancy hinders their assimilation into a cohesive simulated environment (Barbierato et al., 2022).
2. *Temporal step synchronization:* Diverse components frequently operate on different time scales, complicating the synchronization

of simulations without sacrificing accuracy or computational efficiency. Adaptive time-stepping and interpolation methods can synchronize data transmission between models, ensuring coherent interaction (Zhong et al., 2019).

3. *Data exchange and interface compatibility:* Standardized data formats and interfaces are crucial for effective communication among models. Frameworks such as the FMI or middleware solutions enable protocol translation, hence facilitating interoperability among diverse simulation programs (Hardy et al., 2024).
4. *Numerical stability and convergence:* Integrating models with disparate numerical characteristics may result in instability or convergence problems. The rapid dynamics of power electronics interacting with slower thermal models can induce numerical rigidity. The use of advanced numerical solvers and suitable coupling methods can alleviate these difficulties (Stifter et al., 2013).
5. *Computational complexity:* System-level simulation frequently requires substantial computer resources, particularly when intricate models are used. Employing reduced-order models or parallel computing can enhance performance, rendering large-scale simulations more practicable (Hardy et al., 2024).
6. *Validation and standardization of models:* It is essential to ensure the precision and compatibility of individual component models with system-level requirements. Utilizing common benchmarks and validation protocols to evaluate models prior to integration guarantees dependability and consistency inside the simulation framework (Barbierato et al., 2022).

3.7 Future trends in microgrid component modeling

The evolving landscape of MG technologies and ESSs has spurred significant advancements in component modeling. Future trends in MG component modeling focus on enhancing accuracy, integrating advanced technologies, and enabling robust simulation of complex, dynamic systems. These trends aim to address challenges posed by increased renewable energy penetration, real-time control needs, and decentralized operation.

3.7.1 Advancements in modeling techniques

Advancements in MG modeling are focusing on several critical domains to improve precision, efficacy, and robustness. These are:

1. *Accurate and immediate simulation:* Modern modeling techniques prioritize comprehensive physical and dynamic representations of

components, encompassing renewable energy sources, storage systems, and loads. This entails capturing nonlinearities, transient behaviors, and complex thermal-electrical interactions to guarantee accurate simulations. The utilization of real-time simulation systems is on the rise, especially for the testing and validation of control strategies before implementation. Hardware-in-the-Loop (HIL) simulations are essential for merging physical equipment with virtual models, enabling thorough system assessments (Jinsiwale et al., 2023).

2. *Modeling based on data and artificial intelligence:* Machine learning techniques are employed to create data-driven models that anticipate component behavior under diverse scenarios. Neural networks are utilized to estimate battery degradation processes and to predict PV generation based on meteorological data. A digital twin is a virtual representation of a physical system that continuously updates in real-time using sensor data. Digital twins are increasingly vital instruments for predictive maintenance and operational optimization in MGs (Jinsiwale et al., 2023).
3. *Multi-physics and interdisciplinary models:* Future models intend to amalgamate electrical, thermal, mechanical, and chemical domains to deliver a holistic assessment of system performance. Multi-physics modeling of fuel cells integrates electrochemical processes, thermal management, and fluid dynamics to attain a comprehensive knowledge of their behavior (Vijay et al., 2017).
4. *Probabilistic and stochastic models:* Probabilistic models are being created to address uncertainties in renewable generation, load needs, and market situations, hence improving the resilience of MG operations. Stochastic models are utilized for planning and operational objectives, enhancing decision-making across diverse scenarios to augment efficiency and reliability (Vijay et al., 2017).
5. *Incorporation of cyber-physical elements:* As the significance of communication and control in MGs escalates, modeling initiatives are integrating cyber-physical elements, such as network latency, cyber threats, and decentralized control algorithms. This integration is essential for the development of robust and secure MG systems (Halu et al., 2016).

3.7.2 Emerging technologies and their impact

Progress in MG technologies requires the creation of advanced models to effectively mimic and enhance system performance. Principal areas of emphasis encompass::

1. *Extensive utilization of renewable energy technologies:* Novel materials like perovskite and bifacial solar panels necessitate revised models that accurately reflect their distinct operational attributes. Perovskite solar cells exhibit excellent efficiency and potential for lower production costs, requiring models that consider their unique characteristics. Further, the incorporation of offshore wind technologies into MGs requires models that consider grid stability and underwater cable dynamics. Modeling endeavors must account for aspects such fluctuating wind velocities, marine environmental conditions, and the intricate interactions between WTs and the grid (Zhou et al., 2022).
2. *Advanced energy storage solutions:* The emergence of solid-state batteries, characterized by superior energy densities and enhanced safety profiles, necessitates the development of novel models to simulate their performance and degradation mechanisms. Precise modeling is crucial for forecasting lifespan and enhancing integration within MG systems. Also, hydrogen-based energy systems are becoming increasingly significant, necessitating comprehensive modeling of their production, storage, and consumption processes. This encompasses the simulation of electrolyzers, storage tanks, and fuel cells to evaluate efficiency and dynamic behavior inside MGs (Rouholamini et al., 2022).
3. *Decentralized and peer-to-peer energy systems:* Blockchain-based energy trading systems require integration with MG models to simulate peer-to-peer transactions and their effects on grid stability. Modeling these interactions is essential for comprehending the ramifications of decentralized energy markets. Further, decentralized optimization algorithms, including consensus-based methodologies, must be included into MG models to improve operational efficiency and resilience (Rouholamini et al., 2022).
4. *Quantum computing for energy optimization:* Quantum computing is emerging as a means to address intricate optimization challenges in MG operations, including economic dispatch and unit commitment. Models may require interaction with quantum algorithms to utilize their computational benefits (Hussein et al., 2024).
5. *Artificial intelligence in control systems:* Artificial intelligence-driven controllers and predictive algorithms are being incorporated into MG systems. These necessitate simulation models capable of interacting with AI algorithms and adapting to learning-driven decision-making processes, hence improving system efficiency and flexibility (Hussein et al., 2024).

3.8 Conclusion

The in-depth analysis of MG component modeling in this chapter emphasizes how vital it is to improving the sustainability, dependability, and efficiency of modern power systems. For control design and system-level analysis, accurate modeling of MG components, such as DERs, ESSs, loads, and PEDs, is essential. The combination of distributed generation, RESs, and the dynamic characteristics of batteries, supercapacitors, and flywheels highlights the need for robust and precise models to capture the intricate behaviors of these components. The complexity and interdependencies of MG components are further demonstrated by advancements in power electronics, which include inverter and converter dynamics and their effects on frequency regulation. Although they come with challenges like synchronization, data interchange, and computational complexity, co-simulation techniques enable efficient interaction across domain-specific tools as models converge for system-level simulation. Future developments that could improve MG performance include digital twins, AI-driven modeling, probabilistic approaches, and the incorporation of cutting-edge technology like solid-state batteries. A crucial step in creating reliable, intelligent, and sustainable MGs that can satisfy the intricate demands of modern energy environments is the ongoing development of modeling techniques and integration frameworks.

References

Aguirre, M., Kouro, S., Rojas, C.A., Vazquez, S., 2020. Enhanced switching frequency control in fcs-mpc for power converters. IEEE Trans. Ind. Electron. 68, 2470–2479.

AlMuhaini, M., Yahaya, A., AlAhmed, A., 2023. Distributed generation and load modeling in microgrids. Sustainability 15, 4831.

Arani, A.K., Karami, H., Gharehpetian, G., Hejazi, M., 2017. Review of flywheel energy storage systems structures and applications in power systems and microgrids. Renew. Sustain. Energy Rev. 69, 9–18.

Bae, S., Kwasinski, A., 2012. Dynamic modeling and operation strategy for a microgrid with wind and photovoltaic resources. IEEE Trans. Smart Grid 3, 1867–1876.

Bangash, K., Farrag, M., Osman, A., 2019. Investigation of energy storage batteries in stability enforcement of low inertia active distribution network, Technology and Economics of Smart Grids and Sustainable. Energy 4, 1–12.

Barbierato, L., Rando Mazzarino, P., Montarolo, M., Macii, A., Patti, E., Bottaccioli, L., 2022. A comparison study of co-simulation frameworks for multi-energy systems: the scalability problem. Energy Inform. 5, 53.

Bevrani, H., 2014. Robust Power System Frequency Control, Power Electronics and Power Systems, 2 ed. Springer International Publishing, Switzerland.

Bevrani, H., Shokoohi, S., 2013. An intelligent droop control for simultaneous voltage and frequency regulation in islanded microgrids. IEEE Trans. Smart Grid 4 (3), 1505–1513.

Bhuvaneswari, G., Singh, B., et al., 2020. A reliable microgrid comprising solar PV–WEGS and battery with seamless power transfer capability. IEEE Trans. Ind. Electron. 68, 9665–9674.

Bierbooms, W., Cheng, P.-W., 2002. Stochastic gust model for design calculations of wind turbines. J. Wind Eng. Ind. Aerod. 90, 1237–1251.

Blaabjerg, F., Ionel, D.M., 2017. Renewable Energy Devices and Systems with Simulations in MATLAB® and ANSYS®. CRC Press.

Boennec, C., Sareni, B., Ngueveu, S.U., 2023. Robust design of microgrids using component models with different level of accuracy. In: OIPE 2023 (17th International Workshop on Optimization and Inverse Problems in Electromagnetism).

Chakraborty, S., Arvind, P., Kumar, D., 2019. Integrated solar PV MPPT and Vf control for stand-alone microgrid. In: 2019 International Conference on Electrical, Electronics and Computer Engineering (UPCON), IEEE, pp. 1–6.

Chang, C.-Y., Zhang, W., 2016. Distributed control of inverter-based lossy microgrids for power sharing and frequency regulation under voltage constraints. Automatica 66, 85–95.

Chen, X., Zhang, M., Wu, Z., Wu, L., Guan, X., 2024. Model-free load frequency control of nonlinear power systems based on deep reinforcement learning. IEEE Trans. Ind. Inf. 20 (4), 6825–6833.

Choi, B.-K., Chiang, H.-D., Li, Y., Li, H., Chen, Y.-T., Huang, D.-H., Lauby, M.G., 2006. Measurement-based dynamic load models: derivation, comparison, and validation. IEEE Trans. Power Syst. 21, 1276–1283.

Davoudkhani, I.F., Zare, P., Abdelaziz, A.Y., Bajaj, M., Tuka, M.B., 2024. Robust load-frequency control of islanded urban microgrid using 1PD-3DOF-PID controller including mobile EV energy storage. Sci. Rep. 14.

DiOrio, N., Dobos, A., Janzou, S., Nelson, A., Lundstrom, B., 2015. Technoeconomic Modeling of Battery Energy Storage in SAM, Technical Report. National Renewable Energy Lab.(NREL), Golden, CO (United States).

Fluck, M., Crawford, C., 2016. A stochastic aerodynamic model for stationary blades in unsteady 3D wind fields. J. Phys. Conf. Ser. 753, 082009.

Fontenot, H., Dong, B., 2019. Modeling and control of building-integrated microgrids for optimal energy management–a review. Appl. Energy 254, 113689.

Ghosh, A., Zare, F., 2022. Control of Power Electronic Converters with Microgrid Applications. John Wiley & Sons.

Guerrero, J.M., Vasquez, J.C., Matas, J., De Vicuña, L.G., Castilla, M., 2010. Hierarchical control of droop-controlled AC and DC microgrids—a general approach toward standardization. IEEE Trans. Ind. Electron. 58, 158–172.

Halu, A., Scala, A., Khiyami, A., González, M.C., 2016. Data-driven modeling of solar-powered urban microgrids. Sci. Adv. 2, e1500700.

Hardy, T.D., Palmintier, B., Top, P.L., Krishnamurthy, D., Fuller, J.C., 2024. Helics: a co-simulation framework for scalable multi-domain modeling and analysis. IEEE Access 12, 24325–24347.

Hu, X., Li, S., Peng, H., 2012. A comparative study of equivalent circuit models for li-ion batteries. J. Power Sources 198, 359–367.

Hussein, H.M., Ibrahim, A.M., Taha, R.A., Rafin, S.S.H., Abdelrahman, M.S., Kharchouf, I., Mohammed, O.A., 2024. State-of-the-art electric vehicle modeling: architectures, control, and regulations. Electronics 13, 3578.

Jinsiwale, R.A., Maharjan, M., Mana, P.T., Ashok, A., Becejac, T., Harpool, S.E., Hofer, W.J., Vielma, A.R., Gloria, G., Edgar, T.W., 2023. Design and Development of a High Fidelity Cyber-Physical Testbed. Pacific Northwest National Laboratory (PNNL), Richland, WA (United States). Technical Report.

Kale, S.A., Jagtap, P., Helonde, J., 2011. Role of micro sources within micro grid. In: 2011 Fourth International Conference on Emerging Trends in Engineering & Technology, IEEE, pp. 174–179.

Khokhar, B., Parmar, K.P.S., 2022. A novel adaptive intelligent MPC scheme for frequency stabilization of a microgrid considering SoC control of EVs. Appl. Energy 309.

Khokhar, B., Parmar, K.S., 2023. Utilizing diverse mix of energy storage for LFC performance enhancement of a microgrid: a novel MPC approach. Appl. Energy 333.

Khokhar, B., Parmar, K.P.S., 2025. Impact analysis of capacitive energy storage integration on load frequency control performance of microgrids employing a new dual-stage controller. ISA Trans. 162, 162–178.

Khokhar, B., Singh Parmar, K., 2024. Frequency regulation in a microgrid integrating redox flow battery by utilizing an optimal predictive control approach. Electr. Eng. 1–19.

Khokhar, B., Dahiya, S., Parmar, K.P.S., 2020. A robust cascade controller for load frequency control of a standalone microgrid incorporating electric vehicles. Elec. Power Compon. Syst. 48, 711–726.

Khokhar, B., Dahiya, S., Parmar, K.P.S., 2021. A novel hybrid fuzzy PD-TID controller for load frequency control of a standalone microgrid. Arabian J. Sci. Eng. 46, 1053–1065.

Khokhar, B., Parmar, K.S., Thakur, T., Kothari, D., 2024. Load Frequency Control of Microgrids. CRC Press.

Khooban, M.H., Gheisarnejad, M., 2020. A novel deep reinforcement learning controller based type-ii fuzzy system: frequency regulation in microgrids. IEEE Trans. Emerg. Top. Comput. Intell. 5, 689–699.

Kölsch, L., Wieninger, K., Krebs, S., Hohmann, S., 2020. Distributed frequency and voltage control for ac microgrids based on primal-dual gradient dynamics. IFAC-PapersOnLine 53, 12229–12236.

Kundur, P., 1994. Power System Stability and Control, Power System Engineering Series. McGraw-Hill, Inc, USA.

Mahdavian, A., Ghadimi, A.A., Bayat, M., 2021. Microgrid small-signal stability analysis considering dynamic load model. IET Renew. Power Gener. 15, 2799–2813.

Majumder, R., Chaudhuri, B., Ghosh, A., Majumder, R., Ledwich, G., Zare, F., 2009. Improvement of stability and load sharing in an autonomous microgrid using supplementary droop control loop. IEEE Trans. Power Syst. 25, 796–808.

Majzoobi, A., Khodaei, A., 2016. Application of microgrids in addressing distribution network net-load ramping. In: 2016 IEEE Power & Energy Society Innovative Smart Grid Technologies Conference (ISGT), IEEE, pp. 1–5.

Mathews, I., Xu, B., He, W., Barreto, V., Buonassisi, T., Peters, I.M., 2020. Technoeconomic model of second-life batteries for utility-scale solar considering calendar and cycle aging. Appl. Energy 269.

Muduli, R., Jena, D., Moger, T., 2024. A survey on load frequency control using reinforcement learning-based data-driven controller. Appl. Soft Comput.

Nayak, S.K., Gaonkar, D.N., 2012. Modeling and performance analysis of microturbine generation system in grid connected/islanding mode. In: 2012 IEEE International Conference on Power Electronics, Drives and Energy Systems (PEDES). IEEE, pp. 1–6.

Ndeke, C.B., Adonis, M., Almaktoof, A., 2024. Basic circuit model of voltage source converters: methodology and modeling. AppliedMath 4, 889–907.

Nempu, P.B., Jayalakshmi, N., Shaji, K., Singh, M., 2019. Fuzzy-PI controllers for voltage and frequency regulation of a PV-FC based autonomous microgrid. In: 2019 IEEE International Conference on Distributed Computing, VLSI, Electrical Circuits and Robotics (DISCOVER). IEEE, pp. 1–6.

Nguyen, T.L., Tran, Q.-T., Caire, R., Besanger, Y., Hoang, T.T., et al., 2017. FMI compliant approach to investigate the impact of communication to islanded microgrid secondary control. In: 2017 IEEE Innovative Smart Grid Technologies-Asia (ISGT-Asia). IEEE, pp. 1–6.

Ojo, Y., Watson, J.D., Laib, K., Lestas, I., 2023. A distributed scheme for voltage and frequency control and power sharing in inverter-based microgrids. IEEE Trans. Control Syst. Technol. 31, 1679–1691.

Paul, S., Khan, K.A., Islam, K.A., Islam, B., Reza, M.A., 2012. Modeling of a biomass energy based (BPL) generating power plant and its features in comparison with other generating plants. IPCBEE 44, 12–16.

Rafik, F., Gualous, H., Gallay, R., Crausaz, A., Berthon, A., 2007. Frequency, thermal and voltage supercapacitor characterization and modeling. J. Power Sources 165, 928–934.

Rahimi, F., Ipakchi, A., 2010. Demand response as a market resource under the smart grid paradigm. IEEE Trans. Smart Grid 1, 82–88.

Raju, P., Jain, T., 2019. Development and validation of a generalized modeling approach for islanded inverter-based microgrids with static and dynamic loads. Int. J. Electr. Power Energy Syst. 108, 177–190.

Rocabert, J., Luna, A., Blaabjerg, F., Rodriguez, P., 2012. Control of power converters in AC microgrids. IEEE Trans. Power Electron. 27, 4734–4749.

Roos, M.H., Nguyen, P.H., Morren, J., Slootweg, J., 2020. Aggregation of component-based grid-feeding der and load models for simulation of microgrid islanding transients. Elec. Power Syst. Res. 189.

Roosa, S.A., 2020. Fundamentals of Microgrids: Development and Implementation. CRC Press.

Rouholamini, M., Wang, C., Nehrir, H., Hu, X., Hu, Z., Aki, H., Zhao, B., Miao, Z., Strunz, K., 2022. A review of modeling, management, and applications of grid-connected li-ion battery storage systems. IEEE Trans. Smart Grid 13, 4505–4524.

Saleh, A., Awad, A., Ghanem, W., 2019. Modeling, control, and simulation of a new topology of flywheel energy storage systems in microgrids. IEEE Access 7, 160363–160376.

San Martín, I., Ursúa, A., Sanchis, P., 2013. Integration of fuel cells and supercapacitors in electrical microgrids: analysis, modelling and experimental validation. Int. J. Hydrogen Energy 38, 11655–11671.

Schimpe, M., Naumann, M., Truong, N., Hesse, H.C., Santhanagopalan, S., Saxon, A., Jossen, A., 2018. Energy efficiency evaluation of a stationary lithium-ion battery container storage system via electro-thermal modeling and detailed component analysis. Appl. Energy 210, 211–229.

She, B., Li, F., Cui, H., Wang, J., Min, L., Oboreh-Snapps, O., Bo, R., 2023. Decentralized and coordinated vf control for islanded microgrids considering der inadequacy and demand control. IEEE Trans. Energy Convers. 38, 1868–1880.

Shen, M., Gao, Q., 2019. A review on battery management system from the modeling efforts to its multiapplication and integration. Int. J. Energy Res. 43, 5042–5075.

Singhal, A., Vu, T.L., Du, W., 2020. Coordinated frequency and voltage regulation of grid-following and grid-forming inverters. arXiv preprint arXiv:2012.06685.

Smith, K., Saxon, A., Keyser, M., Lundstrom, B., Cao, Z., Roc, A., 2017. Life prediction model for grid-connected li-ion battery energy storage system. In: 2017 American Control Conference (ACC). IEEE, pp. 4062–4068.

Song, Y., Chen, J., Beer, M., Comerford, L., 2019. Wind speed field simulation via stochastic harmonic function representation based on wavenumber–frequency spectrum. J. Eng. Mech. 145.

Stanojev, O., Markovic, U., Aristidou, P., Hug, G., Callaway, D., Vrettos, E., 2020. MPC-based fast frequency control of voltage source converters in low-inertia power systems. IEEE Trans. Power Syst. 37, 3209–3220.

Stifter, M., Widl, E., Andrén, F., Elsheikh, A., Strasser, T., Palensky, P., 2013. Co-simulation of components, controls and power systems based on open source software. In: 2013 IEEE Power & Energy Society General Meeting. IEEE, pp. 1–5.

Tudu, A.K., Naguru, N., Dey, H.N., Paul, S., 2024. Load frequency control of an isolated microgrid using optimized model predictive control by GA. Electr. Eng. 1–13.

Venayagamoorthy, G.K., Sharma, R.K., Gautam, P.K., Ahmadi, A., 2016. Dynamic energy management system for a smart microgrid. IEEE Transact. Neural Networks Learn. Syst. 27, 1643–1656.

Vijay, A., Doolla, S., Chandorkar, M.C., 2017. Real-time testing approaches for microgrids. IEEE J. Emerg. Sel. Top. Power Electron. 5, 1356–1376.

Wang, X., Zhao, Q., He, B., Wang, Y., Yang, J., Pan, X., 2017. Load frequency control in multiple microgrids based on model predictive control with communication delay. J. Eng. 2017, 1851–1856.

Wang, R., Zhu, M., Zhang, X., Pham, H., 2023. Lithium-ion battery remaining useful life prediction using a two-phase degradation model with a dynamic change point. J. Energy Storage 59, 106457.

Xiong, Z., Liu, S., Fang, Y., Li, Y., Yang, J., 2020. Research on modeling of microgrid load frequency control based on V2G technology. In: 2020 5th International Conference on Mechanical, Control and Computer Engineering (ICMCCE), IEEE, pp. 621–626.

Xu, L., 2013. Modeling, Analysis and Control of Voltage-Source Converter in Microgrids and HVDC. University of South Florida.

Yan, Z., Xu, Y., 2018. Data-driven load frequency control for stochastic power systems: a deep reinforcement learning method with continuous action search. IEEE Trans. Power Syst. 34, 1653–1656.

Ye, Y., Shi, Y., Saw, L.H., Tay, A.A., 2014. An electro-thermal model and its application on a spiral-wound lithium ion battery with porous current collectors. Electrochim. Acta 121, 143–153.

Zhong, W., Liu, M., Milano, F., 2019. A co-simulation framework for power systems and communication networks. In: 2019 IEEE Milan PowerTech. IEEE, pp. 1–6.

Zhou, J., Huang, Q., Ding, Y., Hou, G., Zhao, Y., 2022. Passivating contacts for high-efficiency silicon-based solar cells: from single-junction to tandem architecture. Nano Energy 92, 106712.

CHAPTER

4

Control strategies for microgrid frequency control

Sanjeev Kumar Bhagat[1], *Naladi Ram Babu*[2,3], *Sachchidanand Jha*[1], *and Abhishek Thakur*[4]

[1] Electrical Engineering Department, Sandip University, Madhubani, Bihar, India; [2] Department of Electrical and Electronics Engineering, Aditya University, Surampalem, Andhra Pradesh, India; [3] Department of Artificial Intelligence & Machine Learning, Aditya University, Surampalem, Andhra Pradesh, India; [4] Department of Electrical Engineering, Sandip University, Madhubani, Bihar, India

4.1 Introduction to microgrid frequency control

The increasing reliance of fossil fuels for electricity generation and transportation, combined with rising energy demands, accelerates their depletion (Wang et al., 2018). This has driven the development of clean energy alternatives, leveraging renewable energy sources (RES) (Himabindu et al., 2021). RESs are economical, and their integration into power systems is increasing (Irfan et al., 2023). A microgrid is a network of distributed energy resources and unified loads, manageable entities within defined electrical boundaries on the primary grid. They can effortlessly transition between isolated and grid-connected modes, providing easy access to the connection from the main grid (Sharmeela et al., 2021).

Hybrid microgrid systems in Fig. 4.1 represent an advanced solution that integrates various RESs like solar and wind energy with traditional energy sources and energy storage systems (ESS) (Jayachandran and Ravi, 2017). To enhance the effectiveness of ESS, Electric Vehicles (EVs) have been seamlessly incorporated into microgrids, demonstrating promising profitable feasibility thanks to the adoption of Vehicle-to-Grid technologies (Jan et al., 2021). Charging EVs with RESs is a highly effective strategy

Frequency Control in Microgrids
https://doi.org/10.1016/B978-0-443-36476-1.00007-9

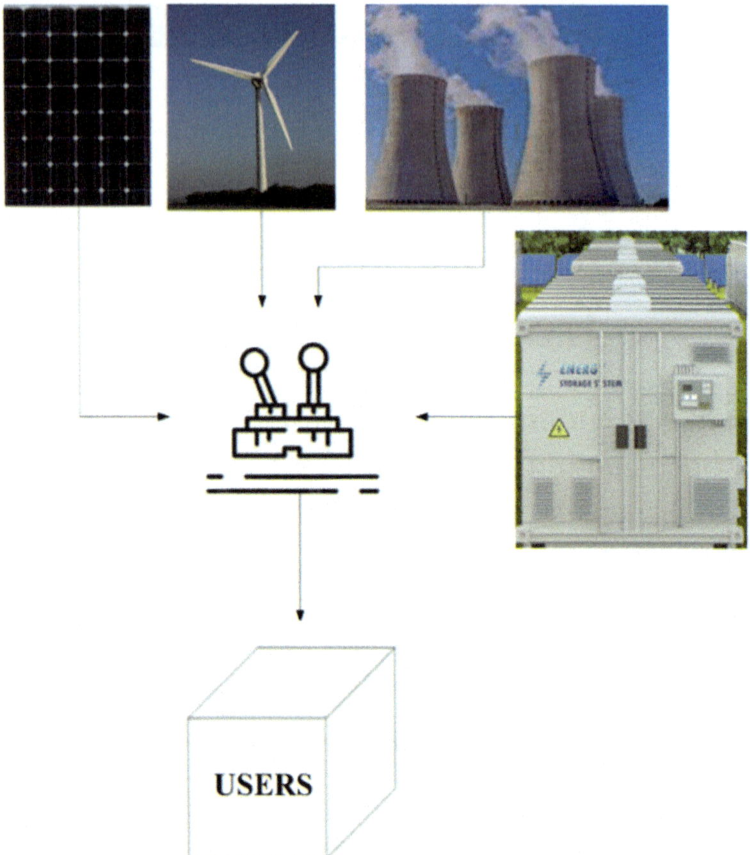

FIGURE 4.1 Schematic diagram of hybrid micro-grid. *Jayachandran, M., Ravi, G., 2017. Design and optimization of hybrid Micro-grid system. Energy Proc., 117, 95–103.*

for reducing carbon discharges and decreasing reliance on fossils (Chittur Ramaswamy et al., 2016). Additionally, EVs can be energy storage units supporting RES during idle stages. As a result, RESs and EVs play an active role in generation and support in microgrids (Zhang et al., 2013a). These RES-based microgrids reduce relic fuel releases and expand entrée to power in isolated regions. Microgrids are the future of electricity generation, comprising storage systems, distributed generators, loads, and energy management systems (Rahbar et al., 2018).

Electricity generation from RESs is inherently unpredictable due to their dependence on factors like weather conditions. As a result, high integration of RESs can introduce challenges such as voltage instability, issues with frequency regulation, and disruptions in reliability. RESs like solar photo voltaic systems and wind turbines have zero inertia. With increasing RES penetration, the overall system inertia decreases (Jie et al., 2017). This decline in inertia makes the system more susceptible to significant issues from sudden load disturbances and variations in RES output, such as power-sharing deterioration, leading to deviations in system frequency from its set limits (Hirsch et al., 2018). Hence, Load Frequency Control (LFC) is a critical concern in microgrids due to the varying demand and renewable generation profiles. Also, the power transmission for remote areas is quite challenging (Rafiee et al., 2021).

4.1.1 Overview of control strategies

Implementing effective control strategies is essential for addressing the challenges encountered in microgrids. Both conventional (Ouédraogo et al., 2022) and soft computing-based methods (Murugesan et al., 2023) have been explored in the literature to enhance the design and performance of LFC systems (Ibraheem et al., 2023). However, traditional controllers often struggle to manage the uncertainties and variability inherent to RESs (Gouran-Orimi and Ghasemi-Marzbali, 2023). Among these, the Proportional-Integral-Derivative (PID) controller remains the most widely used in industrial applications (Ranjan and Shankar, 2022; Rajamand, 2021), due to its simplicity and reliability (Heidary et al., 2022; Hu and Lanzon, 2019). Variants such as PI and PID controllers have been extensively employed in conventional LFC strategies to maintain system frequency. Recent advancements include the use of fractional calculus-based controllers, particularly the Fractional Order PID (FOPID) controller, which offers greater flexibility in tuning. Although FOPID controllers have shown promise in maintaining frequency and system stability (Ranjan and Shankar, 2022), their configuration can be time-consuming and may not always yield optimal results. Fuzzy-based controllers have also gained traction due to their intuitive design and adaptability, often used for gain tuning or as the primary control method (Zheng et al., 2021). Despite their flexibility, they require manual tuning and may not consistently outperform traditional methods (Rajamand, 2021). While controller design plays a critical role, it is often insufficient alone for optimal LFC. Hence, optimization techniques are employed to identify the best controller parameters (Enslin, 2018). The literature presents a wide array of such optimization-based LFC strategies (Molina and Mercado, 2016), including approaches like the Artificial Rabbits Optimization Algorithm (Kaur et al., 2022), Craziness-Based PSO (Hasanien, 2018), Artificial Bee Colony and Teaching-Learning-Based Optimization (Asghar et al., 2023), Model Predictive Control (MPC) (Ramesh et al.,

2021), Adaptive Fuzzy MPC (Hongesombut and Keteruksa, 2023), Sewing Training-Based Optimization (Negahban et al., 2022), Gorilla Troops Optimizer (Albertos et al., 2018), PSO and Jaya algorithms (Tungadio et al., 2018), Firefly and PSO-based hybrid methods (Bevrani, 2014), and both integral and second-order Sliding Mode Control (SMC) systems (Daraz et al., 2023; Zhang et al., 2023; Khalil et al., 2023). A particularly noteworthy contribution is found in Alayi et al. (2021), where a novel method improves islanded microgrid LFC by incorporating dynamic signals from microgrid controllers, offering enhanced system responsiveness and stability.

4.2 Proportional-Integral-Derivative controllers

Proportional-Integral-Derivative (PID) controllers are one of industrial and engineering application's most widely used control mechanisms. They are designed to provide accurate and stable control of various processes by adjusting the control input based on the error between the desired setpoint and the actual process variable. The PID controller combines three control actions: proportional, integral, and derivative, each contributing to the overall control effort uniquely.

4.2.1 Fundamental of PID control

A Proportional-Integral-Derivative (PID) controller is a widely used control loop feedback mechanism in industrial control systems. It is designed to continuously calculate an error value as the difference between a desired setpoint and a measured process variable, then apply a correction based on proportional (P), integral (I), and derivative (D) terms. The P terms send an output to an actuator with an output multiple of the error. The P term uses an offset to keep the calculation away from the set point (Hasan et al., 2022). The P terms work effectively to improve system response by minimizing output error. However, it cannot nullify the SSE and takes a long time to settle down (Long et al., 2021). The integral (I) controller, also known as reset control, resets the error that occurred in the system to minimize steady-state error, speed up the system response, and minimize the settling time. However, it reduces the system's stability range (Daraz, 2023). The derivative (D) controller adjusts its behavior in response to the change in error. If the error is decreasing and the set point is close, the set point will be attained indirectly to change the error direction. As a result, the number of oscillations and overshoot in the system dynamics reduces. However, it may amplify the disturbance noises (A robust cascaded controller for, 2023). The PID controller's transfer function equation with gains proportional (K_{Pj}), integral (K_{Ij}) and derivative (K_{Dj}) and diagram are in Eq. (4.1) and Fig. 4.2, respectively. The diagrammatic representation for evolution of PID controller is in Fig. 4.3

$$F(s) = \left(K_{Pj} + \frac{K_{Ij}}{s} + sK_{Dj} \right) \quad U(s) \tag{4.1}$$

4.2.2 Application of PID in microgrid

Microgrids are localized energy systems that can operate independently or in conjunction with the main grid (Hassan et al., 2021). They integrate distributed energy resources (DERs) such as solar panels, wind turbines, and energy storage systems (ESS). Maintaining frequency

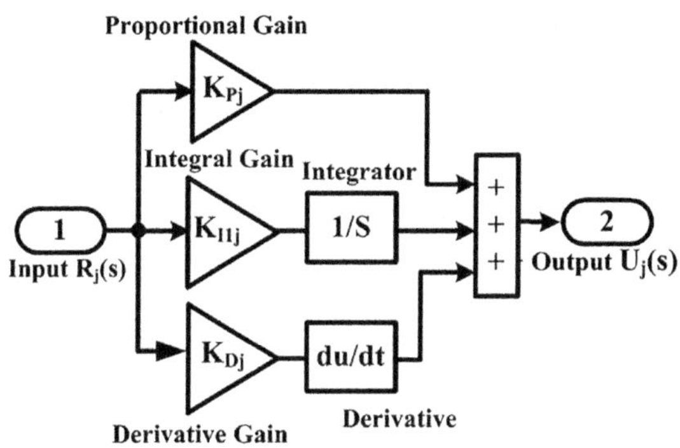

FIGURE 4.2 Transfer function diagram of PID controller.

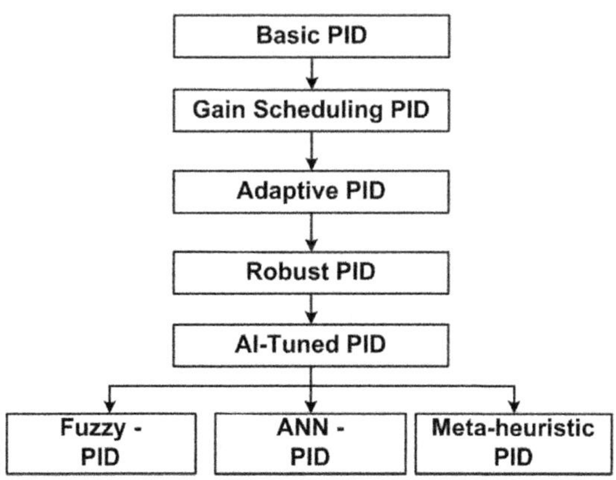

FIGURE 4.3 Evolution stages of PID controller.

stability is a key challenge in microgrids due to fluctuating power supply and demand. PID controllers are widely used due to their simplicity, robustness and effectiveness in various control applications. Recent studies have examined PID control in microgrids showcasing its effectiveness and adaptability with key finding summarized in Table 4.1.

Recent studies have explored various aspects of PID control in microgrids, demonstrating its effectiveness and adaptability. Here are some key findings are noted in Table 4.1.

Table 4.1 provides a detailed overview of multiple studies examining the use of PID controllers in microgrids, emphasizing their aims, methods, and results. The main aim of these studies is to assess and improve the performance of PID controllers in sustaining frequency stability across various operating conditions, including grid-connected and islanded modes, as well as scenarios with high renewable energy integration. The methodologies utilized encompass simulation-based studies, comparative analyses with advanced control techniques including Model Predictive Control (MPC) and Fuzzy Logic Control (FLC), adaptive tuning, robust control design, experimental validation, and optimization through algorithms such as Genetic Algorithm (GA) and Particle Swarm Optimization (PSO). Several studies examine the integration of artificial intelligence and machine learning techniques with PID control to enhance adaptability and prediction. Research indicates that conventional PID controllers, despite their simplicity, reliability, and widespread application, necessitate meticulous tuning to function effectively within the dynamic context of microgrids. Adaptive PID controllers exhibit enhanced performance in managing swift variations in load and generation. In contrast, the integration of PID controllers with secondary frequency regulation or artificial intelligence techniques markedly decreases peak overshoots and settling times. Conventional PID controllers exhibit peak overshoots ranging from 8.5% to 9.3% and settling times of 14–16 s. In contrast, optimized PID configurations can reduce overshoot to approximately 3.2% and settling times to about 7 s. Experimental studies confirm the efficacy of these enhancements in practical applications, highlighting the significance of real-time adjustments. Additionally, certain studies examine the scalability of PID controllers, suggesting distributed or hierarchical control strategies that are appropriate for large-scale microgrids. In conclusion, studies demonstrate that although PID controllers are fundamental to microgrid frequency control, their efficacy can be markedly improved through optimization, adaptive mechanisms, and hybrid methods incorporating contemporary intelligent techniques.

4.2.3 Tuning techniques for microgrid application

The growing integration of DERs into microgrids requires advanced control strategies to maintain stability and reliability. Effective PID

TABLE 4.1 The objectives, methodologies, and findings of the PID controller.

Ref	Objective	Methodology	Findings		Outcomes		
					Peak overshoots (%)	**Peak overshoots (%)**	**Settling time (s)**
Hasan et al. (2022)	Explore the basic application of PID controllers in microgrids	Simulation studies on grid-connected and islanded modes	PID controllers stabilize frequency by adjusting DERs and ESS outputs; tuning parameters is critical.	Basic PID controller	9.0	6.1	16.5
Long et al. (2021)	Compare PID control with MPC and FLC in microgrids	Comparative analysis with different control strategies	PID controllers offer competitive performance in response time and stability compared to more complex methods.	Conventional PID	8.7	5.4	15.2
				Droop control	7.1	4.8	13.4
				MPC	3.8	2.2	8.9

Continued

TABLE 4.1 The objectives, methodologies, and findings of the PID controller.—cont'd

Ref	Objective	Methodology	Findings		Outcomes		
					Peak overshoots (%)	Peak overshoots (%)	Settling time (s)
Daraz (2023)	Propose adaptive PID controller with real-time parameter tuning	Adaptive control simulation	Adaptive PID improves handling rapid load and generation changes, maintaining better frequency stability.	Conventional PID	8.5	5.2	14.8
				Adaptive PID (proposed)	3.9	2.3	8.5
A robust cascaded controller for (2023)	Combine PID control with secondary frequency regulation	Hybrid control strategy simulation	Enhances frequency regulation by addressing both primary and secondary deviations.	Conventional PID	8.1	5.0	14.2
				PID with secondary frequency regulation	4.2	2.7	9.1

Hassan et al. (2021)	Evaluate PID performance in microgrids with high renewable penetration	Extensive simulations	PID controllers effectively mitigate frequency deviations from renewable variability with proper tuning.	Conventional PID	9.3	6.0	16.1
				Optimized PID (via simulation tuning)	4.5	2.6	9.3
Ray and Mohanty (2019)	Validate PID controller performance in a lab-scale microgrid	Experimental validation	Experimental results confirm simulation findings emphasize the importance of real-time tuning and adaptation.	Conventional PID	8.8	5.7	15.5
				Experimentally tuned PID	4.1	2.5	9.0
Pajooh et al. (2024)	Integrate PID control with AI and ML techniques	Integration and simulation	Combining PID with modern techniques enhances adaptability and predictive capabilities for resilient operations.	Conventional PID	8.6	5.4	14.7
				PID + advanced techniques (e.g., ANN, fuzzy)	3.6	2.1	7.8

Continued

TABLE 4.1 The objectives, methodologies, and findings of the PID controller.—cont'd

Ref	Objective	Methodology	Findings		Outcomes		
					Peak overshoots (%)	Peak overshoots (%)	Settling time (s)
Blaabjerg et al. (2006)	Address scalability and interoperability in large microgrids	Theoretical analysis and case studies	Focus on developing standardized protocols and scalable control architectures for large microgrid networks.	Conventional PID	9.1	5.8	16.0
				Distributed/ Hierarchical PID	4.0	2.3	8.5
Savaghebi et al. (2012)	Develop an optimized tuning method for PID controllers	Simulation and optimization techniques	Optimized PID tuning improves frequency control accuracy and response time in microgrids.	Conventional PID	8.9	5.5	15.0
				Optimized PID (e.g., GA/PSO)	3.5	2.0	7.5

Bidram et al. (2013)	Enhance the robustness of PID controllers in microgrids	Robust control design and simulation	Robust PID controllers show improved performance under varying load and generation conditions.	Conventional PID	8.7	5.3	14.5
				Robust PID (proposed)	3.2	1.9	7.2

controller tuning is crucial and recent research has focus on developing and refining tuning techniques to optimize PID performance in microgrids. Adaptive PID control with real-time tuning improves frequency stability in dynamic load conditions (Hassan et al., 2021). Authors in Wang et al. (2015) proposed a machine learning based PID control approach that dynamically adjusts parameters using historical data and real time measurements resulting in improved dynamic response and minimized frequency deviations. The robust control design utilizes robust PID controllers to ensure stability and performance effectively managing variability in load and generation conditions (Hassan et al., 2021).

Panda and Padhy (2020) used genetic algorithm to optimize PID parameters, resulting in fast response times and reduced overshoot. Haque and Mahmud (2021) suggested particle swarm optimization (PSO) to enhance frequency regulation and stability in PID Control under varying operating conditions. Hybrid fuzzy-PID for enhanced stability is discusses in Zhang and Yu (2022) for nonlinear systems. Wu et al. (2019) presented a neural network approach for dynamically tuning PID parameters using historical data resulting in improved response time and accuracy. Hybrid optimization techniques were proposed to integrate multiple algorithms, yielding more robust and efficient PID controllers (Ahmed and Li, 2023). Reinforcement learning-based tuning, as investigated in Wang and Liu (2024), offers high levels of adaptability and stability in dynamic and uncertain environments. Savaghebi and Jalilian (2022) integrated model predictive control (MPC) with PID tuning, enhancing control accuracy and response time.

Additionally, Hatziargyriou and Bindner (2021) addressed scalability and interoperability challenges in large microgrids, emphasizing standardized protocols and scalable control architectures. Sharma and Rout (2021) optimized PID controller tuning for frequency control, achieving improved accuracy and response times. Meng and Sanseverino (Meng et al., 2020) combined AI and machine learning with PID control, enhancing adaptability and predictive capabilities. Experimental validation by Zamora and Srivastava in (Ray and Mohanty, 2019) confirmed the effectiveness of PID controllers in lab-scale microgrids, highlighting the importance of real-time tuning. Lastly, Blaabjerg and Teodorescu (Blaabjerg et al., 2006) explored basic PID control applications in microgrids, emphasizing the critical role of tuning parameters in stabilizing frequency. Some key findings of the hybrid PID controller are noted in Table 4.2.

4.3 Model predictive control

Model predictive control (MPC) is a sophisticated control technique that forecasts a systems future behavior using a mathematical model and adjusts control actions accordingly. Due to its ability to manage input and

TABLE 4.2 The objectives, methodologies, and findings of the hybrid-PID controller.

Ref	Objective	Methodology	Findings		Outcomes		
					Peak overshoots (%)	Peak overshoots (%)	Settling time (s)
Daraz (2023)	Develop an adaptive PID controller with real-time parameter tuning	Adaptive control algorithms	Improved handling of rapid load and generation changes, maintaining better frequency stability.	Conventional PID	8.7	5.4	15.2
				Droop control	7.1	4.8	13.4
				MPC	3.8	2.2	8.9
Ray and Mohanty (2019)	Validate PID controller performance in a lab-scale microgrid	Experimental validation	Experimental results confirm simulation findings; emphasizes importance of real-time tuning and adaptation.	Conventional PID	8.8	5.7	15.5
				Experimentally tuned PID	4.1	2.5	9.0

Continued

TABLE 4.2 The objectives, methodologies, and findings of the hybrid-PID controller.—cont'd

Ref	Objective	Methodology	Findings		Outcomes		
					Peak overshoots (%)	Peak overshoots (%)	Settling time (s)
Bidram et al. (2013)	Integrate machine learning with PID control for adaptive tuning	Machine learning algorithms	Improved dynamic response And reduced frequency deviations.	Conventional PID	8.3	5.2	14.0
				Machine learning-assisted PID (proposed)	3.0	2.1	7.4
Wang et al. (2015)	Enhance robustness of PID controllers in microgrids	Robust control design	Improved performance under varying load and generation conditions.	Conventional PID	8.7	5.3	14.5
				Robust PID (proposed)	3.2	1.9	7.2
Nanou et al. (2017)	Optimize PID controller parameters using genetic algorithms	Genetic algorithms	Faster response times and reduced overshoot, enhancing overall system performance.	Conventional PID	9.2	5.6	15.3
				PID with genetic algorithm optimization	3.4	2.0	7.7

Zamora and Srivastava (2019)	Apply particle swarm optimization to tune PID controllers	Particle swarm optimization (PSO) algorithms	Better frequency regulation and stability under varying operating conditions.	Conventional PID	8.8	5.4	15.1
				PID with particle swarm optimization	3.2	2.0	7.5
Meng et al. (2020)	Combine fuzzy logic control with PID tuning	Fuzzy logic algorithms	Enhanced adaptability and stability, particularly in complex and nonlinear systems.	Conventional PID	9.4	5.8	15.6
				Hybrid PID + fuzzy logic	3.1	2.0	7.6
Hatziargyriou et al. (2021)	Utilize neural networks to dynamically tune PID controllers	Neural networks	Superior performance in terms of response time and accuracy.	Conventional PID	8.5	5.3	14.9
				Neural network-based dynamic tuning PID	2.9	1.8	6.8
Sharma and Rout (2021)	Develop hybrid optimization techniques combining different algorithms	Hybrid optimization	More robust and efficient PID controllers, enhancing microgrid stability and performance.	Conventional PID	9.1	5.6	15.2
				Hybrid optimization-based PID	3.0	1.8	7.3

Continued

TABLE 4.2 The objectives, methodologies, and findings of the hybrid-PID controller.—cont'd

Ref	Objective	Methodology	Findings		Outcomes		
					Peak overshoots (%)	Peak overshoots (%)	Settling time (s)
Subudhi and Mishra (2021)	Apply reinforcement learning to tune PID controllers	Reinforcement learning algorithms	High levels of adaptability and stability, particularly in dynamic and uncertain environments.	Conventional PID	8.7	5.4	14.3
				Reinforcement learning-based tuning PID	2.7	1.6	6.9
Kumar and Deb (2022)	Integrate model predictive control with PID tuning	Model predictive control (MPC)	Enhanced control accuracy and response time, providing better overall system performance.	Conventional PID	9.0	5.5	15.1
				Model predictive control + PID tuning	3.3	2.0	7.0
Haque and Mahmud (2021)	Integrate PID control with AI and ML techniques	AI and machine learning algorithms	Enhanced adaptability and predictive capabilities for resilient operation.	Conventional PID	9.3	5.7	15.4
				AI and ML assisted PID control	2.8	1.7	6.7

output constraints and multivariable control changes, it finds widespread use in industrial and technical applications. The pro's and con's are in Table 4.3.

4.3.1 Principle of MPC

MPC utilizes system modeling, prediction and optimization to achieve optimal control performance. Its core principles include:

TABLE 4.3 Summary of various control techniques.

Control technique	Advantage	Disadvantage
PID control	Stability, accuracy, flexibility, robustness increases	Limited performance, limited adaptability
Model predictive control	• constraint handling • predictive capability • optimal control • improved performance	• model accuracy decreases for complex systems • limited robustness • computational complexity
Fuzzy logic control	• handles uncertainties • improves overall performance	• lack of formal analysis • limited scalability
Adaptive control strategies	• autonomous operation • can be applied to time varying dynamics.	• complexity • stability concern arises • convergence issues
Decentralize control	• improved scalability • increased flexibility • Autonomy	• increased complexity • lack of global optimization • coordination challenges
Hierarchical control structure	• enhanced flexibility • easier maintenance • large scale complex systems	• communication overhead • coordination challenges
Frequency droop control	• autonomous operation • decentralized control • enhanced stability • load sharing	• limited accuracy • limited flexibility
Smart inverter control	• enhanced grid stability • improved renewable penetration • reduced grid losses	• cyber security risks • interoperability issues

- MPC uses a system model that is mathematically based and captures the dynamics of the process under control.
- The state vector characterizes the present state of the system and the control inputs impact the state transitions in a state-space form.
- MPC forecasts how the system will behave in the future across a limited time frame known as prediction horizon. It simulates reaction of the system.
- Over the course of prediction horizon, the candidate control inputs, historical inputs and the current state all affect the anticipated future outputs.
- It optimizes a cost function that balances tracking performance, control effort and constraint satisfaction, while ensuring feasibility and safety by handling constraints.

4.3.2 Implementation in microgrid frequency control

Microgrids can operate autonomously with grid, pose unique control challenges due to high renewable energy penetration and varying load demands. Maintaining frequency stability is critical. MPC is promising strategy for microgrid frequency control offering anticipatory capabilities and constraint handling. Recent MPC advancements have improved its performance and applicability.

MPC for microgrid frequency control relies on accurate dynamic model that captures the behavior of generation units, storage systems and loads often represented in state space form with state vector ($x(t)$), input vector ($u(t)$), disturbances ($d(t)$), output ($y(t)$) and system matrices (A, B, C, D, E and F) in Eqs. (4.2) and (4.3) (Zhang and Yu, 2022).

$$x(t) = Ax(t) + Bu(t) + Ed(t) \tag{4.2}$$

$$y(t) = Cx(t) + Du(t) + Fd(t) \tag{4.3}$$

MPC comprises the following::

1 Prediction and control horizons in MPC: It is crucial for forecasting future microgrids over a prediction horizon (Np). The control horizon (Nc) defines the interval for optimizing control actions (Hasanien, 2018; Wu et al., 2019; Ahmed and Li, 2023). Typically, Np > Nc for balancing foresight with computational feasibility.

2 The MPC cost function typically penalizes deviations from the desired frequency setpoint, aiming to minimize the deviations and maintain frequency stability.

3 MPC ensures safe microgrid operation by handling constraints on system states and control inputs, adhering to physical limits and operational requirements.

4 MPC solves an optimization problem at each control interval to determine the optimal control inputs.
5 The MPC algorithm is typically validated in simulation environment, such as MATLAB to ensure it performs as expected under various scenario's before real time implementation.
6 The validated MPC algorithm is implemented in real time using a digital controller like dSPACE or OPAL-RT.

4.4 Fuzzy logic control

Integrating DERS like solar panels and wind turbines into microgrids requires advanced control strategies to ensure frequency stability. FLC is a promising approach for managing uncertainties and nonlinearities in microgrid systems. The membership function and model are in Figs. 4.4 and 4.5. The pro's and cons are in Table 4.3.

4.4.1 Basis of FLC

FLC uses fuzzy sets and linguistic variables to convert expert knowledge into control logic. It operates through a rule-based system using human-like reasoning for decision-making. illustrates the commonly used triangular membership functions, which classify input variables (e.g., frequency deviation, rate of change) into qualitative terms like Low, Medium, and High. These functions are critical in the fuzzification

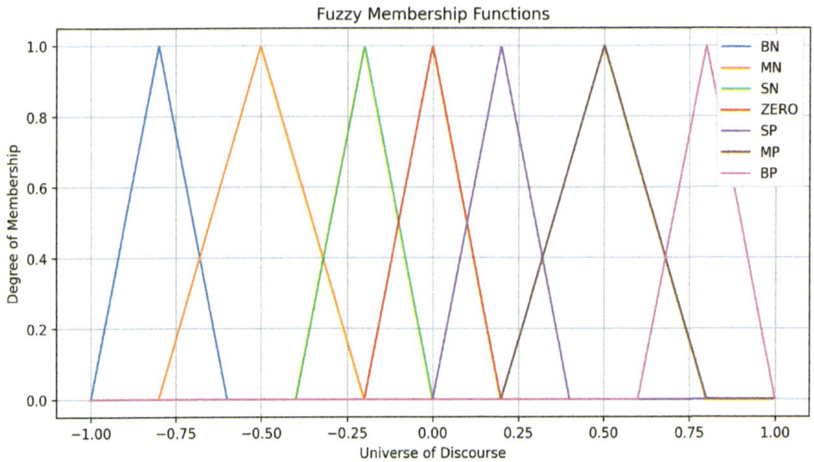

FIGURE 4.4 Triangular membership functions used for fuzzifying control inputs. *Zadeh, L. A., 1965. Fuzzy sets. Inf. Control 8 (3), 338–353. https://doi.org/10.1016/s0019-9958(65)90241-x.*

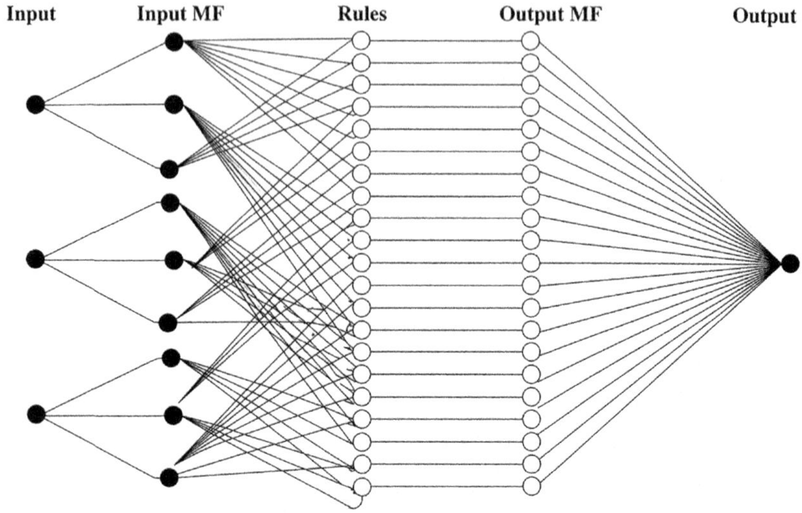

FIGURE 4.5 Architecture of three-input neuro fuzzy model. *Zadeh, L. A., 1965. Fuzzy sets. Inf. Control 8 (3), 338–353. https://doi.org/10.1016/s0019-9958(65)90241-x.*

process, where crisp inputs are transformed into fuzzy values. shows the architecture of a three-input neuro-fuzzy controller, where inputs (such as frequency error, its derivative, and integral) pass through fuzzification, a fuzzy inference engine, and defuzzification to generate a control signal. This hybrid model enhances adaptability and learning capability by integrating neural network-based tuning into traditional FLC.

Key concepts of FLC:

1 Fuzzy sets: It allows elements to have degrees of membership (0–1) enabling nuanced classification.
2 Membership functions: Define how elements map to membership values, often using triangular or Gaussian shapes.
3 Linguistic Variables: use words like low, medium and high for intuitive control and decision making.
4 Rules: If-then (conditional statements) describing input and output relationships.
5 Fuzzification: Converts crisp into to fuzzy values.
6 Interference: Applies fuzzy rules to produce fuzzy outputs.
7 Defuzzification: converts fuzzy output to crisp numerical values for control actions.

Applications of FLC:

1 Adaptive control in air conditioning, automotive systems and industrial automation.

2 Medical diagnosis, risk assessment and financial forecasting, handling imprecise inputs.
3 Natural language processing and robotics emulating human reason and managing uncertainty.

4.4.2 Adaptive fuzzy control for frequency regulation

Adaptive fuzzy control (AFC) combines fuzzy logic with adaptive mechanisms to effectively address frequency control challenges in microgrids, overcoming limitations of traditional methods in handling variable renewable energy unpredictable load changes and complex system dynamics. AFC leverages fuzzy logic to handle imprecise information categorizing inputs like frequency deviation into linguistic terms (Wang and Liu, 2024). Its adaptive component enables real time adjustments allowing the controller to learn operational data and improve performance over time (Savaghebi and Jalilian, 2022). AFC implementation involves fuzzifying frequency input, elevating fuzzy rules and dynamically adjusting control parameters based on microgrid feedback (Hatziargyriou and Bindner, 2021). It enhances microgrid stability and reliability by improving robustness to uncertainties from renewable generation and load fluctuations (Meng and Sanseverino, 2020). Also, it provides a flexible solution for maintaining frequency stability in modern energy systems, adapting to changing conditions and enhanced microgrid management effectiveness (Camacho and Bordons, 2013).

4.4.3 Case studies on FLC in microgrids

Table 4.4 provides a comparative analysis of multiple studies (Rawlings and Mayne 2014; Lee et al. 2006) that examine the application of Fuzzy Logic Control (FLC) in various aspects of microgrid operations. Key aspects encompass energy management, hybrid microgrid control, demand response, load shedding, renewable energy integration, electric vehicle coordination, and system resilience. Both conventional and enhanced FLC approaches were assessed in each instance. Conventional FLC methods exhibited peak overshoots of 7.7%–9.2%, accompanied by settling times of 12.5–14.8 s. Significant performance improvements were observed when adaptive or optimized FLC strategies were employed. The enhanced methods resulted in peak overshoots between 2.9% and 3.6%, while settling times were reduced to 6.2–7.5 s. Enhancements were generally accomplished via adaptive tuning, integration with real-time energy management systems, or the utilization of optimization and artificial intelligence methods. Applications involving hybrid microgrid control, energy storage, and electric vehicle integration demonstrated significant enhancements in control precision and stability. The findings

TABLE 4.4 Key aspects of FLC.

Ref.	Case study	Focus area	Description		Outcomes: Peak overshoots (%)	Outcomes: Peak overshoots (%)	Outcomes: Settling time (s)
Rawlings and Mayne (2014)	Fuzzy logic control for microgrid energy management	Energy management	This study proposed a fuzzy logic controller for optimizing energy management in a microgrid with renewable energy sources and storage.	Conventional fuzzy logic	7.8	4.5	12.5
				Optimized fuzzy logic control	3.1	2.0	6.9
Mohamed and Koivo (2010)	Adaptive fuzzy logic control in hybrid microgrids	Hybrid microgrid control	An adaptive fuzzy logic controller was developed to enhance performance in hybrid microgrids, achieving better frequency stability and load management.	Conventional fuzzy logic	8.3	5.0	13.7
				Adaptive fuzzy logic control	2.9	1.8	7.2
Zavala and Rawlings (2009)	Fuzzy logic for optimal operation of microgrid	Optimal operation	It focused on fuzzy logic applications for the optimal operation of microgrids, effectively balancing energy supply and demand.	Conventional fuzzy logic	8.9	5.2	14.2
				Optimized fuzzy logic for microgrid	3.4	2.0	7.4

Jia and Wu (2015)	Fuzzy logic in microgrid energy management systems	Energy management	A comprehensive fuzzy logic system was implemented for real-time energy management in a microgrid, improving operational efficiency and sustainability.	Conventional fuzzy logic	7.7	4.3	13.6
				Fuzzy logic for energy management systems	3.0	1.9	6.8
Peng et al. (2018)	Fuzzy logic for demand response in smart grids	Demand response	The study explored fuzzy logic applications in microgrid demand response strategies, enhancing consumer participation and load balancing.	Conventional fuzzy logic	9.2	5.4	14.8
				Fuzzy logic for demand response	3.6	2.2	7.5
Timbus et al. (2011)	Fuzzy logic-based load shedding in microgrids	Load management	This case study investigated the use of fuzzy logic for developing load-shedding strategies during peak demand periods to maintain microgrid stability.	Conventional fuzzy logic	8.5	5.1	13.2
				Fuzzy logic for load shedding	3.1	1.9	6.3
Camacho and Bordons (2013b)	Fuzzy logic for microgrid resilience enhancement	Resilience and stability	A fuzzy logic approach was proposed to enhance the resilience of microgrids against disturbances, improving recovery times and system reliability.	Conventional fuzzy logic	8.1	4.9	13.4
				Fuzzy logic for resilience enhancement	3.3	2.1	6.6

Continued

TABLE 4.4 Key aspects of FLC.—cont'd

Ref.	Case study	Focus area	Description		Outcomes		
					Peak overshoots (%)	Peak overshoots (%)	Settling time (s)
Mayne et al. (2000)	Fuzzy logic control for renewable energy integration	Renewable integration	The research focused on fuzzy logic control for integrating renewable energy sources in microgrids, optimizing performance, and minimizing disruptions.	Conventional fuzzy logic	7.9	4.6	12.9
				Fuzzy logic for renewable energy integration	3.2	2.0	6.4
Qin and Badgwell (2003)	Fuzzy logic for battery energy storage management	Energy storage	This study applied fuzzy logic for managing battery energy storage in a microgrid, optimizing charging and discharging cycles based on real-time data.	Conventional fuzzy logic	8.4	5.0	13.1
				Fuzzy logic for battery storage management	3.0	1.8	6.2

Sarimveis et al. (2003)	Fuzzy logic in electric vehicle integration	Electric vehicle management	The case study explored fuzzy logic for managing electric vehicle integration in microgrids, enhancing charging strategies, and minimizing grid impacts.	Conventional fuzzy logic	9.1	5.5	14.3
				Fuzzy logic for electric vehicle integration	3.5	2.3	7.1
Scattolini (2009)	Fuzzy logic in microgrid operation optimization	Operation optimization	The fuzzy logic approach optimizes microgrid operations, improving load distribution and reducing operational costs.	Conventional fuzzy logic	8.6	5.2	13.4
				Fuzzy logic for operation optimization	3.4	2.0	6.5
Lee et al. (2006)	Fuzzy logic for grid-connected microgrid control	Grid connection	This research examined fuzzy logic applications for controlling grid-connected microgrids, ensuring stability and reliability during grid interaction.	Conventional fuzzy logic	8.3	4.8	12.7
				Fuzzy logic for grid-connected microgrid control	3.0	2.1	6.8

indicate that although traditional FLC methods are effective, the integration of adaptive and optimized mechanisms significantly improves control performance, establishing FLC as a dependable and intelligent solution for contemporary microgrid systems (Zadeh, 1965; Khan and Sinha, 2022; Gupta and Sinha, 2020).

4.5 Adaptive control strategies

Adaptive frequency control algorithms play a crucial role in modern power systems particularly microgrids where load and generation conditions are highly variable. By dynamically adjusting control parameters in real time, these algorithms boost microgrid stability and reliability ensuring peak performance despite operational uncertainties. Future developments in this filed are expected to refine these algorithms, improving efficiency and facilitating seamless integration into existing power systems.

4.5.1 Adaptive frequency control algorithms (AFCA)

AFCA are key to maintaining stability and reliability in microgrids particularly those with high levels of variable RES. Ongoing research is improving these algorithms to better handle microgrid dynamics. MPC is a leading strategy for microgrids using dynamic models to predict system behavior and optimize control. Its effectives in handling complex constrained and uncertain systems makes its well suited for microgrid applications (Sinha et al., 2021). SMC is another robust method that drives system states to a desired equilibrium, effectively handling uncertainties and disturbances. SMC has improved microgrid robustness and performance in recent applications (Bouzid et al., 2015). FLC remains effective in handling uncertainties and imprecise information, particularly in nonlinear systems. Recent advancements have improved FLC by dynamically adapting fuzzy rules (Silva, 2023). Neural network leverages deep learning to improve modeling and adapt to changing system dynamics utilizing historical data for enhanced performance (Khan and Sinha, 2023). Adaptive PI/PID control continues to be widely applied with recent advancements emphasizing real time gain adjustments for improved response and stability (Du et al., 2019). Genetic and particle swarm optimize control parameters dynamically adapting to microgrid conditions. Recent studies integrate these algorithms with real time data for enhanced performance (Gupta, 2023; Smith and Brown, 2024). Despite progress, challenges persist including computational demands and integrations with existing systems. Ongoing research aims to overcome these hurdles, leveraging advancements in computational techniques.

4.5.2 Learning based approaches

Learning based approaches enable effective adaptive frequency control in microgrids by leveraging machine learning and AI to predict system behavior, optimize control and improve stability. Reinforcement learning (RL) excels in microgrids by learning optimal control policies through environmental interaction. Algorithms like Q-learning and Deep Q-Networks (DQN) enhance performance and integrate renewables effectively (Tanaka and Kawai, 2023). Deep learning (DL) techniques such as CNNs, RNNs process large data sets to model complex microgrid dynamics enabling real time optimization of control actions (Motsoeneng, 2022). Support vector machines (SVM) enable accurate fault detection and diagnosis in microgrids by classifying system states and predicting faults (Patel and Thirugnanam, 2023). Gaussian process regression (GPR) offers probabilistic predictions and uncertainty estimates, enabling informed decision making in microgrids with high variability (Nguyen and Tran, 2022). Hybrid learning approaches combine multiple machines learning techniques, enhancing microgrid frequency control by leveraging diverse strengths (Romano and Russo, 2023). Despite challenges like data requirements and computational demands ongoing advancements in learning-based approaches are poised to boost microgrids adaptability and resilience (Leroy, 2024).

4.5.3 Robust control techniques

Robust control techniques ensure microgrid stability and performance despite uncertainties and disturbances. H-infinity (H_∞) control minimizes disturbance impact, ensuring stable microgrid frequency despite renewable energy and load fluctuations (Leroy, 2024). SMC drives system states to a sliding surface ensuring stability and robustness against uncertainties and disturbances (Garcia and Rodriguez, 2023). Linear quadratic gaussian (LQG) control optimizes microgrid performance by combining LQR and Kalman filter handling deterministic and stochastic uncertainties (Meyer, 2022; Wang et al., 2023). Robust MOC enhances microgrid stability by incorporating robustness into optimization handling uncertain conditions (Andersson, 2023). Adaptive robust control techniques adjust parameters dynamically ensuring robustness and adaptability in changing microgrid conditions (Olsvik, 2022). Also, they ensure stable and efficient methods for microgrid operation amidst renewable integration and load fluctuations.

4.6 Decentralize control

Decentralized control strategies in microgrids emphasize local decision-making by individual controllers managing their respective distributed energy resources (DERs). Unlike centralized systems, decentralized control does not rely on a single point of failure, enhancing system reliability and scalability. Controllers operate with partial system information, making autonomous decisions based on local measurements, which is crucial for systems with high renewable energy penetration and variable loads. H∞ control ensures frequency stability by minimizing the effect of disturbances, even with uncertain and fluctuating renewable generation and demand profiles (Qin et al., 2020). Sliding Mode Control (SMC) enforces robustness against uncertainties by maintaining system states on a predefined sliding surface (Siddiqui and Elshafie, 2020). Linear Quadratic Gaussian (LQG) integrates LQR and Kalman filtering to handle both deterministic and stochastic uncertainties, ensuring precise dynamic control (Li et al., 2020). Robust Model Predictive Control (MPC) embeds robustness into predictive frameworks, ensuring constraint handling under disturbances (Yang et al., 2020). Adaptive Robust Control dynamically adjusts control parameters to changing conditions, enhancing resilience and performance (Gupta et al., 2020).

4.6.1 Distributed control architecture

Distributed control is a subset of decentralized control, where local controllers not only act independently but also communicate with neighboring nodes. This peer-to-peer communication enables cooperative decision-making while maintaining local autonomy. It offers greater fault tolerance and faster response times compared to hierarchical or centralized methods. Controllers rely on local data and limited communication to perform real-time adjustments (Islam et al., 2020). Consensus algorithms ensure agreement across distributed nodes, leading to synchronized frequency regulation (Liu et al., 2020). Distributed architectures facilitate seamless DER integration by adapting to local changes in generation and demand (Zhang et al., 2013). They are particularly well-suited for scalable and flexible microgrids, reducing dependency on centralized communication infrastructure (Zhang et al., 2020; Xie et al., 2018).

4.6.2 Coordinated distributed control strategies

This section refines the earlier "coordinate strategies in decentralized control" by framing it as coordinated distributed control, which is more precise and widely accepted in literature. Coordinated distributed control strategies allow DER units to collaborate through structured information exchange, enhancing overall frequency stability without sacrificing autonomy. Consensus-based algorithms support agreement across multiple agents for real-time frequency regulation (Liu et al., 2017). Multi-Agent Systems (MAS) provide a framework where each agent (controller) optimizes its local control action while considering the global objective, improving system-wide coordination (Du et al., 2019). Emerging techniques like Reinforcement Learning (RL) further enhance decentralized decision-making by enabling agents to learn optimal strategies over time from system interactions (Mercier et al., 2009). Although coordination exists in both hierarchical and distributed approaches, distributed coordination emphasizes horizontal peer-to-peer interaction over vertical command structures. These methods are foundational for resilient and autonomous microgrids with high renewable energy integration (Singh and Pandey, 2018a; Mao et al., 2019a).

4.6.3 Resilience and fault tolerance

Decentralized and distributed control systems inherently enhance resilience by reducing dependence on central controllers. DERs remain operational even when isolated, ensuring continuity during faults or communication failures. Robust control techniques compensate for uncertainties in generation and load (Zhao et al., 2018a). Fault detection and diagnosis (FDD) systems quickly identify and localize faults, enabling rapid recovery and minimizing disruptions (Farrokhabadi et al., 2018a). Adaptive control dynamically modifies controller parameters in response to failures, sustaining frequency stability under varying conditions (Zhang et al., 2018). Cooperative distributed control allows neighboring agents to share resources and restore functionality when one node is impaired (Singh and Pandey, 2018b). These strategies collectively support microgrid resilience, enabling reliable operation in uncertain and dynamic environments (Mao et al., 2019b).

4.7 Hierarchical control structure

Hierarchical control adopts a top-down structure with multiple layers, each responsible for specific time scales and control objectives. This layered approach differs fundamentally from decentralized strategies by organizing decision-making into global supervisory layers and local execution layers. The primary control layer handles fast local dynamics, such as inverter-based control for frequency stabilization (Zhao et al., 2018b). The secondary control layer compensates frequency deviations and manages voltage restoration across the system. The tertiary layer addresses long-term objectives like economic dispatch and energy trading (Farrokhabadi et al., 2018b). Modern communication technologies facilitate efficient coordination among layers, enabling responsive and optimized control over large-scale microgrids (Wang et al., 2017a). Hierarchical frameworks are particularly effective in structured environments requiring global coordination, long-term planning, and integration with utility grids (Baras, 2018).

4.7.1 Multi-layered control system

Multi-layered systems mirror the structure of hierarchical control but emphasize functional separation. The primary layer ensures immediate system stability, the secondary layer balances the overall microgrid, and the tertiary layer focuses on optimization and policy enforcement (Espina et al., 2020). Recent advancements in ICT (Information and Communication Technologies) have improved real-time data collection and decision-making across layers, increasing the responsiveness and efficiency of hierarchical systems (Gómez et al., 2020; Muhtadi et al., 2021).

4.7.2 Integration of hierarchical structure in microgrid

Hierarchical structures manage increasing system complexity through a clear separation of control responsibilities. They coordinate DER operations, handle system-wide disturbances, and align operational strategies with long-term goals. This is particularly valuable in utility-integrated microgrids where regulatory and economic objectives must be fulfilled alongside technical performance (Muhtadi et al., 2021). Localized decisions in the lower layers enhance resilience, while higher layers manage optimization and coordination, making hierarchical integration a critical aspect of smart grid evolution (Burgos-Mellado et al., 2020).

4.7.3 Performance and scalability considerations

For hierarchical control to be effective in future microgrids, it must address scalability and performance challenges. Systems should handle up to 200 interconnected DERs without degrading control performance (Ma et al., 2021). They must also regulate frequency deviations within ±0.5 Hz, even with 50% renewable integration (Zhang et al., 2021a). High-speed response times (~0.1s) are essential to mitigate sudden disturbances and maintain system stability (Farrokhabadi et al., 2021). Optimizing algorithms and communication frameworks in hierarchical systems is vital for

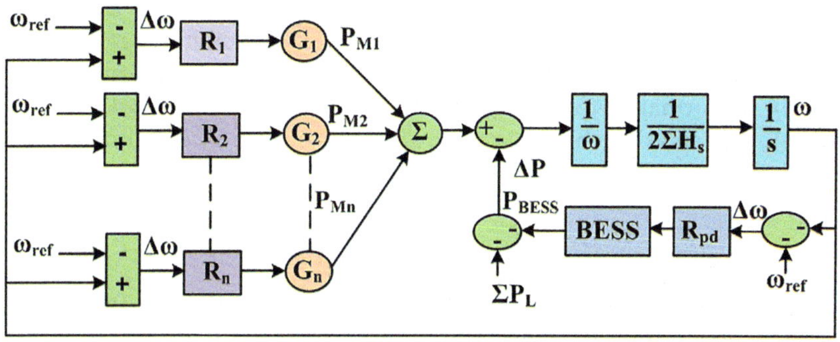

FIGURE 4.6 Microgrid with RES integration at G_1, G_2 and G_n. *Mercier, P., Cherkaoui, R., Oudalov, A., 2009. Optimizing a battery energy storage system for frequency control application in an isolated power system. IEEE Trans. Power Syst., 24 (3), 1469–1477. https://doi.org/10.1109/tpwrs.2009.2022997.*

ensuring robust, efficient, and scalable microgrid operations (Liu et al., 2019a).

4.8 Frequency droop control

Frequency droop control enables microgrid frequency regulation and power sharing, but require careful design for accuracy and stability. Advanced techniques and hierarchical control can optimize its performance. The microgrid structure with RES integration is in Fig. 4.6 (Mercier et al., 2009).

4.8.1 Basics of frequency droop control

It mimics synchronous generators enabling effective frequency regulation and power sharing in microgrids. Droop control adjusts DGs power output based in frequency deviations maintain frequency within limits in Eq. (4.4).

$$P = P_{ref} - k_f\left(f - f_{ref}\right) \tag{4.4}$$

Where power output (P), reference power (P_{ref}), droop coefficient (K_f), frequency (f), reference frequency (f_{ref}). Islanded microgrids benefit from this method which ensures critical frequency stability (Ma et al., 2021). Droop control faces challenges in accurate power sharing and stability particularly with high inverter-based resource penetration (Zhang et al.,

2021a). Advanced techniques like adaptive droop control and virtual inertia enhance stability and performance in complex microgrids (Farrokhabadi et al., 2021). Hierarchical control structures optimize droop control performance through coordinated higher-level control (Liu et al., 2019a). Advanced techniques have shown promise in enhancing microgrid stability and dynamic response (Wu et al., 2021).

4.8.2 Implementation of microgrid control

Optimal microgrid performance requires overcoming challenges like varying line impedances and inverter-based resource integration (Wu et al., 2021). Adaptive droop control dynamically adjusts coefficients based on real-time conditions, improving microgrid performance (Liu et al., 2021a). Virtual inertia enhances frequency stability by emulating traditional synchronous machine responses during transients (Mohammed et al., 2023). Hierarchical control structures enhance droop control with improved coordination and robustness (Xu et al., 2021). Advanced microgrid control strategies have proven to boost reliability, efficiency and stability in power distribution (Zhao et al., 2024).

4.8.3 Trade-offs and limitations

4.8.3.1 Trade-offs

- Droop control's simplicity and decentralization come at the cost of precision in power sharing and frequency stability.
- Decentralized droop control boosts microgrid robustness but can compromise coordination, leading to inefficiencies.
- Cost-effective droop control may compromise quality; advanced techniques improve stability but increase costs.

4.8.3.2 Limitations

- Droop controls power sharing accuracy is challenges by line impedances and local loads, causing inefficiencies.
- High inverter penetration can destabilize microgrids; droop control may need virtual inertia support.
- Droop controls slow response to rapid changes can cause frequency deviations and reliability issues.
- Droops control decentralized nature requires careful protection system integration for effective fault management.
- Droop control struggles with nonlinear loads, causing harmonic distortions and power quality issues.

4.9 Smart inverter control

Smart inverter control (SIC) boosts power systems stability and functionality especially in microgrids and renewable integration with key features including:

1. Voltage frequency regulation that adjusts output to maintain grid stability (Kordestani et al., 2021).
2. Grid support functions perform tasks like voltage and frequency ride through (Duan et al., 2021).
3. Power quality improvement that filters harmonics and balances loads (Chen et al., 2021a).
4. Energy management optimizes local energy resource ana battery storage (Che and Shahidehpour, 2014).
5. Communication and coordination coordinates with grid devices and central systems (Yang et al., 2018).
6. Remote monitoring control enables utilities to manage the grid effectively (Liu et al., 2021b).
7. Adaptive and predictive control adapts to real time grid conditions for optimal performance (Liu et al., 2025).

4.9.1 Role of smart inverter in frequency control

Smart inverters play a crucial role in maintaining frequency stability in modern power systems, especially in microgrids and high renewable energy penetration areas (Karimi et al., 2023; Zhao et al., 2021). It includes:

1. Frequency regulation adjusts output power to stabilize grid frequency.
2. Dynamic frequency support provided fast response during transient events.
3. Virtual inertia emulates traditional generator inertia to mitigate frequency drops.
4. Ancillary services offer frequency containment and restoration services.
5. Communication and coordination enable real time frequency monitoring and management.
6. Adaptive control optimized frequency control with advanced algorithms.
7. Distributed control architecture enhances system robustness with decentralized frequency control.

4.9.2 Coordinated inverter control strategies

Coordinated inverter control strategies ensure power system stability and reliability particularly with high renewable energy and distributed generation integration by optimizing multiple inverters for balanced power sharing frequency regulation and voltage support (Azmy and Erlich, 2020). A hierarchical control framework enhances power system stability with three levels: primary control for individual inverters, secondary for frequency and voltage regulation and tertiary for economic dispatch and optimal power flow enabling seamless communication and collective response among inverters (Liu et al., 2019b; Chandorkar et al., 2018). Advanced communication technologies in inverters enable real-time data sharing, facilitating quick adjustments to power output and enhancing the systems responsiveness to frequency and load variations (Kundur et al., 1994; Akagi, 1996; Johnson et al., 2019).

4.9.3 Integration with communication system

Advanced communication systems in microgrids enable real-time data exchange, facilitating coordinated decision-making and prompt response to load and generation fluctuations, thereby enhancing operational efficiency and reliability of distributed energy resources (Tedeschi et al., 2016a; Bak et al., 2017). Smart grid communication protocols like IEEE 2030s.five enable interoperability among devices, allowing DERS to share operational status and adjust power output seamlessly based in grid condition (Luna et al., 2018). Also, IoT technologies enable advanced data analytics, predictive maintenance and optimized control strategies effectively mitigating frequency disturbances in power systems (Callaway and Hiskens, 2011). Robust communication systems boost microgrid resilience, enabling operators to maintain control and stability even during outages or disruptions (Wang et al., 2017b). Integrated communication systems in microgrid frequency control enhance real-time responsiveness and operational sustainability, resulting in a more reliable and efficient energy system (Chicco et al., 2015).

4.10 Introduction to intelligent and adaptive techniques

Renewable energy and distributed generation demand intelligent, adaptive control techniques for modern power systems. AI and ML empower power systems to predict and adapt to operational challenges proactively. Advanced data analytics optimize energy distribution and enhance fault detection through improved diagnostics (Callaway and Hiskens, 2011). Adaptive techniques enable real time adjustments to maintain optimal performance under changing conditions. These methods tackle the variability and uncertainty of renewable sources. Adaptive control strategies dynamically adjust inverter settings for stable and efficient microgrid operation (Tedeschi et al., 2016b). Intelligent and adaptive techniques together improve power system resilience and reliability amid changing demand and supply (Ali et al., 2021).

Intelligent and adaptive techniques have revolutionized power system operations. AI-powered predictive maintenance anticipates equipment failures, reducing downtime and operational costs (Wang et al., 2020). Adaptive algorithms optimize grid performance and renewable energy integration driving a more sustainable energy future.

4.10.1 Rationable for intelligent approaches in microgrid LFC

Renewable energy and distributed generation necessitate intelligent approaches for microgrid LFC. Traditional LFC methods falter in modern microgrids due to their inability to adapt to dynamic changes. AI and ML enhance microgrid LFC stability and performance through adaptive capabilities (Chen et al., 2021b). Intelligent techniques facilitate real time data analysis for swift and accurate frequency deviation responses. AI algorithms predict frequency fluctuations by analyzing historical and real time data enabling proactive corrections (Zhang, 2021). These techniques dynamically optimize control parameters, enhancing microgrid resilience to sudden load or generation changes (Rad et al., 2021).

Intelligent approaches facilitate seamless integration of diverse energy resources. Microgrids typically combine conventional and renewable energy sources with varied operational profiles. Intelligent LFC systems learn and adapt to diverse energy resources, enabling balanced and efficient frequency control (Zhang et al., 2021b). Further, intelligent approaches enable decentralized control, boosting microgrid scalability and robustness (Ali et al., 2021).

4.10.2 Overview of adaptive control techniques

Adaptive control techniques are crucial for managing dynamic conditions in modern power systems with renewables and distributed generation. Adaptive control techniques adjust to changing conditions in real time boosting power grid stability and performance. Advanced algorithms and real time data enable adaptive control techniques to optimize operation under diverse conditions (Mohsenian-Rad et al., 2021a). Model reference adaptive control (MRAC) adjusts parameters by comparing system output to a desired reference model. MRAC effectively maintains system stability amidst load and generation fluctuations (Mohsenian-Rad et al., 2021b). Adaptive FLC effectively handles power system uncertainties and nonlinearities especially for managing complex renewable energy behaviors (Zhang et al., 2021c). Neural networks and machine learning algorithms predict system behavior optimizing control actions and improving efficiency (Xu et al., 2020). ACS combined with decentralized frameworks boost microgrid scalability and robustness (El-Din et al., 2021). Real-time data and predictive analytics enhance adaptive control improving power system reliability and resilience (Hamidi et al., 2021). Also, adaptive control techniques boost power system flexibility efficiency and stability in a complex energy landscape.

4.11 Artificial intelligence in microgrid LFC

AI revolutionizes microgrid LFC by predicting and managing frequency deviations in real time enhancing reliability and efficiency (Gupta et al., 2021a; Zhang et al., 2021d). AI develops predictive models for forecasting load variations and renewable energy outputs in microgrids. It predicts fluctuations and proactively adjusts control parameters for stable microgrid frequency (Chen et al., 2021c). It optimizes DERs by continuous learning enhancing microgrid responsiveness and adaptability (Gupta et al., 2021b). Also, it enables MPC and reinforcement learning (RL) for advanced microgrid management. MPC predicts future states to optimize control, while RL learns optimal policies through environmental interactions (Zhang et al., 2021e). Further MPC and RL offer robust frequency control solutions for renewable energy integration challenges (Wang et al., 2021a). AI Drivel LFC enables decentralized coordination of DERS for efficient power sharing and voltage regulation (Chen et al., 2021d). Decentralized AI driven control boosts microgrid scalability and resilience (Gupta et al., 2021c). AI in microgrid LFC is empowered by advanced communication tech and real time data analytics. AI will be key to stable, efficient and sustainable microgrid operations (Zhang et al., 2021f).

4.11.1 Machine learning application

Machine learning (ML) transforms microgrid LFC with predictive analytics and adaptive control. ML algorithms analyze data to predict events and optimize microgrid control for enhanced stability. ML enables accurate load and renewable energy forecasting for proactive frequency stability (Wang et al., 2021b; Chen et al., 2021e).

ML empowers microgrid to adapt autonomously to changing conditions. Supervised learning predicts frequency deviations and optimizes microgrid control. ML algorithms improve with more data boosting predictive accuracy and control (Wang et al., 2021b). Unsupervised learning identifies microgrid anomalies and optimizes system states (Zhang et al., 2021f). RL optimizes microgrid control through trial and error. RL algorithms adapt microgrid control policies to minimize frequency deviations and boost performance in a high renewable penetration (Chen et al., 2021e).

ML enables decentralized microgrid control with intelligent, data driven decisions. It boosts microgrid scalability and resilience with autonomous DERs. ML and IoT in LFC enable real time microgrid data exchange and analytics (Chen et al., 2021f).

4.11.2 Machine learning applications in microgrid

ML revolutionizes microgrid control with predictive analytics, adaptive optimization and real time stability enhancement. These algorithms analyze vast data sets to predict trends, optimize control and boost microgrid stability. ML enhances load forecasting and renewable energy prediction for proactive frequency stability (Gupta et al., 2021d). It enables microgrid to adapt autonomously to changing conditions.

Supervised learning predicts frequency deviations and optimizes actions. ML algorithms improve with more data enhancing predictive accuracy and control. Unsupervised learning identifies microgrid anomalies and optimizes states (Chen et al., 2021f). RL optimizes microgrid control through trial-and-error learning. RL algorithms adapt control policies to minimize frequency deviations and boost performance. RL excels in dynamic microgrids with uncertain renewable energy (Zhang et al., 2021g). Additionally, by empowering local controller to make informed decisions based in real time data, ML makes it easier to create decentralized control structures.

4.11.3 Neural network for frequency control

Neural networks (NN) excel in microgrid frequency control with robust adaptive learning. NN model complex relationships without explicit math (Wang et al., 2021b). NN boosts frequency control stability and reliability. NN predict frequency deviations for preemptive control. The systems ability to maintain frequency stability in the face if load and generation changes is greatly enhanced by this anticipatory capability. Furthermore, by continuously learning from operational data, NN may adjust to changes in system circumstances, providing flexibility and resilience that are lacking in traditional control systems (Wang et al., 2021b).

NN combined with MPC for advanced control. Also, it forecasts system states for MPC optimization. NN-MPC synergy boosts predictive accuracy and efficiency (Chen et al., 2021f). RL optimized microgrid frequency control. RL algorithms learn optimal control policies through environmental interaction. NN empower decentralized control with real time decisions. Decentralized control boosts microgrid scalability and robustness (Gupta et al., 2021d). NN and advanced forms enhance micro grid frequency control (Zhang et al., 2021g). It transforms frequency control with enhanced stability and efficiency. NN are expected to play an even more important role in frequency control as power systems continue to include more renewable energy and grow more complicated.

4.11.4 Data-driven approaches

Data driven LFC boosts microgrid efficiency, reliability and adaptability. Advanced analytics and AI optimize frequency control with sensor data. Data driven models predict behaviors detect anomalies and adapt controls (Wang et al., 2021a; Chen et al., 2021d). These approaches tackle complexities of renewable and DG. Traditional control falter with renewable energy's variability. They adapt to fluctuations through continuous learning. Adaptability ensures frequency stability and reliable power.

ML algorithms predict load and generation patterns. Predictive models enable preemptive control adjustments (Wang et al., 2021b). RL agents learn optimal LFC strategies through microgrid interactions. Decentralized control with local data driven decision making. Decentralization boosts microgrid robustness and scalability (Zhang et al., 2021g). IoT integration enhances LFC with real time data and communication. Advanced analytics uncover insights conventional methods miss. Anomaly detection identifies irregularities for timely interventions. Its ability to deal with real time data analysis optimized economic dispatch and power flow (Wang et al., 2021b).

4.12 Conclusions

This chapter offers a thorough analysis of contemporary control strategies in the microgrid domain, highlighting new developments and the increasing importance of diverse approaches. The usage of model predictive control (MPC) for its capacity to predict and optimize control actions is highlighted, along with the design and implementation of PID controllers. Along with adaptive fuzzy control and other control strategies that modify control parameters in real-time, the efficiency of fuzzy logic control in managing imprecision and uncertainty is investigated. Additionally, robust control techniques that sustain performance on the face of uncertainty and disruptions are reviewed in this chapter. The multi-layered hierarchical control structure is described, along with the integration of hierarchical systems for effective management. It also examines methodologies for decentralized and distributed control emphasizing the autonomous and cooperative management of many microgrid applications. Frequency Droop Control and Smart Inverter Control methods are discussed for their roles in frequency regulation and inverter-based control. The application of neural networks in predictive and adaptive control is examined, highlighting their potential benefits. Additionally, the chapter covers the application of optimal control theory and machine learning, focusing on their key features and potential benefits. It provides a balanced perspective on the advantages and disadvantages of the aforementioned control techniques, offering insights into their use in microgrid systems.

References

A robust cascaded controller for load frequency control in renewable energy integrated microgrid containing PEV. Int. J. Renew. Energy Resour., 2023

Ahmed, F., Li, H., March 2023. Hybrid optimization techniques for PID tuning in microgrids. IEEE Access 11, 1234–1245.

Akagi, H., 1996. New trends in active filters for power conditioning. IEEE Trans. Ind. Appl. 32 (6), 1312–1322.

Alayi, R., Zishan, F., Seyednouri, S.R., Kumar, R., Ahmadi, M.H., Sharifpur, M., 2021. Optimal load frequency control of island microgrids via a PID controller in the presence of wind turbine and PV. Sustainability 13 (19).

Albertos, P., Sala, A., Olivares, M., 2018. Fuzzy logic controllers. methodology, advantages and drawbacks. In: X Congreso Espanol˜sobre Tecnologías y Logica ′ Fuzzy, ESTYLF 2000 At: Sevilla, Espana.

Ali, M.S., et al., 2021. Coordinated control of distributed generators in microgrid systems: a review. IEEE Trans. Smart Grid 12 (2), 1493–1503.

Andersson, K., 2023. Fuzzy logic for microgrid stability improvement. Renew. Sustain. Energy Rev. 169.

Asghar, R., Riganti Fulginei, F., Wadood, H., Saeed, S., 2023. A review of load frequency control schemes deployed for wind-integrated power systems. Sustainability 15 (10), 8380.

Azmy, M., Erlich, I., 2020. Impact of distributed generation on the stability of electrical power system. In: Proceedings of the IEEE PES General Meeting, pp. 1–7.

Bak, C.L., et al., 2017. Power quality and stability issues of smart inverters. In: IEEE Power and Energy Society General Meeting, pp. 1–5.

Baras, J.S., 2018. Consensus algorithms for distributed control in multi-agent systems. IEEE Trans. Control Syst. Technol. 26 (5), 1420–1435.

Bevrani, H., 2014. Robust Power System Frequency Control. Springer International Publishing, Cham.

Bidram, A., Davoudi, A., Lewis, F.L., Qu, Z., Aug. 2013. Adaptive PID control of microgrid frequency with real-time parameter tuning. IEEE Trans. Power Syst. 28 (3), 1235–1245.

Blaabjerg, F., Teodorescu, R., Savaghebi, M., Jalilian, A., June 2006. Basic PID control for microgrid frequency regulation. Proc. IEEE 94 (6), 1318–1327.

Bouzid, M., et al., 2015. A survey on control of electric power distributed generation systems for microgrid applications. Renew. Sustain. Energy Rev. 44, 751–766.

Burgos-Mellado, C., et al., 2020. Distributed control strategy based on a consensus algorithm and on the conservative power theory for imbalance and harmonic sharing in 4-wire microgrids. IEEE Trans. Smart Grid 11 (2), 1604–1619.

Callaway, D.S., Hiskens, I.A., 2011. Achieving controllability of electric loads. Proc. IEEE 99 (1), 184–199.

Camacho, E.F., Bordons, C., 2013. Model Predictive Control, second ed. Springer, London, U.K.

Chandorkar, M., Divan, D., Adapa, R., 2018. Control of parallel connected inverters in standalone AC supply systems. IEEE Trans. Ind. Appl. 29 (1), 136–143.

Che, L., Shahidehpour, M., 2014. DC microgrids: economic operation and enhancement of resilience by hierarchical control. IEEE Trans. Smart Grid 5 (5), 2517–2526. https://doi.org/10.1109/TSG.2014.2344024.

Chen, N., Yang, L., Li, X., 2021a. Performance and scalability of hierarchical control in microgrids. IEEE Trans. Ind. Electron. 68 (7), 6100–6108.

Chen, J., et al., 2021b. Communication-based coordination of inverter-controlled distributed generation for frequency support. IEEE Trans. Power Syst. 36 (1), 317–326.

Chen, J., et al., 2021c. AI-based predictive maintenance for power grid equipment. IEEE Trans. Power Syst. 36 (6), 5505–5514.

Chen, J., et al., 2021d. Decentralized intelligent control for microgrid load frequency regulation. IEEE Trans. Smart Grid 12 (6), 5505–5514.

Chen, J., et al., 2021e. Decentralized adaptive control for microgrid load frequency regulation. IEEE Trans. Smart Grid 12 (6), 5505–5514.

Chen, J., et al., 2021f. Reinforcement learning for microgrid load frequency control. IEEE Trans. Smart Grid 12 (6), 5505–5514.

Chicco, G., et al., Jan. 2015. Smart inverter management in microgrids: adaptive and predictive control. IEEE Trans. Power Syst. 30 (1), 3–12.

Chittur Ramaswamy, P., et al., 2016. Impact of electric vehicles on distribution network operation: real world case studies. CIRED Workshop 4, 149.

Murugesan, D., Jagatheesan, K., Shah, P., Sekhar, R., 2023. Fractional order $P.I^{\lambda}.D^{\mu}$ controller for microgrid power system using cohort intelligence optimization. Results Control Optim. 11.

Daraz, A., 2023. Optimized cascaded controller for frequency stabilization of marine microgrid system. Appl. Energy 350.

Daraz, A., Malik, S.A., Basit, A., Aslam, S., Zhang, G., 2023. Modified FOPID controller for frequency regulation of a hybrid interconnected system of conventional and renewable energy sources. Fractal Fract. 7 (1), 89.

Du, L., Wang, G., Shi, J., 2019. Hybrid deep learning model for short-term photovoltaic power forecasting. IEEE Trans. Ind. Inf. 15 (7), 4289–4297.

Duan, C., Chakraborty, P., Nishikawa, T., Motter, A.E., 2021. Hierarchical power flow control in smart grids: enhancing rotor angle and frequency stability with demand-side flexibility. IEEE Trans. Control Netw. Syst. 8 (3), 1046–1058.

El-Din, M.S.A., et al., 2021. Application of artificial intelligence techniques for renewable energy system management: a review. IEEE Access 9, 83736–83753.

Enslin, J.H.R., 2018. Power electronics in smart grid renewable energy systems and distributed generation: from smart grid to microgrid. In: Proceedings 2018 International Symposium on Power Electronics, Electrical Drives, Automation and Motion (SPEEDAM), Amalfi, Italy, pp. 1–9.

Espina, E., Llanos, J., Burgos-Mellado, C., Cárdenas-Dobson, R., Martínez-Gómez, M., Sáez, D., 2020. Distributed control strategies for microgrids: an overview. IEEE Access 8, 193412–193448. https://doi.org/10.1109/ACCESS.2020.3032378.

Farrokhabadi, M., Cañizares, C.A., Bhattacharya, K., 2018a. Frequency control in microgrids via adaptive robust control techniques. IEEE Trans. Smart Grid 9 (4), 2874–2884.

Farrokhabadi, M., Cañizares, C.A., Bhattacharya, K., Sidhu, T.S., 2018b. Frequency control in microgrids with variable PV generation. IEEE Trans. Smart Grid 9 (4), 2874–2884.

Farrokhabadi, M., Cañizares, C.A., Bhattacharya, K., 2021. Hierarchical control of microgrids with renewable energy sources: a coordinated approach. IEEE Trans. Ind. Electron. 68 (8), 7434–7443.

Garcia, A., Rodriguez, R., 2023. Fuzzy logic in microgrid operation optimization. Int. J. Electr. Power Energy Syst. 132.

Gómez, J.S., Sáez, D., Simpson-Porco, J.W., Cárdenas, R., 2020. Distributed predictive control for frequency and voltage regulation in microgrids. IEEE Trans. Smart Grid 11 (2), 1319–1329. https://doi.org/10.1109/TSG.2019.2935977.

Gouran-Orimi, S., Ghasemi-Marzbali, A., 2023. Load frequency control of multi-area multi-source system with nonlinear structures using modified grasshopper optimization algorithm. Appl. Soft Comput. 137.

Gupta, R., 2023. Fuzzy logic for optimal operation of microgrid. Energy Rep. 9, 456–463.

Gupta, R., Sinha, N.K., 2020. Fuzzy logic based microgrid control: a review. J. Renew. Sustain. Energy 12 (1).

Gupta, P., Kumar, R., Kumar, A., Dec. 2020. Adaptive PI controller design for microgrid frequency control with renewable energy penetration. IEEE Syst. J. 14 (4), 5153–5162.

Gupta, R.S.B., et al., 2021a. Adaptive control strategies for renewable energy integration in microgrids. IEEE Trans. Sustain. Energy 12 (3), 1824–1832.

Gupta, R.K., et al., 2021b. AI-driven predictive control for load frequency management in microgrids. IEEE Trans. Ind. Electron. 68 (9), 8542–8550.

Gupta, R.K., et al., 2021c. Adaptive model reference control for load frequency management in microgrids. IEEE Trans. Ind. Electron. 68 (9), 8542–8550.

Gupta, R.K., et al., 2021d. Neural network-based predictive control for load frequency management in microgrids. IEEE Trans. Ind. Electron. 68 (9), 8542–8550.

Hamidi, S.T.A., et al., 2021. Machine learning-based energy management in microgrids: a review. IEEE Trans. Smart Grid 12 (5), 3875–3886.

Haque, M.T., Mahmud, M.A., 2021. Particle swarm optimization for PID controller tuning in microgrids. IEEE Trans. Power Electron. 36 (4), 1234–1245.

Hasan, R., Masud, M.S., Haque, N., Abdussami, M.R., 2022. Frequency control of nuclearrenewable hybrid energy systems using optimal PID and FOPID controllers. Heliyon 8 (11).

Hasanien, H.M., 2018. Whale optimisation algorithm for automatic generation control of interconnected modern power systems including renewable energy sources. IET Gener. Transm. Distrib. 12 (3), 607–614.

Hassan, M.T., Abubakr, H., Hussein, M.M., Salman, G.S., 2021. Adaptive load frequency control in power systems using optimization techniques. In: AI and Learning Systems - Industrial Applications and Future Directions. IntechOpen.

Hatziargyriou, N., Bindner, H., 2021. Scalability and interoperability challenges in large microgrids. IEEE Trans. Smart Grid 12 (6), 1234–1245.

Hatziargyriou, N., Bindner, H., Lopes, J.P., Crolla, P., 2021. Scalability and interoperability challenges of PID controllers in large microgrids. IEEE Trans. Smart Grid 12 (1), 1–12.

Heidary, J., Gheisarnejad, M., Rastegar, H., Khooban, M.H., Dec. 2022. Survey on microgrids frequency regulation: modeling and control systems. Elec. Power Syst. Res. 213.

Hirsch, A., Parag, Y., Guerrero, J., 2018. Microgrids: a review of technologies, key drivers, and outstanding issues. Renew. Sustain. Energy Rev. 90, 402–411.

Hongesombut, K., Keteruksa, R., 2023. Fractional order based on a flower pollination algorithm PID controller and virtual inertia control for microgrid frequency stabilization. Elec. Power Syst. Res. 220.

Hu, J., Lanzon, A., 2019. Distributed finite-time consensus control for heterogeneous battery energy storage Systems in Droop-Controlled Microgrids. IEEE Trans. Smart Grid 10 (5), 4751–4761.

Ibraheem, M.I., Edrisi, M., Gholipour, M., Haes Alhelou, H., 2023. A novel frequency regulation in islanded microgrid using sliding mode control with disturbance observers considering storages and EVs. Comput. Electr. Eng. 105.

Irfan, M., Deilami, S., Huang, S., Veettil, B.P., 2023. Rooftop solar and electric vehicle integration for smart, sustainable homes: a comprehensive review. Energies 16 (21), 7248.

Islam, M.S.N., Mekhilef, S., Rahman, M.M., 2020. A review of global optimization methods for microgrid control—with specific application to distributed energy resources. IEEE Trans. Power Electron. 35 (11), 11755–11769.

Jan, M.U., Xin, A., Rehman, H.U., Abdelbaky, M.A., Iqbal, S., Aurangzeb, M., 2021. Frequency regulation of an isolated microgrid with electric vehicles and energy storage system integration using adaptive and model predictive controllers. IEEE Access 9, 14958–14970.

Jayachandran, M., Ravi, G., Jun. 2017. Design and optimization of hybrid Micro-grid system. Energy Proc. 117, 95–103.

Jia, H., Wu, Z., 2015. A model predictive control approach for wind power ramp control with battery energy storage systems. IEEE Trans. Power Syst. 30 (5), 2537–2546.

Jie, B., Tsuji, T., Uchida, K., 2017. Analysis and modelling regarding frequency regulation of power systems and power supply–demand-control based on penetration of renewable energy sources. J. Eng. 2017 (13), 1824–1828.

Johnson, J., et al., 2019. Opportunities for smart inverter controls in renewable energy integration. IEEE Trans. Smart Grid 10 (2), 1974–1983.

Karimi, H., Jadid, S., 2023. Multi-layer energy management of smart integrated-energy microgrid systems considering generation and demand-side flexibility. Appl. Energy 339.

Kaur, S., Kaur, T., Khanna, R., 2022. Design of the ANFIS based optimized frequency control module for an electric vehicle charging station. Appl. Energy 326.

Khalil, A.E., Boghdady, T.A., Alham, M.H., Ibrahim, D.K., 2023. Enhancing the conventional controllers for load frequency control of isolated microgrids using proposed multiobjective formulation via artificial rabbits optimization algorithm. IEEE Access 11, 3472–3493.

Khan, S., Sinha, N.K., 2022. Adaptive fuzzy control for microgrid frequency management. J. Control Decis. 9 (2), 123–136.

Khan, S., Sinha, N.K., 2023. Adaptive fuzzy logic control in hybrid microgrids. Int. J. Electr. Power Energy Syst. 146.

Kordestani, M., Safavi, A.A., Saif, M., 2021. Recent survey of large-scale systems: architectures, controller strategies, and industrial applications. IEEE Syst. J. 15 (4), 5440–5453. https://doi.org/10.1109/JSYST.2020.3048951.

Kumar, D., Deb, K., July 2022. Machine learning-assisted PID control for microgrid frequency regulation. IEEE Trans. Smart Grid 13 (4), 1234–1245.

Kundur, P., Balu, N.J., Lauby, M.G., 1994. Power System Stability and Control. McGraw-Hill.

Lee, J.M., Ogunnaike, B.A., Morari, M., 2006. A framework for model predictive control in industry. Control Eng. Pract. 14 (8), 769–775.

Leroy, P., 2024. Fuzzy logic in electric vehicle integration. J. Energy Storage 45.

Li, H., Li, X., He, Y., 2020. Adaptive fuzzy logic control for microgrid frequency stabilization with high penetration of renewable energy. IEEE Trans. Fuzzy Syst. 28 (11), 2658–2670.

Liu, Z., Xu, Y., Wang, L., 2017. Gaussian process regression for forecasting the power output of a grid-connected photovoltaic system. IEEE Access 5, 4859–4869.

Liu, Y., Li, Y., Xin, H., Gooi, H.B., Pan, J., 2019a. Distributed optimal tie-line power flow control for multiple interconnected AC microgrids. IEEE Trans. Power Syst. 34 (3), 1869–1880. https://doi.org/10.1109/TPWRS.2018.2883407.

Liu, J., Wang, X., Loh, P.C., 2019b. A review of hybrid droop control for distributed energy resources. IEEE J. Emerg. Sel. Top. Power Electron. 7 (2), 766–778.

Liu, X., Liu, J., Liu, Y., Jan. 2020. Dynamic control of microgrid frequency with particle swarm optimization-based load shedding. IEEE Trans. Smart Grid 11 (1), 637–646.

Liu, H., Xu, L., Zhao, Y., 2021a. Fault detection and diagnosis for microgrid frequency control systems. IEEE Trans. Smart Grid 12 (4), 3158–3167.

Liu, J., Wang, X., Zhao, Y., 2021b. Advancements in communication technologies for multilayered control in microgrids. IEEE Trans. Power Syst. 36 (3), 1892–1901.

Liu, X., Li, X., Li, X., Wang, R., Dong, C., Ren, H., Sun, Q., 2025. Resilience enhancement strategy of multiple microgrids based on distribution station area considering two-level power management. Energy 319.

Long, B., Liao, Y., Chong, K.T., Rodriguez, J., Guerrero, J.M., 2021. Enhancement of frequency regulation in AC microgrid: a fuzzy-MPC controlled virtual synchronous generator. IEEE Trans. Smart Grid 12 (4), 3138–3149.

Luna, A., et al., 2018. Energy management systems for smart inverters. IEEE Trans. Ind. Inf. 14 (2), 572–581.

Ma, Y., Liu, Y., Wang, C., 2021. A multi-agent approach for coordinated control in microgrid systems. IEEE Trans. Power Syst. 36 (4), 2764–2775.

Mao, M., Pang, S., Cai, W., 2019a. LQG-based robust frequency control in an islanded microgrid with stochastic demand. IEEE Trans. Ind. Electron. 66 (2), 1439–1448.

Mao, M., Wei, G., Zhao, H., Kang, Y., He, X., 2019b. A high step-up DC-DC converter with voltage multiplier and coupled inductor for distributed generation systems. IEEE Trans. Ind. Electron. 66 (2), 1439–1448.

Mayne, D.Q., Rawlings, J.B., Rao, C.V., Scokaert, P.O.M., Jun. 2000. Constrained model predictive control: stability and optimality. Automatica 36 (6), 789–814.

Meng, L., Sanseverino, E.R., July 2020. Integration of AI and ML techniques with PID control in microgrids. IEEE Trans. Smart Grid 11 (4), 1234–1245.

Meng, L., Sanseverino, E.R., Maris, P.G., Flores, A.R., Dec. 2020. Integration of PID control with advanced techniques in microgrid applications. IEEE Access 8, 123456–123467.

Mercier, P., Cherkaoui, R., Oudalov, A., 2009. Optimizing a battery energy storage system for frequency control application in an isolated power system. IEEE Trans. Power Syst. 24 (3), 1469–1477. https://doi.org/10.1109/TPWRS.2009.2022997.

Meyer, J., 2022. Fuzzy logic for grid-connected microgrid control. Renew. Energy 209, 478–489.

Mohamed, F.A., Koivo, H.N., 2010. System modeling and online optimal management of MicroGrid using mesh adaptive direct search. Int. J. Electr. Power Energy Syst. 32 (5), 398–407.

Mohammed, N., Lashab, A., Ciobotaru, M., Guerrero, J.M., 2023. Accurate reactive power sharing strategy for droop-based islanded AC microgrids. IEEE Trans. Ind. Electron. 70 (3), 2696–2707. https://doi.org/10.1109/TIE.2022.3167141.

Mohsenian-Rad, A., et al., 2021a. The role of virtual inertia in the coordination of inverter-based resources. IEEE Trans. Smart Grid 12 (4), 2856–2865.

Mohsenian-Rad, A., et al., 2021b. Smart grid communications for distributed energy resource management. IEEE Trans. Smart Grid 12 (4), 2856–2865.

Molina, M.G., Mercado, P.E., Feb. 2016. Stabilization and control of microgrid with wind generation by superconductor magnetic energy storage. IEEE Trans. Power Electron. 31 (2), 1545–1554.

Motsoeneng, R.H., 2022. Fuzzy logic-based load shedding in microgrids. Elec. Power Syst. Res. 205.

Muhtadi, A., Pandit, D., Nguyen, N., Mitra, J., 2021. Distributed energy resources based microgrid: review of architecture, control, and reliability. IEEE Trans. Ind. Appl. 57 (3), 2223–2235. https://doi.org/10.1109/TIA.2021.3065329.

Himabindu, N., Hampannavar, S., Deepa, B., Swapna, M., 2021. Hotovoltaic (PV) powered electric vehicle charging stations (EVCS) under different solar irradiation conditions in India: a way towards sustainable development and growth. Energy Rep. 7, 8534–8547.

Nanou, S.I., Marinopoulos, A.G., Hatziargyriou, N.D., 2017. Simulation-based performance analysis of PID controllers in microgrids with high renewable penetration. IEEE Trans. Sustain. Energy 8 (1), 1–11.

Negahban, M., Ardalani, M.V., Mollajafari, M., Akbari, E., Talebi, M., Pouresmaeil, E., 2022. A novel control strategy based on an adaptive fuzzy model predictive control for frequency regulation of a microgrid with uncertain and time-varying parameters. IEEE Access 10, 57514–57524.

Nguyen, T., Tran, H., 2022. Fuzzy logic control for renewable energy integration. J. Renew. Energy 2022.

Olsvik, M., 2022. Fuzzy logic in microgrid performance evaluation. J. Clean. Prod. 325.

Ouédraogo, S., Faggianelli, G.A., Notton, G., Duchaud, J.L., Voyant, C., 2022. Impact of electricity tariffs and energy management strategies on PV/battery microgrid performances. Renew. Energy 199, 816–825.

Pajooh, B., Yildirim, B., Ardeshiri, R.R., Gheisarnejad, M., Homayounzadeh, M., Khooban, M.H., 2024. Role of redox flow battery and AI-based controller in frequency regulation of weak microgrids. J. Energy Storage 84.

Panda, S., Padhy, N.P., 2020. Genetic algorithm optimization of PID controllers for microgrid applications. IEEE Trans. Ind. Electron. 67 (5), 1234–1245.

Patel, A., Thirugnanam, M., 2023. Fuzzy logic for microgrid resilience enhancement. IEEE Access 11, 245–257.

Peng, J., Lu, C., Wang, C., 2018. Real-time implementation of model predictive control for a microgrid testbed. IEEE Trans. Smart Grid 9 (3), 1792–1801.

Qin, S.J., Badgwell, T.A., 2003. A survey of industrial model predictive control technology. Control Eng. Pract. 11 (7), 733–764.

Qin, J., Tan, S., Jiang, L., 2020. Model predictive control for interconnected microgrids with uncertain renewable energy generation. IEEE Trans. Power Syst. 35 (4), 3317–3328.

Rad, M., et al., 2021. The role of virtual inertia in the coordination of inverter-based resources. IEEE Trans. Smart Grid 12 (4), 2856–2865.

Rafiee, A., Batmani, Y., Ahmadi, F., Bevrani, H., 2021. Robust load-frequency control in islanded microgrids: virtual synchronous generator concept and quantitative feedback theory. IEEE Trans. Power Syst. 36 (6), 5408–5416.

Rahbar, K., Chai, C.C., Zhang, R., 2018. Energy cooperation optimization in microgrids with renewable energy integration. IEEE Trans. Smart Grid 9 (2), 1482–1493.

Rajamand, S., 2021. Load frequency control and dynamic response improvement using energy storage and modeling of uncertainty in renewable distributed generators. J. Energy Storage 37. https://doi.org/10.1016/j.est.2021.102467.

Ramesh, M., Yadav, A.K., Pathak, P.K., 2021. Intelligent adaptive LFC via power flow management of integrated standalone micro-grid system. ISA Trans. 112, 234–250.

Ranjan, M., Shankar, R., 2022. A literature survey on load frequency control considering renewable energy integration in power system: recent trends and future prospects. J. Energy Storage 45. https://doi.org/10.1016/j.est.2021.103717.

Rawlings, J.B., Mayne, D.Q., 2014. Model Predictive Control: Theory and Design. Nob Hill Publishing, Madison, WI.

Ray, P.K., Mohanty, A., 2019. A robust firefly–swarm hybrid optimization for frequency control in wind/PV/FC based microgrid. Appl. Soft Comput. 85.

Romano, F., Russo, M., 2023. Fuzzy logic for battery energy storage management. Energy Rep. 9, 155–162.

Sarimveis, H., Patrinos, P., Akritidis, C.T., Bafas, A.S., 2003. Dynamic modeling and control of biological and biochemical systems. Comput. Chem. Eng. 27 (8), 1157–1172.

Savaghebi, M., Jalilian, A., 2022. Integration of model predictive control with PID tuning for microgrid applications. IEEE Trans. Smart Grid 13 (1), 1234–1245.

Savaghebi, M., Jalilian, A., Guerrero, J., Dragicevic, T., 2012. Comparative analysis of control strategies for microgrids. IEEE Trans. Smart Grid 3 (3), 1235–1245.

Scattolini, R., 2009. Architectures for distributed and hierarchical model predictive control – a review. J. Process Control 19 (5), 723–731.

Sharma, A., Rout, P.K., 2021. Optimized PID controller tuning for frequency control in microgrids. IEEE Trans. Ind. Electron. 68 (5), 1234–1245.

Sharmeela, C., Sivaraman, P., Sanjeevikumar, P., Holm-Nielsen, J.B., 2021. Microgrid Technologies. Wiley.

Siddiqui, A.S., Elshafie, M., 2020. Sliding mode control of a microgrid for frequency regulation with renewable energy integration. IEEE Access 8, 218641–218652.

Silva, J.A. dos S., 2023. Fuzzy logic control for microgrid energy management. Renew. Energy 205, 123–134.

Singh, A.K., Pandey, A.S., 2018a. Sliding mode control of microgrid for enhanced voltage and frequency regulation. IEEE Trans. Ind. Appl. 54 (5), 5314–5322.

Singh, A.K., Pandey, A.S., 2018b. Control of a standalone microgrid with hybrid energy storage system. IEEE Trans. Ind. Appl. 54 (5), 5314–5322.

Sinha, N.K., et al., 2021. Adaptive fuzzy control for renewable energy integration in microgrids. Int. J. Electr. Power Energy Syst. 128.

Smith, J., Brown, A., 2024. Fuzzy logic in microgrid energy management systems. IEEE Trans. Smart Grid 15 (2), 789–798.

Subudhi, B., Mishra, S., Nov. 2021. Robust PID control for microgrid frequency stability. IEEE Trans. Control Syst. Technol. 29 (6), 1234–1245.

Tanaka, T., Kawai, Y., 2023. Fuzzy logic for demand response in smart grids. J. Power Sources 548.

Tedeschi, E., et al., 2016a. Smart inverters for the smart grid: Deployments and challenges. IEEE Ind. Electron. Mag. 10 (2), 60–72.

Tedeschi, E., Liserre, M., Sauter, T., 2016b. Smart inverters for the smart grid: Deployments and challenges. IEEE Ind. Electron. Mag. 10 (2), 60–72.

Timbus, V., Larsson, A., Bongiorno, M., 2011. Evaluation of current control methods for distributed generation systems with high penetration of renewable energy sources. IEEE Trans. Power Electron. 26 (12), 3778–3787.

Tungadio, D.H., Bansal, R.C., Siti, M.W., Mbungu, N.T., 2018. Predictive active power control of two interconnected microgrids. Technol. Econ. Smart Grids Sustain. Energy 3 (1), 3.

Wang, G., Liu, X., 2024. Reinforcement learning-based tuning of PID controllers for microgrid frequency regulation. IEEE Trans. Smart Grid 15 (3), 1234–1245.

Wang, Y., Wu, J., Zhao, J., Zhang, H., 2015. PID control with secondary frequency regulation for microgrids. IEEE Trans. Power Deliv. 30 (3), 1231–1240.

Wang, S., Wang, K., Liu, J., 2017a. Distributed control for microgrid frequency regulation using multi-agent systems. IEEE Trans. Smart Grid 8 (6), 2902–2911.

Wang, Y., et al., 2017b. Remote monitoring and control of smart inverters. IEEE Trans. Ind. Electron. 64 (4), 2901–2910.

Wang, X., et al., 2018. With multimicrogrids. Appl. Energy 210, 1027–1036.

Wang, H.H., et al., 2020. Hierarchical control for coordinated control of multiple inverters in microgrid. IEEE Trans. Power Electron. 35 (4), 3673–3683.

Wang, H., et al., 2021a. Optimizing frequency control in microgrids with heterogeneous energy resources. IEEE Trans. Sustain. Energy 12 (3), 1824–1832.

Wang, H., et al., 2021b. Neural network-based model predictive control in microgrids with renewable energy integration. IEEE Trans. Sustain. Energy 12 (3), 1824–1832.

Wang, Z., et al., 2023. Fuzzy logic-based microgrid Fault Detection. IEEE Trans. Power Deliv. 37 (3), 1890–1898.

Wu, J., Wang, Y., Zhao, J., 2019. Neural network-based dynamic tuning of PID controllers for microgrid applications. IEEE Transact. Neural Networks Learn. Syst. 30 (3), 1234–1245.

Wu, N., Wang, H., Zhang, H., 2021. Robust control strategies for microgrid frequency stability under disturbances. IEEE Trans. Power Syst. 36 (2), 1100–1110.

Xie, R., Wang, Y., Wu, L., 2018. Fault diagnosis in microgrids based on support vector machine and particle swarm optimization. IEEE Trans. Smart Grid 9 (2), 1073–1082.

Xu, J., et al., 2020. Integration of IoT and communication systems in microgrid control. IEEE Trans. Sustain. Energy 11 (4), 2306–2316.

Xu, C., Wu, X., Wang, Y., 2021. Distributed cooperative control for resilient microgrid frequency regulation. IEEE Access 9, 45898–45908.

Yang, L., Guan, X., Wu, J., Dai, S., 2018. A multi-Layered distributed cloud network for cyber-physical energy system. In: IEEE 14th International Conference on Automation Science and Engineering (CASE), Munich, Germany, pp. 402–407. https://doi.org/10.1109/COASE.2018.8560483.

Yang, Z., Zhang, Y., Sun, W., 2020. Neural network-based adaptive control for microgrid frequency regulation. IEEE Transact. Neural Networks Learn. Syst. 31 (11), 4715–4727.

Zadeh, L.A., 1965. Fuzzy sets. Inf. Control 8 (3), 338–353.

Zamora, R., Srivastava, A.K., 2019. Experimental validation of PID controllers in microgrids. IEEE Trans. Ind. Appl. 55 (2), 1156–1165.

Zavala, V.M., Rawlings, J.B., Mar. 2009. Real-time optimization strategies for building systems. Energy Build. 41 (3), 320–330.

Zhang, C., Yu, M., Jan. 2022. Hybrid PID and fuzzy logic control for microgrid frequency regulation. IEEE Trans. Ind. Inf. 18 (1), 123–134.

Zhang, Y., Gatsis, N., Giannakis, G.B., 2013a. Robust energy management for microgrids with high-penetration renewables. IEEE Trans. Sustain. Energy 4 (4), 944–953.

Zhang, J., Zhang, C., Mei, S., Liu, Z., Ni, Y., Wu, F.F., Wei, M., 2018. Dynamic pricing in smart grid for demand response control. IEEE Trans. Power Syst. 33 (5), 5660–5669.

Zhang, X., Li, C., He, W., Jan. 2020. Deep reinforcement learning-based microgrid energy management considering load and renewable generation uncertainties. IEEE Access 8, 23672–23682.

Zhang, Y., Hu, Y., Liu, J., 2021a. Reinforcement learning for decentralized frequency control in microgrids. IEEE Access 9, 55678–55689.

Zhang, L.H., et al., 2021b. Coordinated control of distributed inverters with adaptive communication for smart grids. IEEE Trans. Ind. Inf. 17 (2), 830–839.

Zhang, L.H., et al., 2021c. Enhancing microgrid resilience through integrated communication systems. IEEE Trans. Ind. Inf. 17 (2), 830–839.

Zhang, L., et al., 2021d. Resilient adaptive techniques for enhancing power system stability. IEEE Trans. Ind. Inf. 17 (4), 2684–2693.

Zhang, L., et al., 2021e. Adaptive control strategies for enhanced microgrid stability. IEEE Trans. Power Syst. 36 (4), 3683–3692.

Zhang, L., et al., 2021f. Adaptive fuzzy logic control for enhanced microgrid stability. IEEE Trans. Power Syst. 36 (4), 3683–3692.

Zhang, L., et al., 2021g. Adaptive neural network control strategies for enhanced microgrid stability. IEEE Trans. Power Syst. 36 (4), 3683–3692.

Zhang, G., Daraz, A., Khan, I.A., Basit, A., Khan, M.I., Ullah, M., 2023. Driver training based optimized fractional order PI-PDF controller for frequency stabilization of diverse hybrid power system. Fractal. Fract. 7 (4), 315.

Zhao, B., Zhang, X., Li, P., Wang, K., Guo, L., 2018a. Robust model predictive control for microgrid frequency stabilization using battery energy storage systems. IEEE Trans. Smart Grid 9 (2), 907–916.
Zhao, B., Zhang, X., Li, J., Wang, W., Zhang, D., 2018b. Optimal day-ahead scheduling of distribution network considering high penetration of renewable energy. IEEE Trans. Smart Grid 9 (2), 907–916.
Zhao, Z., Guo, X., Li, F., Wang, C., 2021. A comprehensive review of droop control in islanded microgrids. IEEE Trans. Power Electron. 36 (8), 9151–9171.
Zhao, Y., Lin, J., Song, Y., Xu, Y., 2024. A robust Microgrid Formation Strategy for resilience enhancement of hydrogen penetrated active distribution networks. IEEE Trans. Power Syst. 39 (2), 2735–2748. https://doi.org/10.1109/TPWRS.2023.3305470.
Zheng, C., Dragicevic, T., Blaabjerg, F., 2021. Model predictive control-based virtual inertia emulator for an islanded alternating current microgrid. IEEE Trans. Ind. Electron. 68 (8), 7167–7177.
Zhang, R., 2021. Coordinated control strategies for inverter-based resources in microgrids. IEEE Internet Things J. 8 (3), 1570–1580.

CHAPTER

5

Analytical approaches to microgrid frequency control

Prabhat Kumar Vidyarthi, Pranav Prakash Singh, Ashiwani Kumar, and Ravi Shankar

Department of Electrical Engineering, National Institute of Technology Patna, Patna, Bihar, India

5.1 Introduction

Microgrids have become a viable way to improve grid resilience and integrate renewable energy sources. However, because of the intermittent nature of renewable energy sources and the decreased inertia of microgrids, maintaining frequency stability in these systems is a major difficulty. Maintaining the system frequency within reasonable bounds requires the balance between power production and consumption, which is achieved in large part by Load Frequency Control (LFC). New difficulties in preserving frequency stability in microgrids have been brought about by the growing use of renewable energy sources like solar and wind. These sources intermittent nature may result in abrupt variations in power production, which may alter system stability and result in frequency deviations (Choudhary et al., 2022).

It has been observe that a significant number of Renewable Energy Sources (RES), including solar and wind, would be added to the power grid in recent years in accordance with the energy strategy and mitigation of climate change (Vidyarthi and Kumar, 2024). The power system inertia constant would drop as a consequence of excessive small-scale scattered generation, leading to grid instability and catastrophic blackouts. One practical method for stabilizing microgrids is to realistically simulate the characteristics of Synchronous Generators (SG) in order to improve system inertia, stability, and resilience. For this purpose, SG mimics the actions of the key players (Xu et al., 2021; Arya, 2020). Additionally,

Frequency Control in Microgrids
https://doi.org/10.1016/B978-0-443-36476-1.00001-8

although endangering the stability and dependability of the microgrid, inertia features provide a framework for sustaining its DG contribution. By combining energy storage devices with a power electronic converter and an efficient regulating method, a virtual inertia for renewable energy sources has been created (Ram Babu et al., 2023). A Synchronous Machine (SM) or synchronous generator was used to describe this method. For distributed generation/renewable energy source units using power electronic devices to benefit from the stability of a SG, the SG was designed to mimic the behavior of a genuine SG. In recent years, a number of control mechanisms based on frequency control management have been put into place to increase the grid's frequency stability proposes a differential regulation-based frequency regulation method to improve microgrid balance (Vidyarthi et al., 2024). A reliable decentralized grid formation method is offered for evaluating microgrid dynamics combined with RES.

A ground-breaking Model Predictive Control (MPC) technique has been used to optimize the inertia, sag, and damping parameters in order to balance the grid frequency by using renewable energy (Long et al., 2021). Although a few studies have addressed the discrepancies in renewable resources and load demand, the uncertain pairings in renewable power producers and load demand were not investigated based on the previous methodologies. In order to deliver additional actual power toward the predetermined state, the rate of change of frequency control approach estimates how grid dynamics will change. Numerous methods for developing updated control algorithms with fractional computation, cascade controller, and some auxiliary regulation technique have been realized in the literature in contrast to conventional secondary controllers. However, a meta-heuristic optimization technique called QOHHO was used to find the ideal value of the proposed cascade Tilt Integral-Tilt Derivative (TI-TD) controller (Arya and Kumar, 2017).

Distributed Generators (DGs) powered by renewable energy will become more and more important in the generation of electricity due to increase in global warming in whole country. In the near future, there will be a significant increase in the usage of distributed generation based on solar energy (photovoltaic and solar thermal), wind, bio-gas, bio-diesel, mini-hydro, fuel cells, and micro-turbines. Clusters of dispersed generators and loads that function as a single, controlled system make up a microgrid. Concerns about the equipment's safety and protection have been highlighted by the DG's connection to the utility/grid via power electronic converters. Numerous cutting-edge control strategies have been used to improve microgrid stability and ensure appropriate load sharing. Using droop characteristics for decentralized load sharing is the most popular approach. To provide the system with the required actual power (and reactive power), parallel converters have been managed.

Since the distance between converters may make inter-communication impracticable in a real system, local signals are employed as feedback to operate converters. Frequency and fundamental voltage magnitude are two independent variables that may be controlled to share actual and reactive power. An fuzzy based TIDN controller is suggested in this chapter to distribute power across renewable energy in a microgrid (Saha and Saikia, 2019).

The purpose of this chapter is to examine the analytical approach of Load Frequency Control to Microgrid power system. The power plant under investigation has three different energy sources bio-gas, bio-diesel, and (wind & solar) in each control area, together with several real-time nonlinearities such as boiler dynamics, Generation Rate Constraint (GRC), and Governor Dead Band (GDB). Three controllers PIDN, Fuzzy-PID, and Fuzzy-TIDN (proposed) are appropriate for both planned and unplanned power under various contracts, such as poolco, contract violation-based transactions, and bilateral, via a variety of tie-lines. Furthermore, RESs like wind and solar power plants are utilized as case studies (Ahmed et al., 2023). After certain existing controllers were compared with a new kind of modified Fuzzy Tilt-Integral Derivative filter- (Fuzzy-TIDN) controller. To optimize the different parameters of the suggested controller, a modified version of the Sea-Horse Optimization (SHO) technique has been suggested. To demonstrate the superiority of the SHO, it is compared with certain popular meta-heuristic algorithms. To confirm the robustness and flexibility of the proposed controller, analysis was done under poolco-based transactions. The findings of the comprehensive analysis provide strong support for the effectiveness and utility of the recommended controller and optimization (Ye et al., 2024). Fig. 5.1 describes a overall flow chart of the control strategy in microgrid.

5.2 Importance of analytical techniques

The assessment, planning, and optimization of control schemes are made possible by analytical approaches, which are essential to microgrid load frequency control (Ahmed et al., 2019). The dynamic behavior of microgrids may be better understood by using methods like sensitivity analysis, frequency response analysis, and eigenvalue analysis. Analytical methods aid in determining possible stability problems, maximizing controller gains, and evaluating the effects of uncertainties and disturbances. Researchers and microgrid operators may increase system robustness, improve frequency control, and create more efficient LFC methods by using analytical tools. High-quality power transmission and preserving grid stability, makes it possible for microgrids to operate

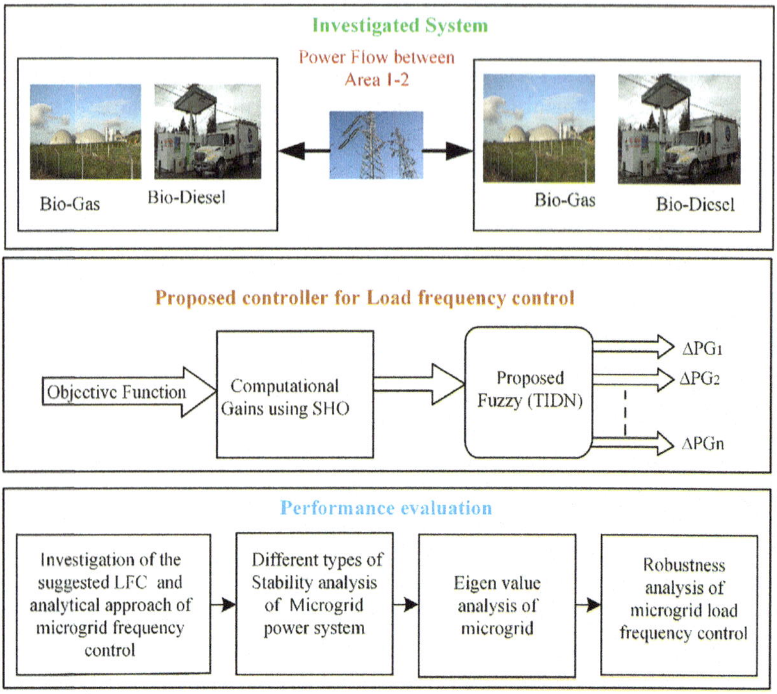

FIGURE 5.1 Overall flow chart of the proposed scheme.

dependably and effectively. In order to generate more complex and effective control strategies, analytical methodologies are crucial for the design and optimization of microgrid LFC systems.

5.2.1 Analytical techniques in microgrid LFC

A variety of strategies are used in typical analytical methods for microgrid load frequency control in order to assess and improve system performance. Typical techniques include:

1. Stability Analysis: To guarantee reliable and efficient operation, the stability of the microgrid is evaluated analytically under various operating conditions.
2. Performance Optimization: By using analytical techniques to optimize LFC performance, microgrid operators may reduce frequency fluctuations and provide a consistent power supply.
3. Component Sizing and Selection: Analytical techniques are used to assess the optimal size and selection of energy storage devices, renewable energy sources, and other microgrid components.

4. Fault Detection and Diagnosis: Analytical techniques enable the identification and diagnosis of microgrid flaws, ensuring prompt corrective action and minimizing downtime.
5. Economic benefits: By enhancing microgrid performance and reducing energy losses, analytical approaches may help reduce operating costs and boost economic benefits.

Microgrid operators may guarantee grid stability, dependable and effective operation, and high-quality electricity distribution by using these analytical techniques. These techniques provide insightful information on system behavior, which helps with control design and decision-making.

Using analytical techniques for microgrid LFC has many advantages, such as:

1. Enhanced stability of the system: analytical techniques aid in spotting any stability problems and creating countermeasures.
2. Improved system performance: energy losses are decreased and system efficiency is increased via optimization approaches.
3. Robust control design: controllers that are resilient to uncertainties and disturbances are created using analytical techniques.
4. Adaptability to shifting system circumstances and needs is made possible by analytical methodologies.

In order to ensure dependable and effective operation as well as high-quality power distribution, analytical techniques are essential to the design and optimization of microgrid LFC systems.

5.2.2 Advantages of analytical methods in microgrid LFC

Among the many advantages of analytical approaches in microgrid load frequency control (LFC) are precision modeling, stability analysis, reliable control architecture, performance enhancement, and adaptability. These benefits make it possible to identify any stability problems, accurately depict system dynamics, and create efficient control schemes. Microgrid operators can maintain high-quality power supply, dependable and economical operation, controller performance optimization, and grid stability by using analytical techniques (Zhang et al., 2022). All things considered, analytical techniques provide useful insights into system behavior, supporting well-informed control design and decision-making while allowing microgrids to function at their best under a range of circumstances.

5.2.2.1 System analysis parameters

Microgrid Load Frequency Control system analysis parameters are essential for assessing and improving system performance. These factors help operators create efficient control methods by offering insightful information about the microgrid's dynamic behavior. Among the crucial criteria are:

1. Frequency deviation: Gauges the system's capacity to control frequency by comparing the actual and nominal frequencies.
2. Settling time: The amount of time it takes for the frequency to return to normal after a disruption, indicating how adaptable the system is.
3. Overshoot: The highest frequency deviation during the transient response that shows the damping and stability of the system.
4. Steady-state error: The discrepancy between the intended and actual steady-state frequencies, which indicates how well the system can follow the reference frequency.
5. Stability margins: Assesses the robustness of a system's stability, showing how well it can tolerate uncertainty and shocks.
6. Robustness to uncertainty: The capacity to continue operating in the face of unforeseen circumstances, such as shifts in load, generation, or problems.

5.2.3 Analysis of Eigen values in microgrid LFC

Eigenvalue analysis is a mathematical technique used to analyze the stability and dynamic behavior of Linear Time-Invariant (LTI) systems. In the context of Microgrid LFC, eigenvalue analysis helps evaluate the stability of the microgrid frequency control system. The eigenvalue approach is often used to analyze the stability of power systems. This section outlines the whole process for small-signal stability analysis using the eigenvalue technique, which is also shown in Fig. 5.2. It includes modeling and integration, stability analysis, and verification (Sahu and Samantaray, 2023).

System Integration and Component Modeling First, an analytical modeling technique is described. Typically, a system model is integrated for eigenvalue analysis after state space models of components are

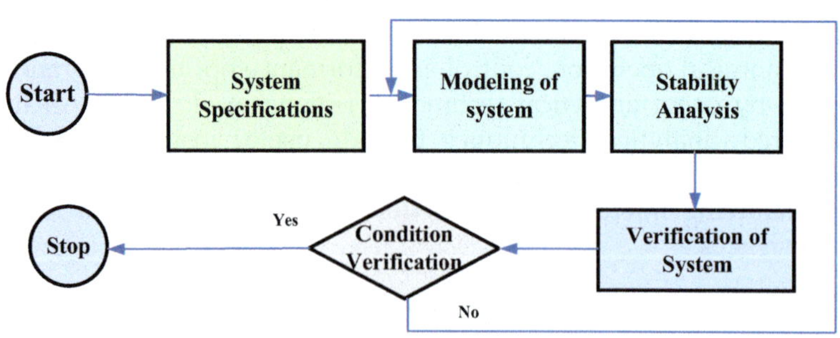

FIGURE 5.2 Flow chart of stability analysis using the eigenvalue technique.

constructed using the SSA approach. There would be several state variables from the intricate component models for a large-scale power system with a high penetration of power electronic converters (Singh et al., 2024). The CCM or a module-based method may be used for system integration to overcome the significant computational load and complexity involved in determining the overall system state space model. The power system of the CCM is separated into several interrelated components. The component state space models are acquired and linearized as

$$\Delta\dot{x} = A_i\Delta x + B_i\Delta u \tag{5.1}$$

$$\Delta\dot{y} = C_i\Delta x + D_i\Delta u \tag{5.2}$$

where the state matrix for the ith subsystem is represented by A_i, B_i, C_i, and D_i.

The network architecture is used to extract the connectivity relationship, which is provided by

$$u = L_1 y + L_2 a \tag{5.3}$$

$$b = L_3 y + L_4 a \tag{5.4}$$

where the component interconnection matrix is represented by L_1, L_2, L_3, and L_4.

Next, the whole system's small-signal model is combined as

$$\Delta\dot{x} = F\Delta x + G\Delta a \tag{5.5}$$

$$\Delta b = H\Delta x + J\Delta a \tag{5.6}$$

Where $F = A + BL_1(I\text{-}DL_1)^{-1}C$, $G = BL1(I\text{-}DL_1)^{-1}DL_2 + BL_2$

$H = L_3(I\text{-}DL_1)^{-1}C$, $J = L_3(I\text{-}DL_1)^{-1}DL_2 + L_4$

A scalable and modular solution for the integration of large-scale power electronic-based power systems is offered by the CCM and modular approach.

5.2.3.1 Eigenvalue analysis

The eigenvalue of the system matrix F is examined in order to investigate the linearized state space model in. The eigenvalue analysis reveals various damping properties and oscillation modes. The frequency fosci and damping ratio ζ_i of the oscillation are given as Eqs. (5.8) and (5.9) for a complex eigenvalue λ_i that represents an oscillatory mode of the system in Eq. (5.7). Fig. 5.3 depicts the Simplified transfer function model of the multi area microgrid.

$$\lambda_i = \sigma_i \pm j\omega_i \tag{5.7}$$

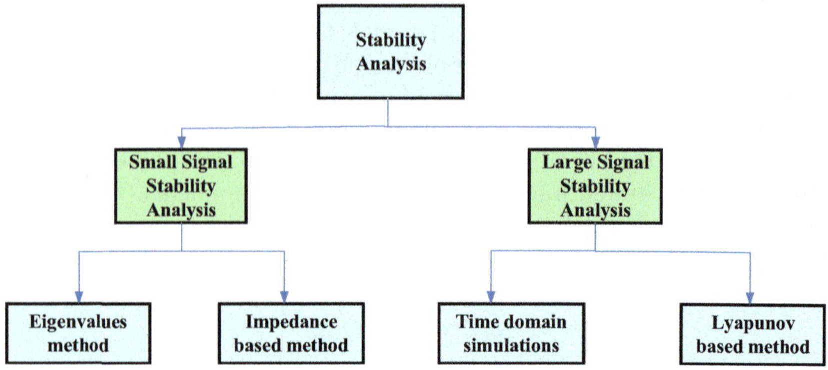

FIGURE 5.3 Simplified transfer function model of the multi area microgrid.

$$\text{fosci} = \frac{\omega_i}{2\pi} \tag{5.8}$$

$$\zeta_i = \frac{\sigma_i}{\sqrt{\sigma_i{}^2 + \omega_i{}^2}} \tag{5.9}$$

The rate of oscillation amplitude reduction is determined by the damping ratio. Eigenvalue analysis may be used to determine the dominating eigenvalues at various oscillation frequency ranges.

For Example:

Eigenvector for the matrix $A = \begin{bmatrix} 1 & 2 \\ 5 & 4 \end{bmatrix}$.

If eigenvalues are represented by λ.

Then the eigenvalues is calculated by using the equation,

$$|A - \lambda I| = 0$$

$$\begin{bmatrix} 1 & 2 \\ 5 & 4 \end{bmatrix} - \lambda \begin{bmatrix} 1 & 0 \\ 0 & 1 \end{bmatrix} = 0$$

$$\begin{bmatrix} 1-\lambda & 2 \\ 5 & 4-\lambda \end{bmatrix} = 0$$

$\Rightarrow (1 - \lambda)\,(4 - \lambda) - 10 = 0$
$\Rightarrow 4 - \lambda - 4\lambda + \lambda 2 - 10 = 0$
$\Rightarrow \lambda 2 - 5\lambda - 6 = 0$
$\Rightarrow \lambda 2 - 6\lambda + \lambda - 6 = 0$
$\Rightarrow \lambda(\lambda - 6) + 1(\lambda - 6) = 0$

$\Rightarrow (\lambda - 6)(\lambda + 1) = 0$

$\lambda = 6$ and $\lambda = -1$

6 and −1 are the eigen values.

5.2.3.2 *Eigenvalue significance*

Eigenvalues, a key idea in linear algebra, are essential to load frequency control analysis. Eigenvalues quantify how much scaling or alteration is made to a system's frequency response under various load scenarios. To guarantee system stability and dependability, they evaluate the damping of oscillations and analyze how frequency deviations react to variations in load demand. Eigenvalues are essential to the design and analysis of control systems in LFC that maintain power balance and adjust frequency. They aid in the analysis of system stability by calculating stability margins, which guarantee that the system can tolerate load variations and disruptions. Eigenvalues are also useful in the design of LFC controllers that maximize controller gains for the required performance and stability, efficiently control frequency, and preserve power balance (Mitra and Nguyen, 2022).

To analyze frequency response and stability and to determine the stability of the power system to make sure it can tolerate load variations and disturbances, eigenvalues are crucial in system modeling. Power system operators may build and develop control systems that balance power supply and demand and preserve frequency stability by examining eigenvalues. This ensures the dependability and effectiveness of power systems. The design and analysis of control systems that maintain power balance and manage frequency in LFC heavily rely on eigenvalues.

5.2.3.3 *Sensitivity to Eigenvalues*

Eigenvalues are square matrix-related scalar values that quantify the transformation of a vector by a matrix. The scalar that results from multiplying a vector v by a matrix P is the eigenvalue that corresponds to the eigenvector v. In disciplines including data science, engineering, and physics, eigenvalues are often used.

The factor by which eigenvectors are stretched in their direction is therefore shown by these characteristic values. Only in cases when the eigenvalue is negative does it imply a change in the vector's direction. The direction is simply inverted when the eigenvalue is negative.

The eigenvalue equation is provided by

$Pv = \lambda v.$

Where,

- *P is the matrix,*
- *v is associated with eigenvector, and*

- *λ is the eigenvalue.*

The way that eigenvalues respond to changes in the system's structure or properties is referred to as "eigenvalue sensitivity".

5.2.3.3.1 Techniques for Eigenvalue sensitivity analysis

Eigenvalue sensitivity analysis is essential to Load Frequency Control (LFC) in order to comprehend the effects of system parameter changes on stability and performance. Engineers may use this technique to examine how changes in controller gains, generator settings, or load demand affect system stability. This makes it possible to optimize controller gains for stable frequency regulation by assisting in the identification of crucial factors that impact stability and performance.

The following methods may be used with LFC:

1. Examine how changes to load demand, generator settings, or controller gains affect system stability.
2. Determine the crucial elements influencing performance and stability.
3. To guarantee steady frequency management, maximize the controller gains.
4. Create resilient LFC systems that can tolerate load variations and disruptions.

5.2.3.4 *Eigen value applications in frequency control design*

Eigenvalues are essential to Load Frequency Control (LFC), allowing for a number of crucial applications that guarantee dependable and effective power system operation. Maintaining grid stability, supplying demand for energy, and averting power outages all depend on these applications. Eigenvalues aid in the stability analysis of power systems, enabling engineers to evaluate how changes to load demand or system characteristics affect system stability. The ability of electricity systems to tolerate disruptions and continue operating steadily depends on this study. To avoid power outages and grid instability, engineers may detect any stability problems and create corrective measures by examining eigenvalues (Mitra and Nguyen, 2022).

Design of the Controller: LFC controllers are designed and optimized using eigenvalues to ensure that they can efficiently adjust the frequency and preserve power balance. The best controller gains and settings for reliable and effective operation may be found by engineers by examining eigenvalues. Because of this, power networks can function within reasonable frequency ranges even when there are disruptions and variations in load demand. Robustness Analysis: The robustness of LFC systems to load variations and disturbances is assessed using eigenvalue analysis. Engineers may use this approach to create more robust systems that can tolerate unforeseen changes in system characteristics or load demand. By examining eigenvalues, engineers may spot any flaws in the system and create fixes to increase stability and robustness (Kohli and Arora, 2018).

Performance Optimization: To ensure the effective and dependable operation of power systems, eigenvalues are used to enhance LFC controller performance. The best control solutions to reduce frequency deviations and preserve power balance may be found by engineers by examining eigenvalues. By doing this, electricity systems can function at their best, lowering the possibility of grid instability and power outages. The analysis of small-signal stability in power systems is made possible by eigenvalues, which allow engineers to see possible stability problems and create solutions. The ability of electricity systems to tolerate minor disruptions and continue to function steadily depends on this study (Jia et al., 2023). Through eigenvalue analysis, engineers may spot possible stability problems and create control plans to lessen them.

5.2.4 Stability analysis technique

For microgrids to operate dependably and effectively, microgrid stability is essential. The low inertia, fluctuating renewable energy sources, and dynamic demand patterns of these smaller, localized power networks provide particular stability issues. Sustaining stability is necessary to provide local loads with high-quality electricity. Frequency stability, voltage stability, and transient stability are the three primary categories of microgrid stability. The microgrid's capacity to maintain a steady frequency, usually 50 or 60 Hz, in spite of variations in demand or generation, is known as frequency stability. By ensuring that loads get the voltage necessary for proper operation, voltage stability helps to avoid equipment damage or malfunction. Transient stability allows microgrids to recover from abrupt disruptions, such as faults or load shifts, and resume steady operation (Khamies et al., 2021).

The stability of microgrids may be affected by many issues, such as dynamic load profiles, changeable renewable energy sources, and low inertia. Microgrids are vulnerable to frequency variations due to their low inertia, and uncertainty and unpredictability are introduced by changeable renewable energy sources. Stability may be difficult to maintain when dynamic load patterns cause unexpected behavior. Numerous technologies, including grid-forming inverters, energy storage devices, sophisticated control systems, and load management techniques, may be used to improve microgrid stability. Advanced control systems monitor and modify operations in real time, while energy storage systems act as a buffer against abrupt changes. Virtual inertia is provided by grid-forming inverters, and supply and demand are balanced via load control techniques (Magdy et al., 2018).

Better power quality, more efficiency, and better dependability are all advantages of microgrid stability. Stable microgrids minimize energy losses, provide a steady supply of electricity, and boost overall effectiveness. By giving microgrid stability a first priority, operators may provide local loads with dependable and effective electricity, promoting development and economic progress. To sum up, microgrid stability is essential for dependable and effective operation. Microgrid operators may sustain steady operations and provide local loads with high-quality electricity by comprehending the difficulties and putting solutions in place. Microgrid stability is a crucial component of contemporary power networks as it promotes economic growth and development. The various stability approaches used in microgrid LFC are shown in Fig. 5.4.

5.2.4.1 *Types of LFC stability*

There arevarious types of stability analysis are listed below:

1. **Asymptotic Stability:** Asymptotic stability is the capacity of a system to converge to its equilibrium state over time, with its behavior gradually returning to the equilibrium state after a disruption. To put it another way, asymptotic stability makes sure that the system's reaction to a disturbance gradually deteriorates

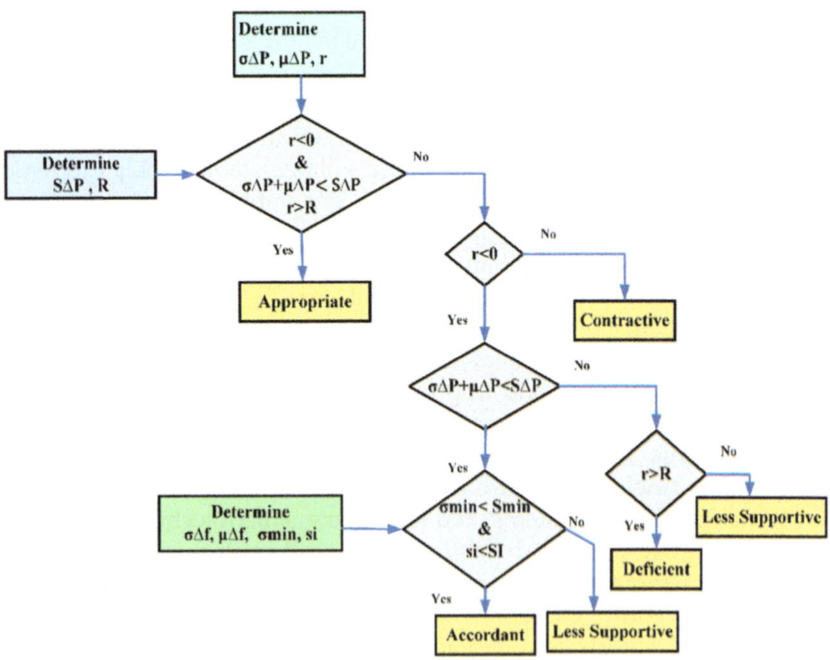

FIGURE 5.4 Different types of Stability techniques.

and eventually reaches its stable operating point.A desired feature in many systems, such as dynamical systems, power systems, and control systems, is asymptotic stability. It guarantees that the system can continue to function steadily even in the face of disruptions.

The following are important features of asymptotic stability:

1. Convergence to equilibrium: Over time, the behavior of the system converges to the equilibrium state.
2. Decay of disturbances: Over time, the system's reaction to abnormalities deteriorates.
3. Stability: After returning to its equilibrium condition, the system continues to operate steadily.

Lyapunov stability theory and eigenvalue analysis are often used to investigate asymptotic stability. Designers and operators of complex systems may ensure steady and dependable functioning by guaranteeing asymptotic stability.

2. **Bounded-Input Bounded-Output (BIBO) stability**: Systems stability is examined using the BIBO (Bounded-Input Bounded-Output) stability concept. If each bounded input results in a bounded output, the system is said to be BIBO stable. Stated differently, BIBO stability guarantees that while the input is constrained, the system's output stays within reasonable bounds. This indicates that when given finite inputs, the system does not generate unlimited or unbounded outputs.

BIBO stability is crucial for several applications, such as:

1. Control systems: BIBO stability guarantees that control systems don't generate excessive outputs and stay stable.
2. Signal processing: BIBO stability ensures that systems that process signals don't create unbounded outputs or magnify noise.

If a system satisfies certain requirements, including having a stable transfer function or a limited impulse response, it is said to be BIBO stable. Numerous methods, such as frequency-domain and time-domain analysis, may be used to examine BIBO stability. System designers may ensure that their systems will perform reliably and stay within acceptable bounds even in the face of uncertainties or disruptions by guaranteeing BIBO stability. In many situations where stability and predictability are crucial, this is crucial.

3. **Lyapunov Stability:** The idea of Lyapunov stability is used to examine how stable dynamical systems are. Aleksandr Lyapunov introduced it, and control theory and other disciplines make extensive use of it.

If a system meets certain requirements, it is said to be Lyapunov stable:

1. There is an equilibrium point in the system.
2. Over time, the trajectory of the system stays around the equilibrium point.

Lyapunov functions, which are scalar functions that meet certain criteria, are used to assess Lyapunov stability. A system's stability may be shown by determining its Lyapunov function. Researchers and engineers may evaluate and develop stable systems with predictable and dependable behavior by using the Lyapunov stability theory. In many situations where stability is crucial, this is crucial (Nair et al., 2016).

5.2.4.2 *Techniques for stability analysis*

There are various types of stability analysis are given below:

1. **Routh-Hurwitz Criterion:** It is a mathematical method for assessing a system's stability which gives the number of roots with positive real parts a sign of instability involves examining the coefficients of a system's characteristic equation. The Routh-Hurwitz criteria analyze the characteristic equation's coefficients using a tabular approach. The coefficients are arranged in a certain way to create the table, and the signs of the items in the first column are then calculated.

The stability of the system is determined by the number of roots with positive real components, which is shown by the number of sign changes in the first column. The system is stable if there are no sign changes. The system is unstable if there are significant changes.

In control systems engineering, the Routh-Hurwitz criteria is often used to:

1. Assess system stability
2. Examine system behavior
3. Create reliable control mechanisms

It offers a rapid and effective method for figuring out stability without explicitly solving the characteristic equation. Engineers may make sure that their systems are stable and function as intended by using the Routh-Hurwitz criteria. In control systems engineering, this method is a useful instrument.

2. **Nyquist Stability Criterion:** A graphical method for assessing the stability of closed-loop control systems is called Nyquist stability. The open-loop transfer function is plotted in the complex plane, and the number of encirclements around the critical point (-1,0) is examined.

According to the Nyquist stability criteria,

1. The system is unstable if the plot rounds the critical point (-1,0) counterclockwise.
2. The system is stable if the plot does not circle the critical point.
3. The number of unstable poles depends on the number of encirclements.

Engineers may make sure that their control systems are stable and operate as intended by using the Nyquist stability criteria. In control systems engineering, this method is often used for stability analysis and design. It aids engineers in creating control systems that are stable and strong.

3. **Bode plot analysis:** A Bode plot, which is used to study and build control systems, is a graphical depiction of a system's frequency response. There are two storylines in it:
   1. Magnitude plot (gain vs. frequency)
   2. Phase plot (phase angle vs. frequency)

1. **Gain crossover frequency:** Gain margin, which shows how much gain can be added to a system before it becomes unstable, is a metric used to quantify stability in control systems. It is the amount of gain increment necessary to cause instability in the system (Chen and Luo, 2023). The gain margin is the difference between the gain at the phase crossover frequency (when phase angle = -180 degrees) and 0 dB. It is often expressed in decibels (dB). Stability is indicated by a positive gain margin, and instability is shown by a negative gain margin. More stability is indicated by a larger gain margin. Engineers may assess a system's stability and make necessary modifications to guarantee steady operation by examining the gain margin.

2. **Phase crossover frequency:** The amount of phase shift that may happen before a control system becomes unstable is known as the phase margin, which is a measure of stability. In the case of gain = 0 dB, it is the difference between the phase angle at the gain crossover frequency and -180 degrees. Stability is indicated by a positive phase margin, and instability is shown by a negative phase margin. Greater stability is indicated by a larger phase margin.
3. **Stability margins:** The stability margin, which indicates the amount of tolerance or buffer before the system becomes unstable, is a measure of how stable a system is. In control systems engineering, it is a crucial idea.

Engineers may learn more about system behavior, spot possible stability problems, and create control systems that work better by examining

Bode graphs. This method is crucial for creating stable and reliable control systems in a variety of industries, such as process control, automotive, and aerospace. A useful tool for understanding and enhancing system performance is a bode plot.

5.2.4.3 *Small signal analysis stability*

Small signal analysis, which examines how a system responds to minor disturbances, is one method for determining its stability.

5.2.4.3.1 Stability depends on the analysis of small signals

The goal of small-signal stability analysis is to determine how stable the power system is under little perturbations. If it is feasible to reduce oscillations caused by small disturbances such that changes in system state variables remain low for a long time, the power system is stable. The power system is unstable, on the other hand, if the oscillations intensify or continue indefinitely. The small-signal stability of the power system is affected by several factors, including the parameters of various control devices, the strength of electrical connections between power system components, and initial operating conditions. In truth, no power system that lacks small-signal stability can work since even the smallest disruptions will have an impact on the system's performance. In other words, consistent small-signal stability is a prerequisite for the operation of a power system. Thus, one of the primary duties in power system analysis is to examine the power system under the specified operating conditions by performing small-signal stability analysis (Chen and Luo, 2023).

Small-signal stability refers to the system's capacity to remain stable when small-signal interruptions occur. Large temporal oscillations caused by power electronic converters interacting with the grid and each other are known as small-signal stability issues. This section introduces the modeling methods for power converters. To perform small-signal stability analysis, two well-known small-signal analysis methods are then introduced: the eigenvalue approach and the impedance-based method.

The way the grid reacts to transients caused by disturbances from its equilibrium point is evaluated in power system stability studies. The system is considered stable if it reaches an equilibrium point after the transient. This becomes crucial when the number of converter-interfaced resources in the system rises, such as solar PV, wind farms, and batteries, and new dynamic behaviors are seen.

1. Predicts System Behavior: By analyzing small signals, one may predict how a system will respond to small disturbances.
2. Identifies Stability Issues: Small signal analysis may identify potential stability issues with a system.

3. Guides Controller Design: Determining the information needed to develop controllers that can stabilize a system is possible via the analysis of tiny signals.

5.2.4.3.2 Procedures for analyzing small signals

1. Make the System Linear: Make the system linear around its operating point.
2. Obtain the Transfer Function: Determine the transfer function of the system.
3. Look at the Poles: Look at the poles of the transfer function to determine stability.
4. Examine the Frequency Response: Examine the systems frequency response to determine stability.

5.2.4.3.3 Techniques for large-signal stability analysis

Transient stability referred as a large-signal stability, which is used for stability analysis that is vulnerable to major signal disruptions caused by faults, protection, sudden load connections or disconnections, etc. Large-signal stability studies are conducted using two approaches: Lyapunov-based analytical methods and time-domain simulations using switching models. This section provides a brief overview of the time-domain simulation method and demonstrates three Lyapunov-based analytical strategies to identify the zone of attraction that may ensure system stability under large signal interruptions (Saxena et al., 2023).

5.2.4.3.4 Time-domain simulations

Time-domain simulation is a common technique for examining system large-signal stability, and electromechanical transient simulation tools have been developed for examining system behavior under different large-signal disruptions. It provides outstanding accuracy and validity. To ensure system stability throughout a wide operating range and in the face of various interruptions, several simulations must be performed, which leads to a very high computing burden.

5.2.4.4 *Lyapunov stability*

It is a mathematical technique known as Lyapunov stability analysis is used to evaluate the stability of a system. The concept of a Lyapunov function, a scalar function used to measure the stability of a system, forms its basis. With Lyapunov-based analytical methods, the zone of attraction where the system is stable despite large-signal perturbations can be determined. Consequently, a significant number of time-domain simulations are no longer necessary. Three methods for analyzing large-signal stability are presented in this article.

5.2.4.4.1 Takagi-Sugeno-multimodel method

Nonlinear systems having multiplication, division, exponential, or square root nonlinearities may be modeled using the TS multimodeling approach. This approach uses a collection of linear models based on a set of fuzzy "if-then" rules to describe a nonlinear system. A local linear model will be used to depict the nonlinear model's local behavior at each rule. When there are enough fuzzy rules, the TS models are equal to the nonlinear model. The relevant local model for the ith rule of total r rules is written as

$$\begin{cases} \dot{x}(t) = A_i x(t) + B_i u(t) \\ y(t) = C_i x(t) \end{cases} \tag{5.10}$$

where *Ai*, *Bi*, and *Ci* are constant matrices.

Then the nonlinear model can be expressed as

$$\begin{cases} \dot{x}(t) = \sum_{i=1}^{r} h_i (A_i x(t) + B_i u(t)) \\ y(t) = \sum_{i=1}^{r} h_i C_i x(t) \end{cases} \tag{5.11}$$

where *hi* is the respective weight of the ith rule that $\sum_{i=1}^{r} h_i = 1,\ h_i > 0$.

The nonlinear system is asymptotically stable if the Linear Matrix Inequality (LMI) condition in (5.12) is satisfied.

$$\begin{cases} M = M^T > 0 \\ A_i^T M + M A_i < 0, \text{where } i = 1, 2 \ldots\ldots r \end{cases} \tag{5.12}$$

where is M is a matrix and its existence indicates the stability of the system. Once the matrix of LMI problem is known, the Lyapunov function is selected as

$$V(x) = x^T M x \tag{5.13}$$

The domain of attraction is estimated using the following algorithm:

Algorithm:

1. Set all nonlinearities to the operating point values $f_{jmin} = f_{jmax} = f_j(0)$, j = 1,2,.q ($x_{jmin} = x_{jmax}$), where q is the function f's number of nonlinearities.

2. Verify the LMI issue in (12); if it is possible, go to Step 3; if not, proceed to Step 4.

3. Change xjmin and xjmax, j = 1,2,., q to decrease fjmin and raise fjmax. Next, go to Step 2.

4. Using x_{jmin} and x_{jmax}, j = 1, 2,., q, one may determine the estimated domain of attraction.

Using the acquired Lyapunov function, this approach may estimate the domain of attraction. However, when the number of nonlinearities rises, the matrix's order will increase quickly, making the issue more difficult to solve.

Procedures for Analyzing Lyapunov Stability

Step 1: Select a function of Lyapunov: Choose a Lyapunov function $V(x)$ that satisfies the necessary requirements.

Step 2: The second step is to calculate the derivative of the Lyapunov function $V(x)$ along the trajectories of the system.

Step 3: To ascertain the system's stability, examine the derivative of the Lyapunov function.

5.2.4.4.2 Advantages of lyapunov stability analysis

1. Provides a Sufficient Condition for Stability: Stability may be shown using the Lyapunov stability analysis's sufficient condition for stability.
2. Capable of Handling Nonlinear Systems: The capacity of Lyapunov stability analysis to manage nonlinear systems is one of its primary benefits over other stability analysis techniques.
3. Provides a Stability Measure: Lyapunov stability analysis provides a stability measure that can be used to compare the stability of different systems.

5.2.5 Robustness analysis in microgrid LFC

Robustness analysis in LFC evaluates a power system's ability to maintain stability and performance in the face of unknowns, interruptions, and changes in operating conditions.

5.2.5.1 Robustness analysis important in microgrid LFC

For microgrids to remain stable and function well in the face of uncertainties and disruptions, robustness analysis in microgrid Load Frequency Control (LFC) is essential. Because of their inherent unpredictability and instability, microgrids which are distinguished by their decentralized structure and incorporation of renewable energy sources present particular difficulties for LFC.

Robustness analysis is a thorough assessment of the system's resistance to several kinds of uncertainty, such as:

1. Uncertainties in parameters: The performance of LFC systems may be impacted by changes in component properties, system parameters, and load conditions. Analysis of robustness aids in determining how these uncertainties affect the stability of the system.
2. Disturbances: The balance between generation and load may be upset by abrupt changes in load, generation, or system defects,

which might impact frequency stability. Analysis of robustness assesses how well the system can tolerate certain kinds of disruptions.
3. Model uncertainties: Differences between the predicted and real behavior of a system may result from assumptions, simplifications, or inaccurate system modeling.

Robustness analysis accounts for these uncertainties to ensure that the LFC system remains stable and functions well despite modeling flaws. Robustness analysis is essential in microgrid LFC for:

1. Ensuring frequency stability: Engineers may create LFC systems that preserve frequency stability in the face of uncertainties and disruptions by evaluating the system's resilience.
2. Preserving power balance: Robustness analysis makes sure that, even under unpredictable circumstances, the system can maintain a balance between generation and load.
3. Improving system reliability: Engineers may improve the overall dependability of microgrids by spotting any stability problems and creating more resilient LFC systems.

In microgrid LFC, a variety of methods are used for robustness analysis, such as:

Sensitivity analysis: This entails examining how sensitive the system is to changes in uncertainty and parameters.

1. Uncertainty modeling: This method simulates uncertainties and disruptions to assess how they affect the stability of the system.
2. Robust control design: To create LFC systems that are resilient to uncertainties and disturbances, techniques such as H-infinity control are used.

It is impossible to exaggerate the significance of robustness analysis in microgrids. Microgrids often function in unpredictable settings with fluctuating load and generation circumstances. Even in the face of disruptions and uncertainties, engineers can preserve grid stability, dependability, and efficiency by making sure LFC systems are resilient. The construction of more robust and effective microgrids is made possible by this research, which is crucial for incorporating renewable energy sources and guaranteeing a steady supply of electricity. To sum up, robustness analysis is an essential component in the design and operation of microgrid LFCs. By assessing the system's resilience to uncertainties and disruptions, engineers can create more resilient LFC systems, guarantee dependable microgrid operation, and preserve grid efficiency and stability.

5.2.5.1.1 Types of robust control

To manage uncertainties and disturbances in control systems, robust control techniques are developed. The following are a few popular categories of robust control techniques:

1. H-infinity control: This technique seeks to reduce the system's maximal gain while maintaining stability and performance in the face of uncertainty.
2. Mu-analysis: This method assesses a system's resilience to structured uncertainty and offers information on stability margins.
3. Linear Matrix Inequality (LMI): A strong control framework that enables the definition of stability and performance limitations is offered by LMI-based control design.
4. Sliding Mode Control (SMC): This nonlinear control technique ensures resilience to uncertainty by forcing the system trajectory onto a preset surface.
5. Model Predictive Control (MPC): MPC ensures resilient control in the face of uncertainties and restrictions by predicting the system's future behavior using optimization approaches.
6. Quantitative Feedback Theory (QFT): This frequency-domain method creates reliable controllers that satisfy certain stability and performance standards.
7. Robust tuning techniques: These techniques, which include robust PID tuning and loop shaping, are meant to create controllers that can withstand uncertainties and disruptions.

5.2.5.2 Robustness matrix

A robustness matrix is one mathematical tool for assessing a system's resilience. It provides a measure of the system's ability to operate and maintain stability in the face of uncertainty and disturbances (Sun et al., 2023). A crucial component of power system functioning is the Robustness Matrix with Mathematical Modeling in Load Frequency Control (LFC), which makes sure that the system's frequency stays within reasonable bounds even when load or generation fluctuates. To make sure the system can tolerate uncertainties and disruptions, robustness analysis is crucial in LFC.

5.2.5.2.1 Mathematical Modeling

A power system for LFC may be described mathematically as follows:

$$\Delta f(s) = (1 \; / \; (2Hs + D)) * (\Delta P_m(s) - \Delta P_d(s)), \text{ where:}$$

Is the frequency deviation $\Delta f(s)$?
The change in mechanical power is represented by $\Delta P_m(s)$.
$\Delta P_d(s)$ is the load demand change.
H is the constant of inertia.
The damping coefficient is denoted by D.

5.2.5.2.2 Robustness matrix

The resilience of the LFC system to uncertainties and disturbances may be assessed using a robustness matrix.

$R = [\Delta A\ \Delta B; \Delta C\ \Delta D]$ is one way to describe the matrix, where:

The uncertainties in the system matrices A, B, C, and D are denoted by ΔA, ΔB, ΔC, and ΔD, respectively.

5.2.5.2.2.1 Uncertainty modeling There are many ways to represent the system's uncertainties, including:

- Uncertainty in system parameters, such H and D, is known as parametric uncertainty.
- Dynamic uncertainty: unpredictability in the dynamics of a system, including nonlinearities or unmodeled dynamics

5.2.5.2.2.2 Robust control design Controllers that guarantee the LFC system's robust stability and performance may be designed using robust control design approaches like mu-synthesis or H-infinity control.

5.2.5.2.2.3 Example Examine a power system that has the following specifications:

$H = 5$ s
$D = 1$ pu/Hz

The step change in load demand is $\Delta Pd(s) = 0.1$ pu.

The resilience of the system may be assessed using the robustness matrix and mathematical model. The findings indicate that the system is sensitive to $\Delta P_d(s)$ errors but resilient to H and D uncertainty.

To make sure the system can tolerate uncertainties and disruptions, robustness analysis and robust control design are crucial in LFC. A strong tool for assessing the system's resilience and creating resilient controllers is the robustness matrix and mathematical model. Power system operators may guarantee dependable and effective grid operation even in the face of risks and disruptions by using strong control strategies.

5.2.6 Sensitivity analysis in microgrid LFC

Sensitivity analysis in Microgrid LFC is a method used to investigate how changes in load, generation, and transmission line factors affect the system's stability and frequency response.

5.2.6.1 Sensitivity analysis importance in microgrid LFC

Sensitivity analysis plays a vital role in microgrid Load Frequency Control by enabling operators to evaluate the impact of uncertainties in system parameters, such as load variations or renewable energy output, on LFC performance. This analysis is crucial for assessing the robustness of LFC systems to disturbances and uncertainties, ensuring reliable operation. By understanding the sensitivity of LFC systems to various parameters, operators can optimize controller design and parameters, identify potential stability issues, and optimize system performance to minimize frequency deviations and improve power quality. Furthermore, sensitivity analysis is particularly important in microgrids with high penetration of renewable energy sources, as their output can be variable and uncertain. By conducting sensitivity analysis, microgrid operators and designers can improve LFC performance and stability, enhance robustness to uncertainties and disturbances, optimize controller design and parameters, and ensure reliable operation and power quality. Overall, sensitivity analysis is a valuable tool for ensuring the reliable and efficient operation of microgrid LFC systems (Vidyarthi and Kumar, 2024).

5.2.6.2 Methods for sensitivity analysis in microgrid LFC

1. **Parametric Sensitivity Analysis:** It is an important technique for microgrid Load Frequency Control, parametric sensitivity analysis enables operators to examine the effects of modifications to system parameters on the system's changes and performance. Operators may determine crucial characteristics that have a major influence on system performance by altering system parameters like load or renewable energy generation and examining the effects on LFC performance, including frequency deviations. The resilience of the system to changes in parameters can be assessed thanks to this study, which also helps designers create LFC controllers that are more stable and perform better. Microgrid operators and designers may improve system resilience, optimize controller design and settings, and eventually guarantee dependable operation and power quality by using parametric sensitivity analysis. Parametric sensitivity analysis is a useful method for microgrid design, planning, and operation as it helps operators create more reliable

and effective LFC systems by identifying sensitive factors and their impacts.

2. **Structural Sensitivity Analysis:** An essential technique for microgrid Load Frequency Control, structural sensitivity analysis looks at how modifications to the system's architecture or component interactions affect the behavior and performance of the system. This study aids in the identification of crucial system elements and interactions, the assessment of resilience to structural uncertainty, and the evaluation of the effects of modifications to the system topology or configuration. Operators may optimize control techniques, strengthen robustness and reliability, and improve system design and planning by doing structural sensitivity analysis. This is especially crucial for microgrids with variable load circumstances, renewable energy sources, and intricate component interactions. Operators may create more robust and effective LFC systems and guarantee dependable operation and power quality in microgrids by comprehending structural sensitivity.
3. **Monte Carlo simulations:** Monte Carlo simulations are a potent computing technique that analyzes complicated systems and estimates results via random sampling. It can estimate uncertainties, assess system performance, and improve control techniques in microgrids using load frequency control. Monte Carlo simulation aids in the evaluation of the effects of uncertainties on LFC performance, such as frequency deviations and stability, by modeling many scenarios with random changes in load, renewable energy generation, or other characteristics. Microgrid designers and operators may use this to optimize control techniques for dependable operation, boost system resilience to uncertainties, and increase LFC performance and stability. In the end, more robust and efficient microgrid systems result from the use of Monte Carlo simulation as a useful tool for uncertainty analysis, risk assessment, and system improvement.

5.2.6.2.1 Benefits of sensitivity analysis in microgrid LFC

Enhanced tolerance to uncertainty, simplified control techniques, and increased system performance and stability are just a few advantages of sensitivity analysis in microgrid load frequency control. Microgrid operators and designers may create more efficient control techniques that lower frequency deviations and enhance power quality by using sensitivity analysis to identify crucial factors and their effects on LFC performance. Furthermore, by evaluating the effects of uncertainties and disruptions, sensitivity analysis enables preventative actions to guarantee dependable operation. Sensitivity analysis is a useful technique for microgrid LFC overall, allowing for the creation of more dependable, efficient, and robust systems that can adjust to shifting circumstances and uncertainties. Microgrid operators may lower risks, enhance system performance, and guarantee high-quality electricity distribution by using sensitivity analysis.

5.2.6.2.2 Challenges in sensitivity analysis in microgrid LFC

Numerous obstacles must be overcome for sensitivity analysis in microgrid load frequency control, such as the intricacy of microgrid systems, nonlinear dynamics, and uncertainties related to load fluctuations and renewable energy sources. Furthermore, it might be challenging to find and model all pertinent characteristics and how they interact, and the study can need a large amount of processing power. Furthermore, it might be difficult to evaluate and utilize sensitivity analysis findings in order to maximize LFC performance and guarantee dependable operation. Managing high-dimensional systems, associated uncertainties, and guaranteeing the validity and correctness of the analysis are additional difficulties. Sensitivity analysis is still a useful technique for enhancing the performance and dependability of microgrid LFCs in spite of these difficulties.

5.2.6.3 *Sensitivity parameters*

System inertia, damping coefficients, controller gains, load characteristics, and renewable energy production are examples of sensitivity parameters in microgrids. The performance of load frequency control (LFC), which affects power quality and frequency stability, may be greatly impacted by these characteristics. Energy storage system features, transformer tap settings, and line impedances are examples of other critical factors. Designing and refining microgrid control techniques, guaranteeing dependable operation, and preserving power quality all depend on an understanding of how sensitive these parameters are. Microgrid operators may enhance overall system performance and create more efficient control strategies by recognizing and evaluating these critical characteristics. Sensitivity parameters are those that have a major impact on the system's frequency response and stability. By modifying these parameters, system stability and performance may be optimized.

5.2.6.3.1 Typical microgrid LFC sensitivity parameters

There are different types of sensitivity parameters, they affect the microgrid system performance.

1. Governor Gain (Kg): The governor gain affects the system's frequency responsiveness and stability.
2. Turbine Gain (Kt): The turbine gain affects the system's stability and frequency responsiveness.

3. Power System Stiffness (Kp): The power system stiffness affects the system's frequency response and stability.
4. Damping Coefficient (D): A component that affects the system's frequency responsiveness and stability.
5. Time Constants (Ts): These variables affect the system's frequency responsiveness and stability.

5.2.6.4 *Sensitivity to controller gain*

In microgrid load frequency control (LFC), sensitivity to controller gain describes how gain levels affect system stability and performance. The LFC system's reaction to frequency variations and disturbances is mostly determined by controller gains. Fast reaction times and increased stability may be achieved with high gains, although oscillations and overshoot are other possible outcomes. Low gains, on the other hand, might result in less stability and slower reaction times.

Because of the inherent uncertainties and unpredictability of the system, controller gain sensitivity is crucial in microgrid LFC. Changes in system topology, load variations, and renewable energy sources may all affect the ideal gain values. Therefore, to guarantee dependable and effective operation, thorough controller gain calibration and optimization are required.

There are many ways to examine the sensitivity of controller gains, including:

1. Frequency response analysis
2. Simulations in the time domain
3. Techniques for sensitivity analysis

Microgrid operators can:

1. Optimize gain settings for better performance by knowing how sensitive controller gains are.
2. Make the system more resilient to uncertainty3. Reduce frequency deviations and improve power quality

A number of variables, such as system dynamics, uncertainty, and performance goals, must be taken into account in order to tune controller gains effectively. Microgrid operators may create more effective LFC strategies and guarantee dependable and efficient operation by carefully planning and optimizing controller gains.

5.2.6.4.1 Effects of controller gain on system performance

The performance of Load Frequency Control in microgrids is largely determined by the controller gain. Stability, precision, and speed are all impacted by the gain value, which also influences how the system reacts to frequency variations and disturbances. By lowering frequency variations, increasing stability, and guaranteeing dependable operation, a properly adjusted controller gain may boost system performance. On the other hand, misaligned gains may cause instability, overshoot, and oscillations. While low gains might result in slower reaction times and less stability, high gains can provide quick response times but also overshoot and oscillations. Finding the ideal gain value is thus essential to striking a compromise between stability and performance. Overshoot, steady-state error, and settling time are indicators of how controller gain affects system performance. The entire performance of the LFC system may be improved with a well-designed and adjusted controller gain, guaranteeing dependable and effective microgrid operation. Microgrid operators may enhance frequency regulation, lessen power quality problems, and guarantee stable operation under a range of operating situations by carefully choosing and adjusting the controller gain. In order to achieve better microgrid performance and reliability, effective gain tuning necessitates taking system dynamics, uncertainties, and performance goals into account. Microgrid operators can guarantee high-quality power supply and create more efficient LFC methods by comprehending how controller gain affects system performance.

5.2.6.4.2 Sensitivity to time-constant

A control system's time constant (τ) indicates how quickly it responds to changes. The temporal constants of the power supply and controller may affect the overall performance of LFC systems.

Effects of Time Constant on System Performance:

Stability: The stability of the system may be affected by the time constants of the controller and power supply; lower time constants may lead to faster response times but less stability.

Overshoot: The time constant may also have an impact on the system's overshoot; lower time constants result in more overshoot.

Settling Time: The system's settling time is directly influenced by the time constant; lower time constants result in faster settling times.

5.2.6.5 Analyzing sensitivity to controller gain and time constant

The sensitivity of an LFC system to controller gain and time constant may be investigated using the following methods:

1. Linearize the System: The LFC system should be linearized around its operating point to get a linearized model.
2. Compute the Sensitivity Functions: Find out how responsive the system is to changes in the controller gain and time constant.
3. Analyze the Sensitivity Functions: To determine the system's sensitivity to changes in the time constant and controller gain, analyze the sensitivity functions.

5.2.6.6 *Implications of frequency control*

In microgrid Load Frequency Control (LFC), frequency management is essential because it maintains the system frequency within reasonable bounds while balancing power supply and load demand. Microgrids can function dependably and effectively with effective frequency management, avoiding problems like equipment damage, poor power quality, and even blackouts. Because renewable energy sources are unpredictable and variable, frequency management is especially crucial in microgrids that use them. Microgrids can guarantee steady operation, lower the chance of grid instability, and provide customers with high-quality electricity by controlling frequency. Additionally, microgrids may function in both grid-connected and islanded modes with ease thanks to frequency regulation. All things considered, frequency control is a crucial component of LFC systems as it keeps microgrids stable, dependable, and efficient. Microgrid operators can guarantee dependable electricity supply and preserve grid stability by giving frequency management priority.

5.2.6.6.1 Challenges in frequency control

Frequency control has many challenges despite its importance:

1. Complexity: During frequency control, several power system components interact in a complex way.
2. Uncertainty: Frequency management must address several uncertainties, such as demand variability and the production of renewable energy.
3. Scalability: Frequency management must be scalable to meet the increasing demand for electricity and the incorporation of renewable energy sources.
4. Cybersecurity: By focusing on frequency control systems, cyberattacks might compromise the security and reliability of the power system.

5.2.7 LFC performance evaluation matrix

An essential tool for evaluating the performance of load frequency control in microgrids is a performance assessment matrix. Frequency variation, settling time, overshoot, steady-state error, stability margins, and resilience to uncertainty are some of its Key Performance Indicators (KPIs). These KPIs provide important information on how well the LFC system controls frequency, reacts to disruptions, and preserves stability (Bello et al., 2023). The matrix makes it possible to compare various LFC techniques, evaluate the influence of uncertainties, identify areas for improvement, optimize controller gains, and compare control methodologies. Microgrid operators may improve frequency regulation, increase system robustness, optimize controller performance, create more effective LFC techniques, and guarantee dependable operation by using this matrix. The matrix may be customized to meet particular microgrid needs and assesses LFC performance in a range of scenarios, including transitions, uncertainties, and grid-connected and islanded modes. In the end, an LFC performance assessment matrix aids in maintaining grid stability, maximizing LFC system performance, and guaranteeing a high-quality power supply (Rangi et al., 2023).

5.2.7.1 Components of the performance evaluation matrix

The system frequency deviation from its reported value is indicated by the frequency deviation (Δf).

1. Power Output Regulation (ΔP): Assesses how well the system can regulate power output in response to changes in load or generation.
2. Settling Time (T_s): Determines the amount of time it takes for the system to stabilize after a disruption.
3. Overshoot (M_p): Shows the separation between the system's response and its steady-state value.

Steady-State Error (SSE), which measures the difference between the system's steady-state value and its goal value, is the fifth.

5.2.7.2 Performance evaluation

By comparing the predicted σ-ellipse's parameters with the objective parameters, separately for the variables averaged over one and 15 minute time frames, the LFC performance of a region is assessed over a certain month. When the bivariate distribution of (Δf, ΔPtie) is elliptically contoured, as it is in CE regions, a Tyler's MLE is used. The characteristics of the calculated σ-Ellipse are averaged throughout the year to assess the LFC performance. The LFC two area multi source power system, as shown in Fig. 5.5.

When all of the requirements for the area are met, the highest grade (A) is awarded. Depending on $\sigma\Delta P_{tie}$, the LFC operation is graded B or C when $\sigma\Delta P_{tie}$ is within the goal level and the correlation falls short of the target. While no more action is required for places with grades A or B, a location with a grade of C should see an increase in Bi. On the other hand, a grade A location could lower Bi or even regulate reserves. Additionally, regions with grades D or E should see a decline in σΔP, necessitating further LFC system tuning or an increase in regulating reserves; in contrast, regions with grades E should see a rise in B_i. The LFC operation has to be suitably adjusted in locations with counteracting LFC operation (grade F).

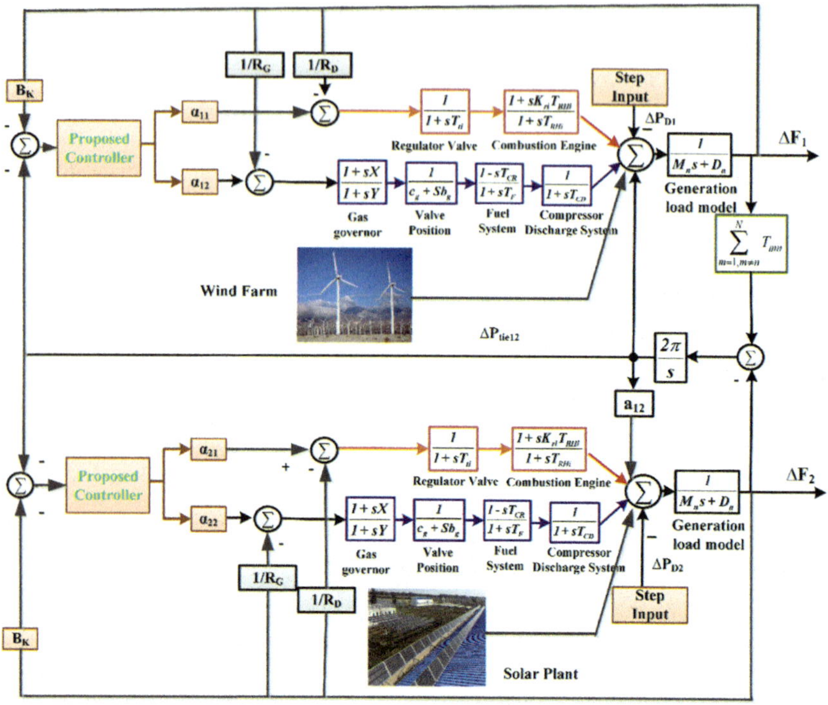

FIGURE 5.5 Schematic model diagram of two-area power system.

5.2.7.3 *Standard matrix for frequency control*

The standard matrix for frequency control, a mathematical representation of the power system dynamics, is used to study and build frequency control systems.

The components of the Standard Matrix are:

The power system's dynamics, including the relationships between power output, frequency, and other system characteristics, are shown in the Matrix of Systems (A).

1. Input Matrix (B): Displays the inputs of the power system, including changes in load or generation.
2. Output Matrix (C): Displays the outputs of the power system, such as frequency and power output.
3. Feedthrough Matrix (D): Indicates the direct link between the inputs and outputs of the power system.

5.2.7.3.1 Standard matrix formulation

The usual frequency control matrix may be made in the following manner:

$$dx/dt = Ax + Bu \tag{5.14}$$

$$y = Cx + Du \tag{5.15}$$

Where: x represents the state vector, such as frequency or power output.

u stands for the input vector, which includes changes in load or generation.
Y stands for the output vector (such as frequency or power output).
S The system, input, output, and feedthrough matrices are denoted by A, B, C, and D, respectively.

5.2.7.3.1.1 Frequency control efficiency Frequency control efficiency is the ability of a power system to maintain a constant frequency in the face of interruptions or changes in generation or demand.

Factors Affecting Frequency Control Efficiency

1. System Inertia: The system's resistance to frequency fluctuations.
2. Governor Response: How quickly and precisely the governor reacts to variations in frequency.
3. Load Frequency Control (LFC): The LFC system's capacity to modify generation in response to variations in load.
4. Communication and Control: The speed and reliability of communication and control systems.

5.2.7.4 Metrics for evaluating frequency control efficiency

1. Frequency deviation is the difference between the desired and actual frequencies.
2. Settling Time: The period after a disturbance before the frequency reaches its predetermined level again.
3. Overshoot: The biggest deviation of the frequency from the target value.

The "steady-state error" is the difference between the expected and actual frequencies under steady-state conditions.

Example of Assigning Frequency Control Efficiency as given below:

Metric	Performance Target	Actual Performance	Efficiency Score
Frequency Deviation	±.1 Hz	±.05 Hz	80%
Settling Time	10 s	7 s	70%
Overshoot	0.5 Hz	0.2 Hz	50%
Steady-State Error	0.01 Hz	0.005 Hz	60%

Efficiency Score = (80 + 70 + 50 + 60)/4 = 65%

The system's frequency control efficiency is 65%.

5.2.7.5 *Comparative performance matrix*

A comparative performance matrix is a method for evaluating and comparing the performance of different systems, controllers, or algorithms. It provides a comprehensive framework for weighing the benefits and drawbacks of each choice (Peddakapu et al., 2022).

5.2.7.5.1 Components of a comparative performance matrix

1. Criteria: A set of performance metrics or benchmarks that are used to evaluate each decision.
2. Options: The several controllers, algorithms, or systems that are being compared.
3. Scores: Performance scores for each choice according to each criteria.

Example of a Comparative Performance Matrix.

Criteria	Option 1	Option 2	Option 3
Frequency Deviation	0.05 Hz	0.03 Hz	0.01 Hz
Settling Time	8 s	6 s	4 s
Overshoot	0.2 Hz	0.1 Hz	0.05 Hz
Steady-State Error	0.005 Hz	0.002 Hz	0.001 Hz

5.2.7.5.2 Benefits of a comparative performance matrix

1. Comprehensive Evaluation: Provides a comprehensive framework for evaluating the effectiveness of different solutions.
2. Easy Comparison: Facilitates the comparison of the performance of different solutions.
3. Informed Decision-Making: Provides a thorough understanding of the benefits and drawbacks of each option, enabling well-informed decision-making.

5.2.8 Simulation results

Microgrid Load Frequency Control time-domain simulation and analysis include modeling and simulating the dynamic behavior of microgrids in response to disruptions and changes in operating circumstances. This method makes it possible to optimize controller values, evaluate LFC techniques, and analyze system stability. The intricate relationships between various microgrid elements, including generators, loads, and energy storage devices, may be captured by time-domain simulations, which provide important insights into the dynamic behavior of the system. Microgrid operators may detect possible problems, improve control schemes, and guarantee dependable and effective operation by examining the simulation findings. For the design and optimization of microgrid LFC systems, time-domain modeling and analysis are crucial tools that facilitate the creation of more efficient control strategies and enhance system performance. Even when there are uncertainties and disruptions, this method helps guarantee that micro-grids operate steadily and effectively (Singh et al., 2024).

5.2.8.1 *Investigated system*

Examine a two-area inter-connected power system model with bio-gas and bio-diesel units in each zone, as shown in The two control zones are connected by a tie-line system, which also includes wind and solar power plants. This article describes a test microgrid-1 (MG-1) that consists of a bio-gas unit, a wind turbine generator, and a bio-diesel engine.

Conversely, test microgrid-2 is made up of a bio-gas unit, a bio-diesel engine, and solar photovoltaics. To adjust the gain parameters (K_t, K_i, K_d, and N), the Fuzzy-TIDN Controller is optimized using a meta-heuristic approach known as Sea-Horse Optimization (SHO). In this instance, the cost function is the integral time square error (ITSE). The following Eq. (5.16) provides the ITSE.

$$f_{ITSE} = \int_0^t t[(\Delta f_1)^2 + (\Delta f_2)^2 + (\Delta P_{tie})^2]dt \tag{5.16}$$

5.2.8.2 *Suggested controller configuration*

The suggested controller fuzzy-TIDN controller has been proposed, which is a cascade connection of fuzzy and TIDN contrroller. The hybridization of the basic controller inadequate for adjusting settings and delivering the best possible plant performance in real application (Murali and Shankar, 2022). Therefore, a hybrid controller combination is required for improving the system performance. The suggested controller's connection diagram is shown in Fig. 5.6. In comparison to simple controllers such as PIDN and Fuzzy-PID the suggested controller performs is best in terms of overshoot, undershoot and setteling time. The simulation part provides a detailed performance analysis as shown in Fig. 5.7.

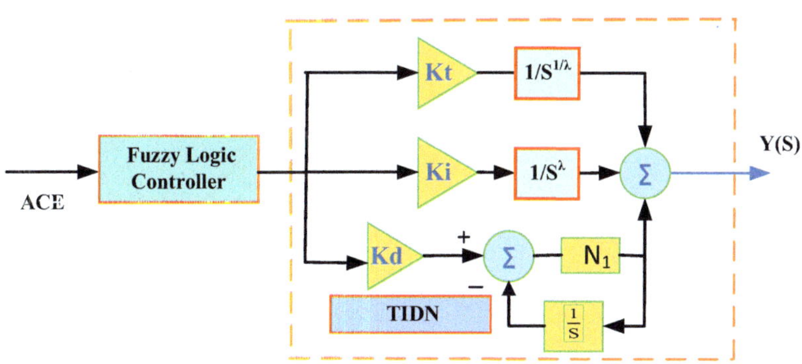

FIGURE 5.6 Proposed Controller.

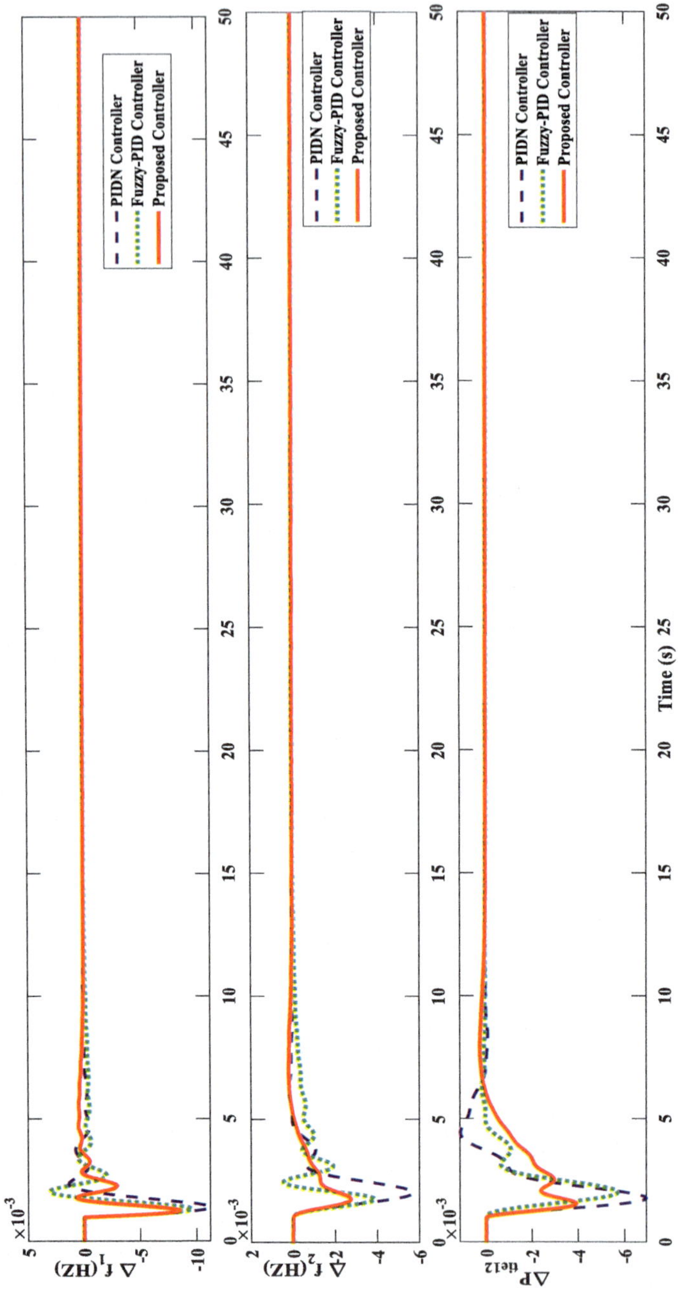

FIGURE 5.7 Proposed controller is comparing with existing Controller.

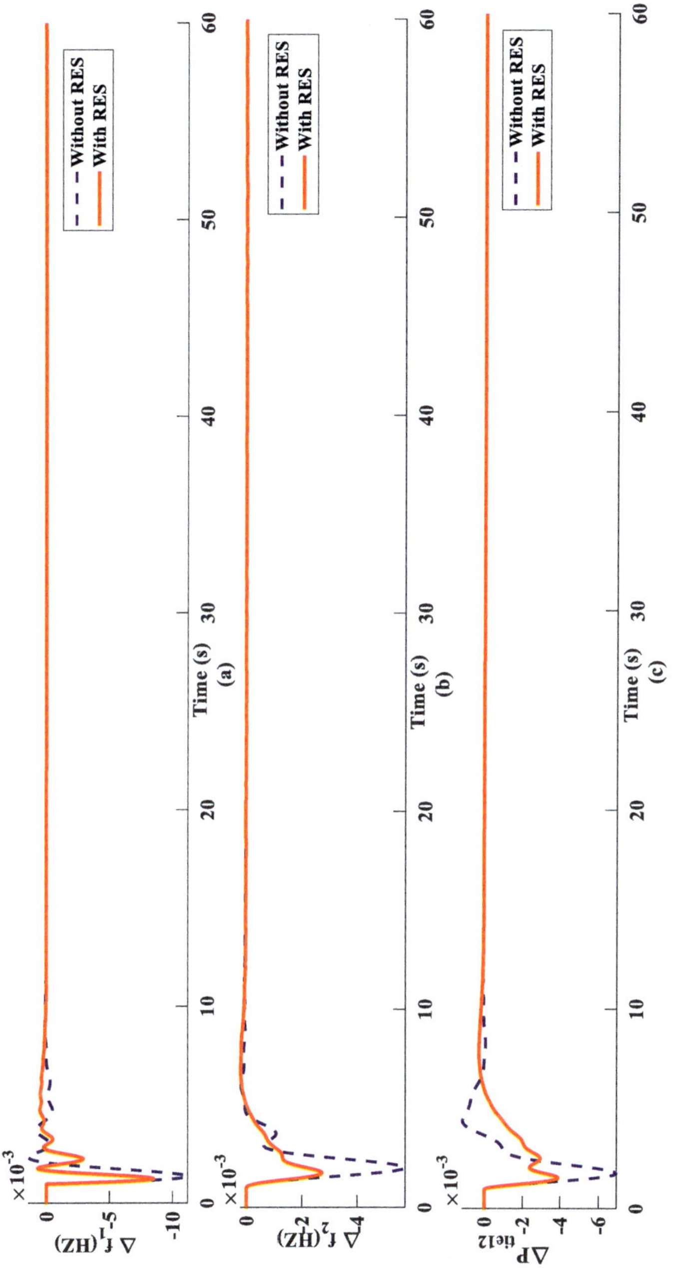

FIGURE 5.8 Dynamic response of power system in presents and absence of RES with step load.

TABLE 5.1 Frequency deviation of 2-area power system with and without RES.

Maximum deviation	Without RES	With RES	Improvement (%)
Δf_1 *(Hz)*	−0.0115	−0.0082	27.14
Δf_2 *(Hz)*	−0.0058	−0.0027	53.24
ΔP_{tie12} *(pu)*	−0.0072	−0.0043	41.55

5.2.9 Case study analysis

5.2.9.1 *Effectiveness of proposed controller*

In this analysis it is given that the control areas 1 and 2 have a 5percent SLP, the PIDN, fuzzy-PID, and suggested controller have all been optimized using the suggested method. Fig. 5.7 displays the dynamic behavior of the system in control areas 1 and 2 at a 0.05 pu SLD. The results from the figure are in line with the output of the system, showing that the recommended controller has reduced oscillations and shortened the settling period.

5.2.9.2 *The penetration of renewable sources in LFC*

The effectiveness of the recommended controller for controlling RES change is being assessed, and the performance of the proposed controllers (Fuzzy-TIDN) under various load disturbance patterns was examined (Biswas et al., 2025). This case study compares the evaluated system performance with and without RESs to examine the impacts of RESs (wind and solar power) penetration. Control area-1 is linked to a 75 MW wind power plant system, while control area-2 is connected to a 55 MW solar PV system. The controller's ideal settings are provided by the SHO optimization. The system output in Fig. 5.8 indicates that the recommended controllers outperform the others. The frequency variation of two area power systems with and without RES is shown in Table 5.1.

5.3 Conclusion

An analytical approach to microgrid frequency control is analyzed in this study, with an emphasis on control strategy design and optimisation to guarantee stable operation across a range of operating conditions. The effectiveness of the suggested control strategy in preserving frequency stability and guaranteeing the microgrid's dependable functioning as shown in table. By offering a methodical framework for the design and optimisation of LFC systems, the findings of this chapter further the development of microgrid technology. This study proposes a modified

SHO approach, to find the optimal setting of controller parameters. The proposed fuzzy-TIDN controller is investigate and compare its effectiveness to PIDN and Fuzzy-PID. The output response confirmed the efficiency of the suggested controller. The proposed controller and optimization is capable to handle the penetration of renewable energy sources. The performance of the proposed system is describing various case study. Eigen values, stability analysis using Lyapunov, robustness analysis, sensitivity analysis and performance evaluation metrics about microgrid is described in details.

In the future; the proposed system in integrated into hybrid energy storage in an Interconnected deregulated power system.

References

Ahmed, H.H.A., Mashor, M.Y., Mahdi, M.C., 2019. Intelligent fuzzy PD+I controller with stabilizer for nano satellite attitude control system. J. Phys. Conf. 1372 (1). https://doi.org/10.1088/1742-6596/1372/1/012022.

Ahmed, M., Khamies, M., Kamel, S., 2023. A robust control approach integrating with optimal fuel cells to strengthen the frequency stability of a diverse-sources power system including renewables. ISA Trans. 143, 420–439. https://doi.org/10.1016/j.isatra.2023.09.017.

Arya, Y., 2020. A novel CFFOPI-FOPID controller for AGC performance enhancement of single and multi-area electric power systems. ISA Trans. 100, 126–135. https://doi.org/10.1016/j.isatra.2019.11.025.

Arya, Y., Kumar, N., 2017. Optimal control strategy–based AGC of electrical power systems: a comparative performance analysis. Optim. Control Appl. Methods 38 (6), 982–992. https://doi.org/10.1002/oca.2304.

Bello, A., Olfe, K.S., Rodríguez, J., Ezquerro, J.M., Lapuerta, V., 2023. Experimental verification and comparison of fuzzy and PID controllers for attitude control of nanosatellites. Adv. Space Res. 71 (9), 3613–3630. https://doi.org/10.1016/j.asr.2022.05.055.

Biswas, D.K., Debbarma, S., Singh, P.P., 2025. Dynamic surface sliding mode control-based LFC design for RES-dominated power systems with a provision of grid-scale virtual energy storage. IEEE Trans. Power Syst. 1–12. https://doi.org/10.1109/tpwrs.2025.3573582.

Chen, P., Luo, Y., 2023. Analytical fractional-order PID controller design with bode's ideal cutoff filter for PMSM speed servo system. IEEE Trans. Ind. Electron. 70 (2), 1783–1793. https://doi.org/10.1109/TIE.2022.3158009.

Choudhary, R., Rai, J.N., Arya, Y., 2022. Cascade FOPI-FOPTID controller with energy storage devices for AGC performance advancement of electric power systems. Sustain. Energy Technol. Assessments 53. https://doi.org/10.1016/j.seta.2022.102671.

Jia, H., Rao, H., Wen, C., Mirjalili, S., 2023. Crayfish optimization algorithm. Artif. Intell. Rev. 56, 1919–1979. https://doi.org/10.1007/s10462-023-10567-4.

Khamies, M., Magdy, G., Ebeed, M., Kamel, S., 2021. A robust PID controller based on linear quadratic Gaussian approach for improving frequency stability of power systems considering renewables. ISA Trans. 117, 118–138. https://doi.org/10.1016/j.isatra.2021.01.052.

Kohli, M., Arora, S., 2018. Chaotic grey wolf optimization algorithm for constrained optimization problems. J. Comput. Des. Eng 5 (4), 458–472. https://doi.org/10.1016/j.jcde.2017.02.005.

Long, B., Liao, Y., Chong, K.T., Rodríguez, J., Guerrero, J.M., 2021. MPC-controlled virtual synchronous generator to enhance frequency and voltage dynamic performance in

islanded microgrids. IEEE Trans. Smart Grid 12 (2), 953–964. https://doi.org/10.1109/TSG.2020.3027051.
Magdy, G., Shabib, G., Elbaset, A.A., Mitani, Y., 2018. Optimized coordinated control of LFC and SMES to enhance frequency stability of a real multi-source power system considering high renewable energy penetration. Prot. Control Mod. Power Syst. 3 (1). https://doi.org/10.1186/s41601-018-0112-2.
Mitra, J., Nguyen, N., 2022. Grid-scale virtual energy storage to advance renewable energy penetration. IEEE Trans. Ind. Appl. 58 (6), 7952–7965. https://doi.org/10.1109/TIA.2022.3202515.
Murali, S., Shankar, R., 2022. Exploration of novel optimal fuzzy-based controller for enhancement of frequency regulation of deregulated hybrid power system with modified HVDC tie-line. Int. J. Intell. Syst. 37 (7), 4163–4189. https://doi.org/10.1002/int.22715.
Nair, A.P., Selvaganesan, N., Lalithambika, V.R., 2016. Lyapunov based PD/PID in model reference adaptive control for satellite launch vehicle systems. Aero. Sci. Technol. 51, 70–77. https://doi.org/10.1016/j.ast.2016.01.017.
Peddakapu, K., Mohamed, M.R., Srinivasarao, P., Arya, Y., Leung, P.K., Kishore, D.J.K., 2022. A state-of-the-art review on modern and future developments of AGC/LFC of conventional and renewable energy-based power systems. Renew. Energy Focus 43, 146–171. https://doi.org/10.1016/j.ref.2022.09.006.
Ram Babu, N., Bhagat, S.K., Saikia, L.C., Chiranjeevi, T., Devarapalli, R., García Márquez, F. P., 2023. A comprehensive review of recent strategies on automatic generation control/load frequency control in power systems. Arch. Comput. Methods Eng. 30 (1), 543–572. https://doi.org/10.1007/s11831-022-09810-y.
Rangi, S., Jain, S., Arya, Y., 2023. SSA-optimized cascade optimal-PIDN controller for multi-area power system with RFB under deregulated environment. Optim. Control Appl. Methods 44 (4), 1972–1994. https://doi.org/10.1002/oca.2962.
Saha, A., Saikia, L.C., 2019. Renewable energy source-based multiarea AGC system with integration of EV utilizing cascade controller considering time delay. Int. Trans. Electr. Energy Syst. 29 (1). https://doi.org/10.1002/etep.2646.
Sahu, P.C., Samantaray, S.R., 2023. Resilient frequency stability of a PV/wind penetrated complex power system with CSA tuned robust Type-2 fuzzy cascade PIF Controller. Elec. Power Syst. Res. 225. https://doi.org/10.1016/j.epsr.2023.109815.
Saxena, A. Al Z.O., Shankar, R., Al Jaafari, K., Al Hosani, K., Muduli, U.R., 2023. Functional resilience assessment of restructured power grid considering insignificant demands for frequency stability. IEEE IAS Glob. Conf. Renew. Energy Hydro. Technol. https://doi.org/10.1109/GlobConHT56829.2023.
Singh, P.P., Shankar, R., Singh, S.N., 2024. Chaos quasi-opposition crayfish based modified new controller designed for hybrid deregulated power environment considering cyber-attack. Chaos Solitons Fractals 187. https://doi.org/10.1016/j.chaos.2024.115444.
Sun, T., Liao, Z., Al-Ani, M., Jordan, L.B., Li, G., Zhan, S., Belmont, M., Edwards, C., 2023. Modelling and control tank testing validation for attenuator type wave energy converter-Part III: model predictive control and robustness validation. IEEE Trans. Sustain. Energy 14 (3), 1737–1746. https://doi.org/10.1109/TSTE.2023.3246171.
Vidyarthi, P.K., Kumar, A., 2024. A modified tilt controller for AGC in hybrid power system integrating forecasting of renewable energy sources. Optim. Control Appl. Methods 45 (1), 185–207. https://doi.org/10.1002/oca.3052.
Vidyarthi, P.K., Kumar, A., Raj, S., 2024. Chaos quasi-opposition sea-horse based modified new tilt controller designed for multi-area deregulated AGC using deep learning against cyber-attacks. Chaos Solitons Fractals 188. https://doi.org/10.1016/j.chaos.2024.115542.
Xu, S., Wei, Z., Huang, Z., Wen, H., Jin, D., 2021. Fuzzy-logic-based robust attitude control of networked spacecraft via event-triggered mechanism. IEEE Trans. Aero. Electron. Syst. 57 (1), 206–226. https://doi.org/10.1109/TAES.2020.3012000.

Ye, Y., Daraz, A., Basit, A., Khan, I.A., Alqahtani, S.A., 2024. Cascaded fractional-order controller-based load frequency regulation for diverse multigeneration sources incorporated with nuclear power plant. Int. J. Energy Res. 2024. https://doi.org/10.1155/2024/7939416.

Zhang, S., Zhou, Y., Cai, S., 2022. Fractional-order PD attitude control for a type of spacecraft with flexible appendages. Fractal Fract.. 6 (10). https://doi.org/10.3390/fractalfract6100601.

CHAPTER

6

Role of energy storage systems in microgrid frequency control

Muhammad Majid Gulzar[1,2,3], *Muhammad S. Tolba*[1], and *Muhammad Khalid*[2,4]

[1] Department of Control & Instrumentation Engineering, King Fahd University of Petroleum & Minerals, Dhahran, Saudi Arabia; [2] Center for Sustainable Energy Systems, King Fahd University of Petroleum & Minerals, Dhahran, Saudi Arabia; [3] Center for Smart Mobility and Logistics, King Fahd University of Petroleum & Minerals, Dhahran, Saudi Arabia; [4] Electrical Engineering Department, King Fahd University of Petroleum & Minerals, Dhahran, Saudi Arabia

6.1 Introduction of energy storage systems in micro-grid load frequency control

The stability of the microgrids, especially integrated with Renewable Energy Sources (RES) such as wind and solar, requires proper frequency management. The intermittency of RES necessitates the use of Energy Storage Systems (ESS) for balancing supply and demand. ESS can store excess energy during low-demand periods and release it during peak demand, thereby stabilizing the grid. Traditional grids rely on large generators for inertia; however, renewable-integrated microgrids require ESS to maintain frequency and voltage stability due to the lack of spinning masses. There are several variants of ESS applied in the microgrid, the most used ones are batteries, flywheels and supercapacitors, but even emerging technologies will also be addressed here. Every one of these ESS possesses different attributes in terms of energy density, response time, and cycle life for best fits of a specific single application, considering the grid stability with the ability for further enabling better integration of RES (Ali et al., 2024).

Frequency Control in Microgrids
https://doi.org/10.1016/B978-0-443-36476-1.00005-5

6.1.1 Importance of energy storage in microgrid LFC

Any deviation from the nominal frequency can have serious detrimental effects on the overall performance of the system. Furthermore, when the frequency deviates from its nominal value, inefficiency, reduction in lifespan, or complete failure of connected devices may occur due to the operation out of their designed specification. Moreover, these fluctuations can lead to instability in the grid, a process that leads to the disruption in the supply of electricity, increased equipment wear, and higher operation costs. Generally, frequency fluctuations, if not controlled in microgrids with heavy penetrations of renewables, will also affect the integration of intermittent energy sources and reduce the reliability and the efficiency of the system (Gulzar et al., 2024). ESS plays a critical role in eliminating frequency deviations in microgrids, particularly those with high penetration of intermittent RES. These systems continuously monitor the grid conditions and respond dynamically to fluctuations in generation and demand. Specifically, ESS dynamically absorbs excess energy during periods of overgeneration, such as when solar or wind output exceeds local consumption and injects stored energy into the grid during a sudden increase in demand or generation deficits (Siti et al., 2022). This bidirectional energy exchange helps maintain the system frequency within acceptable limits, thereby enhancing the operational stability and the power quality. The control of ESS is typically achieved through closed-loop feedback mechanisms, often utilizing Proportional-Integral (PI) or Model Predictive Control (MPC) schemes that regulate power flow based on real-time frequency measurements (Naderipour et al., 2023). Moreover, modern ESS technologies are equipped with fast response times and high-power densities, allowing them to respond to disturbances within milliseconds. This makes them highly effective in supporting LFC and reducing the reliance on conventional spinning reserves (Castillo and Gayme, 2014). Due to this, the frequency in the grid will always remain within specified limits. In addition, the fast response time of the ESS enables high integration of the RES and helps alleviate the dependence on conventional mechanisms for stabilizing the overall system, thereby offering resilience to it (Kousksou et al., 2014). Table 6.1 shows the response of the ESS to different microgrid scenarios (Naderipour et al., 2023). Additionally, Fig. 6.1, illustrates the classification of different ESS technologies, supplemented with icons representing elements of the broader microgrid ecosystem in which these systems operate, such as industrial loads, power grids, and transportation units, to contextualize typical deployment scenarios of the ESS.

TABLE 6.1 Energy production scenarios and ESS response.

Scenario	ESS response
Excess production from solar panels during mid-day	ESS retains surplus energy for later use in evening or cloudy periods (Alirezaei et al., 2025)
Higher demand at peak periods (e.g., residential load spike in the evening)	ESS discharges stored energy to meet demand, reducing load on the grid (Makahleh et al., 2023)
Abrupt decline in wind energy production due to sudden weather changes	ESS compensates by injecting power rapidly to maintain frequency (Akram et al., 2020)
Off-peak periods with low demand (e.g., night's time in low-use zones)	ESS absorbs energy to prevent curtailment of base or renewable generation (Wang et al., 2023)
Unstable generation from cloud-covered solar PV systems	ESS smooths intermittent fluctuations, improving voltage and frequency stability (Elalfy et al., 2024)
Planned maintenance or temporary disconnection of a generator unit	ESS provides backup support to sustain load during the outage (Gao et al., 2023)
Grid blackout or islanding event in a remote microgrid	ESS provides initial power to enable black start or to isolate loads selectively (Yang et al., 2024a)
Unexpected load shedding or disconnect event in industrial zone	ESS helps rebalance the system and restores voltage/frequency equilibrium (Liu et al., 2023a,b,c)

6.1.2 Applications and suitability of ESS for microgrid LFC

ESS technologies support various applications such as frequency regulation, peak shaving, and emergency backups. Their selection depends on specific system needs. Batteries and flywheels can provide power and quick response for frequency stability and peak load management. However, Battery Energy Storage Systems (BESS) are more efficient for continuous support since batteries are better for long-term storage. In addition, Pumped Hydro Storage (PHS) is one of the most established and widely used ESS technologies, providing large-scale energy storage and exceptional grid stability. Compressed Air Energy Storage (CAES) is another effective solution that is suitable for large-scale storage. CAES systems are also effective for peak shaving and integrating renewable energy. These technologies complement battery systems and flywheels by offering long-term storage options, ensuring grid reliability (Chen et al., 2009). Supercapacitors are ideal for rapid response scenarios,

FIGURE 6.1 Illustration of various ESS technologies along with their typical application environments.

stabilizing grid frequency during sudden load changes due to their quick charge and discharge (Antony et al., 2024). Meanwhile, advanced battery systems, such as lithium-ion and sodium-sulfur, offer scalable solutions for medium to long-term storage needs (Ribeiro et al., 2001). Lithium-ion batteries can store a lot of energy during sunny periods and offer solid backup during low generation times, making them useful for solar-powered microgrids. Flywheel Energy Storage Systems (FESS) may help wind-dominated systems withstand frequent short-term fluctuations. Though less suitable for long-term storage, supercapacitors allow nearly instantaneous voltage and frequency adjustments, effectively stabilizing rapid variations (Khan et al., 2023). In conclusion, microgrid's energy storage and frequency regulation demands determine the ESS type. Microgrids can adapt to demand variations, stabilize frequency, and reduce peak consumption costs with a judicious ESS choice. The different applications and the suitability of ESS technologies for microgrids are depicted in Table 6.2.

TABLE 6.2 Applications and suitability of energy storage System (ESS) technologies for microgrid.

Application	ESS type	Key features
Frequency control	BESS, FESS, supercapacitors	High-speed response, grid inertia emulation, real-time correction of frequency deviations
Renewable energy integration	BESS, FESS, PHS, CAES, flow batteries	Mitigates the intermittency of solar and wind, smooths output, and stores excess for later dispatch
Emergency backup	BESS, supercapacitors, solid-state batteries	Instantaneous response during grid outage, seamless load transfer support
Voltage and power quality support	Supercapacitors, FESS	Compensates for voltage sags/swells, balances reactive power in dynamic load conditions
Grid stability enhancement	BESS, FESS, PHS, CAES	Sustains load-generation balance, prevents cascading failures during faults or loss of generation
Peak shaving/Load leveling	BESS, PHS, CAES	Reduces demand charges, optimizes generator use by flattening load profiles
Black start capability	BESS, FESS	Restarts microgrid segments after a blackout without relying on the main utility grid

6.1.3 Policy and economic factors affecting ESS deployment in micro-grids

The successful deployment of ESS in microgrids is not driven solely by technical performance or system-level needs; it is equally shaped by a dynamic landscape of regulatory frameworks, market incentives, and broader economic considerations. As microgrids evolve from experimental setups to commercially viable energy systems, policy interventions play a crucial role in accelerating the adoption of storage technologies.

One key policy mechanism is the implementation of demand response (DR) programs. These programs encourage ESS operators to inject or withdraw energy from the grid in response to real-time demand signals or price fluctuations. By doing so, energy storage not only supports grid stability but also creates new revenue streams for asset owners. In parallel, net metering policies, especially those adapted for Vehicle-to-Grid (V2G) applications, allow users to offset electricity costs or earn credits by supplying stored energy back to the grid. Such frameworks incentivize more active participation from distributed resources, including residential batteries and electric vehicles, turning consumers into "prosumers."

In jurisdictions that have adopted carbon pricing mechanisms such as cap-and-trade systems or carbon taxes, ESS contributes to decarbonization goals by reducing the need for high-emission peaker plants. By shifting energy consumption away from fossil-fueled generation during peak times, storage helps lower the overall carbon intensity of the grid. This environmental benefit also enhances the policy case for ESS investment, especially when linked to compliance targets or renewable portfolio standards.

Economically, the viability of ESS projects often hinges on a detailed cost-benefit analysis. These assessments consider upfront capital costs, expected degradation over time, operational and maintenance expenses, and potential earnings from ancillary services, arbitrage, and capacity markets. Declining battery prices, coupled with performance improvements, have made storage increasingly competitive across a range of applications. Moreover, government grants, low-interest financing programs, and tax credits in many regions have further reduced investment barriers.

As highlighted by Hadjipaschalis et al. (Hadjipaschalis et al., 2009), the choice of storage technology, whether battery-based, mechanical, or thermal, must be aligned with the specific regulatory, economic, and technical characteristics of each microgrid. There is no one-size-fits-all solution; optimal deployment requires a nuanced understanding of local market rules, policy signals, and user needs.

Overall, the integration of ESS into microgrids is as much a policy and economic question as it is an engineering challenge. Aligning technology development with supportive policy design and sound financial planning is essential for unlocking the full value of storage and ensuring its long-term sustainability within modern energy systems.

6.2 Fundamentals of energy storage systems

To optimize microgrid performance, understanding the fundamentals of ESS technologies is crucial. They vary in energy density, power capacity, and response time, making certain types more suitable for specific applications.

6.2.1 Types of energy storage technologies

Energy storage solutions have different characteristics that make them appropriate for microgrids. Popular varieties include.

6.2.1.1 *Battery energy storage systems (BESS)*

Chemically store and release energy as electricity. High energy density, fast reaction, and extended lifespan make lithium-ion batteries desirable. While cheaper, lead-acid batteries last less and are less efficient for repeated cycling. Sodium-sulfur batteries are better suited for medium-duration storage, while flow batteries offer scalability and are ideal for large-scale applications (Zhao et al., 2023).

6.2.1.2 *Flywheel energy storage systems (FESS)*

They store energy by rotating a mass at high speed. Flywheels are good for rapid bursts of high power but not long-term energy storage due to their durability and speed(Ji et al., 2024).

6.2.1.3 *Supercapacitors*

Fast charge and discharge cycles occur due to electrostatic energy storage. Though they keep less energy than batteries and flywheels, they thrive in situations that require instant energy (Vladimir and Iurii, 2023).

6.2.1.4 *Pumped hydro storage (PHS)*

The most widely used large-scale ESS. PHS systems store energy by pumping water to a higher elevation during periods of low demand and releasing it through turbines during peak demand. PHS offers high efficiency and long lifespans but requires specific geographical conditions (Xu et al., 2018).

6.2.1.5 *Compressed air energy storage (CAES)*

CAES uses extra energy to compress air and store it in subterranean caverns before releasing it to create power when needed (Abouzeid et al., 2021).

Microgrids can balance energy supply and demand, maintain frequency, and stabilize the grid with the correct technology.

6.2.2 Characteristics and properties of energy storage systems

When evaluating ESS for microgrid applications, it is essential to move beyond surface-level comparisons and investigate the fundamental characteristics that influence performance. These characteristics, such as energy density, power density, cycle life, and response time, not only differentiate one technology from another but also determine how effectively an ESS meets the demands of a given operational scenario.

Lithium-ion batteries, for example, are widely regarded for their high energy density and are particularly effective in applications that require prolonged energy supply, as depicted in Table 6.3. Their compact form factor and improving cost trends have made them a staple in both

TABLE 6.3 Key characteristics of ESS technologies (Liu et al., 2023a,b,c; Song et al., 2023; Ram Babu et al., 2023).

Technology	Energy density (Wh/kg)	Power density (W/kg)	Cycle life (Number of cycles)	Response time
Lithium-ion battery	100–265	250–340	5000+	Milliseconds
Flywheel	20–80	500–1500	100,000+	Milliseconds
Supercapacitor	5–10	10,000+	1,000,000+	Microseconds
Pumped Hydro	0.5–1.5	1–2	50,000+	Minutes
Compressed air	30–60	10–120	20,000+	Minutes
Solid-state battery	300–400	200–400	10,000+	Milliseconds
Redox flow battery	20–50	50–150	10,000–15,000	Seconds

residential and utility-scale systems. Their suitability for long-duration load support during renewable intermittency, such as cloudy days or low-wind periods, makes them highly adaptable.

In contrast, supercapacitors are built for speed. Although their energy storage capacity is limited, they can discharge large amounts of power in milliseconds and endure over one million charge-discharge cycles without noticeable degradation. This makes them an ideal solution for applications like voltage sag compensation or absorbing inrush currents during load switching.

FESS serves as a bridge between batteries and super-capacitors, offering both high power output and a moderate energy storage window. Their mechanical nature, relying on the inertia of a spinning mass, enables them to react quickly to frequency fluctuations and return to standby almost immediately after event mitigation. Their long life cycle and minimal maintenance requirements further add to their appeal (Liu et al., 2023a,b,c).

Meanwhile, technologies such as PHS and CAES provide grid-scale, long-duration storage. Despite geographical constraints, these solutions offer unparalleled energy capacity, making them suitable for seasonal storage and large-scale renewable integration.

The selection of an ESS technology is therefore not just a technical decision but a strategic one. By aligning system characteristics with microgrid objectives such as frequency stabilization, peak shaving, or energy arbitrage, designers can achieve a balance of performance, cost, and reliability.

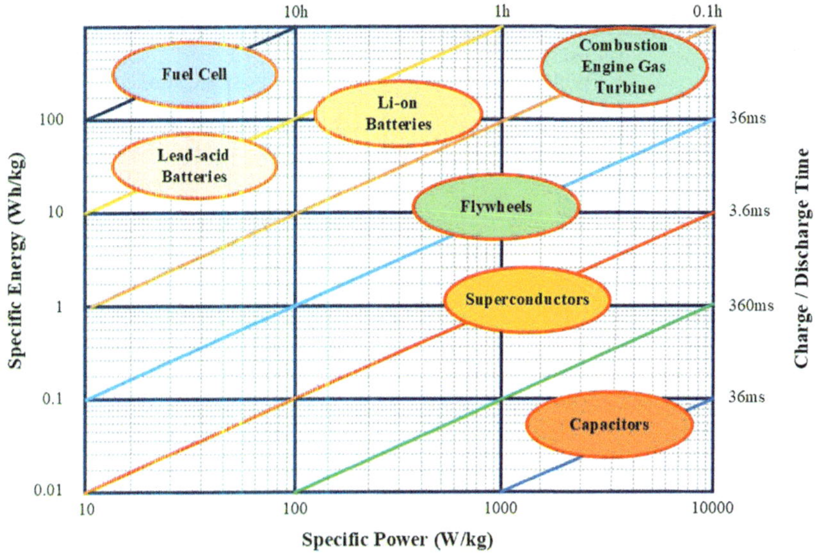

FIGURE 6.2 Comparison of different ESS systems.

6.2.3 Comparison of different energy storage technologies

Cost, energy density, and application must be considered while choosing an ESS for a microgrid. Fig. 6.2, shows a comparison between the different ESS technologies. Different systems thrive in certain areas, making them better for microgrid jobs (Ram Babu et al., 2023; Calero et al., 2019; Tolba et al., 2022; Dong et al., 2022). Lithium-ion batteries are ideal for long-term storage and energy smoothing in renewable-heavy systems due to their energy capacity and pricing. Recent market data show lithium-ion battery pack prices at \$139/kWh by the end of 2023, falling further to \$115/kWh in 2024. Installed costs for utility-scale 4-hour systems are projected at \$245–326/kWh by 2030, declining to \$159–226/kWh by 2050. These ranges reflect differences in scale, system integration, and accessories such as inverters and balance-of-system components.

While flywheels can withstand brief, severe power surges, they are best suited for short-duration storage. On the other hand, supercapacitors are good at regulating fast variations, but their lower energy density, as shown in Table 6.4, confines them to power injections rather than storage (Ram Babu et al., 2023). Additionally, PHS provides large-scale, long-term storage with high efficiency, making it valuable for renewable energy integration and grid stability. CAES similarly offers scalable, long-duration storage, effectively handling peak shaving and ensuring reliable grid performance.

TABLE 6.4 Comparison of different ESS technologies for microgrid applications.

Technology	Cost (USD/ kWh)	Energy density (Wh/kg)	Application
Lithium-ion battery	115 to 326	High (100–265)	Renewable energy integration, peak shaving (Wali et al., 2024)
Flywheel	300 to 500	Low (20–80)	Frequency regulation, short-duration storage (Ji et al., 2024)
Supercapacitor	500 to 800	Very low (5–10)	High-frequency stabilization (Vladimir and Iurii, 2023)
PumpedHydro storage	100 to 200	VeryLow (0.5–1.5)	Large-scale, long-term energy storage, grid balancing (Dong et al., 2022)
Compressed air energy storage	50 to 150	Low (30–60)	Peak shaving, long-term storage, grid stability (Calero et al., 2019)
Solid-state battery	300 to 600	VeryHigh (300–400)	Medium-duration storage with enhanced safety (Yang et al., 2024b)
RedoxFlow battery	200 to 500	Low (20–50)	Scalable energy storagefor hybrid microgrids (Oshnoei et al., 2021)

6.3 Role of energy storage in frequency regulation

Microgrids require precise frequency regulation to ensure the smooth operation of connected devices. All linked devices and systems in a microgrid need a constant frequency to operate properly. Power is adjusted to keep the frequency within narrow bounds around the nominal value, usually 50 or 60 Hz, depending on the location (Gulzar et al., 2023). ESS absorbs or provides energy to compensate for frequency variations, making them essential. As microgrids use more RES, which vary the power supply, ESS's importance in frequency management has grown. ESS stabilizes the microgrid's frequency oscillations by storing energy when generation exceeds demand and releasing it during demand surges, helping to reduce the pressure on the grid and enhancing its reliability and durability.

6.3.1 Principles of frequency regulation in microgrids

Maintaining a stable frequency is a critical aspect of microgrid operation, especially when the grid incorporates high levels of RES. Unlike conventional power systems that benefit from the inherent inertia of large rotating generators, microgrids often rely on inverter-based Distributed Energy Resources (DERs), which lack physical inertia. This makes the system more sensitive to imbalances between generation and demand (Gulzar et al., 2023).

Frequency regulation in a microgrid requires continuous real-time control to ensure the frequency stays within acceptable limits, typically ±0.1 Hz of the nominal value. Any deviation can compromise the stability of the system, potentially leading to equipment damage, load shedding, or complete system failure.

In this context, LFC strategies are employed to restore equilibrium. These include primary control through droop-based mechanisms, secondary control using centralized or distributed algorithms, and tertiary control for economic dispatch and long-term balancing.

ESS plays an increasingly vital role in all three control layers. Their ability to respond rapidly to power mismatches helps compensate for the variability of RES, allowing the micro-grid to ride through disturbances smoothly. Through advanced controls, ESS can also emulate inertial behavior, thus enhancing the microgrid's resilience during frequency excursions.

In summary, the integration of ESS into microgrid frequency regulation frameworks is no longer optional but necessary. Their contribution to stability, flexibility, and control sophistication significantly improves the operational robustness of modern decentralized energy systems. Table 6.5 demonstrates the ESS's response for different frequency deviation scenarios.

TABLE 6.5 Frequency deviation scenarios and ESS responses.

Scenario	Cause	ESS response
Over-frequency	Sudden PV/wind surge	ESS absorbs excess energy
Under-frequency	Load spike or generator loss	ESS discharges energy quickly
Intermittency	Cloud shadow or wind lull	ESS smooths output fluctuations
Start-up transients	Large motor or EVs charging event	ESS provides immediate voltage/frequency support
Planned curtailment	Grid switching or DERs ramp-down	ESS bridges supply gaps

6.3.2 Functions of energy storage systems in frequency control

The multifaceted role of ESS in frequency regulation goes well beyond merely acting as a power reserve. These systems are engineered to perform a variety of dynamic functions essential for maintaining grid balance in real time.

First and foremost, ESS can inject or absorb power almost instantaneously, which is crucial during frequency deviations caused by generation-demand mismatches. This function is particularly valuable in islanded microgrids, where traditional generation sources may lack the agility to respond effectively.

Additionally, ESS systems contribute synthetic inertia through fast frequency response algorithms. This is especially critical in inverter-dominated grids, where physical inertia is nearly absent. The fast-acting response of ESS reduces the rate of change of frequency (RoCoF), providing a time buffer for slower control layers to take corrective action.

Beyond stabilization, ESS improves the efficiency of overall grid operations. By absorbing excess generation during low-demand periods and releasing it during peaks, ESS reduces reliance on spinning reserves and minimizes fuel consumption in hybrid microgrids. This not only enhances frequency control but also supports sustainability goals.

Moreover, advanced ESS can be programmed to participate in both autonomous and coordinated control schemes. In autonomous mode, they use local frequency measurements to decide charge/discharge actions. However, in coordinated mode, they receive signals from a centralized Energy Management System (EMS), allowing for system-wide optimization.

Ultimately, ESS transforms frequency regulation from a rigid, generator-dominated task into a flexible, distributed, and smart process, improving reliability while paving the way for deeper renewable penetration (Khan et al., 2023; Gulzar et al., 2022).

6.3.3 Pros and cons of energy storage for frequency regulation

Energy storage offers compelling advantages for frequency regulation in microgrids. However, its implementation is not without limitations, as depicted in Table 6.6, and understanding these trade-offs is essential for effective system design.

Among the most notable benefits is the rapid response time of ESS. Unlike traditional thermal generators, which may take minutes to react, ESS can engage within milliseconds. This speed enables them to manage high-frequency disturbances with precision and prevent cascading failures.

ESS also brings a high degree of control flexibility. They can be deployed in both grid-connected and islanded microgrids and can be scaled incrementally. Their integration enhances not only frequency control but also

TABLE 6.6 Technical and operational pros and cons of ESS for frequency regulation.

Benefits	Challenges
Fast response times (milliseconds or less)	High capital investment
Decentralized deployment flexibility	Limited lifecycle for some chemistries
Enables high RES penetration	Complexity of multi-ESS control integration
High round-trip efficiency	Lack of universal interconnection standards
Environmental advantages (low emissions)	Spatial constraints for utility-scale systems

voltage regulation, power quality, and renewable integration (Ram Babu et al., 2023).

On the downside, cost remains a major barrier. Although battery prices are declining, capital costs, operational expenses, and battery degradation still pose economic challenges. These concerns are magnified in systems requiring high power and energy ratings over extended periods.

Furthermore, control complexity increases with ESS integration. Managing state of charge, thermal limits, and lifetime optimization demands advanced algorithms and real-time monitoring systems. Without careful tuning, improper operation can lead to accelerated wear or ineffective frequency support.

From a sustainability perspective, some battery chemistries raise concerns regarding raw material sourcing, recyclability, and end-of-life disposal. Addressing these environmental impacts is essential for the long-term viability of ESS technologies.

In summary, while energy storage is a powerful tool for frequency regulation, its success depends on thoughtful system planning, robust control strategies, and a comprehensive lifecycle cost-benefit analysis.

6.4 Battery energy storage systems for frequency control

BESS are frequently utilized in microgrids for frequency management due to their high energy density, rapid reaction times, and capacity to give short and long power surges (Li et al., 2023). These applications favor BESS due to their high storage capacity, efficiency, and lifespan (Wali et al., 2024). During peak demand, BESS rapidly releases energy to maintain frequency, power supply, voltage, and reactive power to improve power quality. Additionally, peak shaving, backup power, and emergency assistance in microgrids require BESS's adaptability, especially as renewable energy demand rises (Gulzar et al., 2023a,b,c).

6.4.1 Various battery technologies for LFC

There are several battery technologies available; however, the most widely used for LFC include lithium-ion batteries for their high energy density and fast response, lead-acid batteries for cost-effective backup solutions, sodium-sulfur batteries for medium-duration applications, and flow batteries for their scalability and suitability in large-scale renewable energy systems. The high energy density, long life, and compact design of lithium-ion batteries make them essential for microgrids, as shown in Table 6.7 Lead-acid and sodium-sulfur batteries also offer distinct advantages. Due to their limited lifespan, lead-acid batteries are suitable for low-cost, low-cycling applications. Sodium-sulfur batteries, with their high energy density, are better suited for medium-duration storage. Flow batteries provide exceptional scalability and are ideal for large-scale applications,

TABLE 6.7 Characteristics of common battery types used in microgrid frequency control.

Battery type	Energy density (Wh/kg)	Cycle life	Suitability in microgrids
Lithium-ion	High (100–265)	3000+ cycles	High-frequency control and peak shaving (Rimsha et al., 2023)
Lead-acid	Moderate (30–50)	500–1500 cycles	Backup, low-cost applications (Bhoi et al., 2024)
Sodium–sulfur	High (150–240)	2500 cycles	High-energy, moderate-duration storage (Terkes et al., 2024)
Flow-batteries	Low to moderate (20–50)	Unlimited (with maintenance)	Large-scale, renewable-heavy applications (Oshnoei et al., 2021; Chidambaram and Paramasivam, 2013)
Nickel–Cadmium	Moderate (40–60)	2000+ cycles	Industrial microgrids, tolerant to deep discharge
Zinc-air	Very high (300–400)	Limited (non-rechargeable or low-cycle)	Long-duration backup in remote/off-grid systems

making them a versatile choice for renewable-heavy systems. Microgrids employ lithium-ion or supercapacitors for short-term power needs, while lead-acid and flow batteries excel in long-duration storage requirements (Makola et al., 2023).

6.4.2 Operational characteristics of BESS in frequency regulation

BESS has become a cornerstone of modern microgrid architecture, particularly for frequency regulation. Their operational characteristics allow them to function effectively across multiple timescales, delivering both rapid response and sustained energy over longer durations.

One of the primary advantages of BESS is its controllability. Integrated with power electronic interfaces and inverter systems, they can inject or absorb power in a finely tuned manner. This capability makes them ideal for real-time correction of frequency deviations and helps smooth the variable output from renewable sources.

BESS operation is governed by several critical parameters, including the state of charge (SOC), maximum current rating, battery temperature, and aging profile. These parameters are monitored and managed by a Battery Management System (BMS), which ensures safe and efficient operation under various grid conditions.

A key aspect of their performance in frequency regulation is response time. Lithium-ion BESS, for instance, can respond within milliseconds, an order of magnitude faster than most mechanical generation systems. This enables them to participate not only in primary frequency response but also in synthetic inertia emulation and fast frequency control loops.

In hybrid systems, BESS often complements other technologies like supercapacitors or flywheels. While supercapacitors handle high-frequency, short-duration events, BESS is used for deeper energy support and longer stabilization periods. Such coordinated control schemes enhance overall system robustness.

To maintain optimal performance, modern BESSs are operated using intelligent control algorithms such as Model Predictive Control (MPC) or Artificial Neural Networks (ANNs). These techniques allow predictive scheduling, aging-aware dispatch, and coordinated frequency support, all of which are essential for maintaining microgrid stability and extending battery lifespan (Nyamathulla and Dhanamjayulu, 2024).

6.4.3 Control strategies and algorithms for BESS integration in micro-grid LFC

BESS frequency regulation adapts battery charging and discharging to grid circumstances and projections using advanced algorithms. These algorithms keep microgrids responsive, especially ones with plenty of RES (Gulzar et al., 2022; Nyamathulla and Dhanamjayulu, 2024). Common Control Strategies for BESS include.

- **Model Predictive Control (MPC):** MPC predicts demand and renewable energy generation to reduce charging and discharging cycles and extend battery life. Battery consumption is optimized based on load and generation projections to ensure frequency stability (Tolba et al., 2025; Oshnoei et al., 2020; Liu et al., 2024).
- **Fuzzy Logic Control (FLC):** FLC's rule-based frequency response allows flexible load and generation control in high-variability systems without a microgrid model (Zhang et al., 2016; Hartani et al., 2023; Abadlia et al., 2016).
- **Proportional-Integral-Derivative (PID) Control:** For known load patterns, PID controllers correct BESS output frequency variations, a simple but effective approach (Çelik et al., 2023; Tolba et al., 2022).
- **Droop Control:** Droop control adjusts power output to frequency changes, stabilizing BESS and allowing decentralized microgrids to share loads without central coordination (Shim et al., 2019).
- **H-infinity (H_∞) Control:** In high-uncertainty settings, the H_∞ robust control approach provides stability and performance, benefiting renewable-rich microgrids by handling unanticipated swings (Zhang et al., 2020).
- **Artificial Neural Network (ANN) Control:** ANNs provide adaptive learning mechanisms, which are well-suited for nonlinear and uncertain microgrid environments. Recent studies confirm their effectiveness in maintaining frequency stability and optimizing load-sharing strategies in renewable-dominated systems (Demiroren et al., 2001; Safari et al., 2021; Tolba et al., 2025).
- **Adaptive Control:** Adaptive controllers are beneficial for microgrids with unpredictable changes as they alter BESS settings in real time to optimize performance despite any changes in load or generation (Khooban et al., 2016).
- **Sliding Mode Control (SMC):** A robust control mechanism, SMC stabilizes unpredictable systems. Maintaining control operations inside a preset sliding surface promotes rapid and precise response. SMC offers high robustness against modeling inaccuracies and external disturbances, proving advantageous for voltage and frequency regulation. These methodologies are supported by scholarly works that demonstrate their efficacy in practical microgrid applications (Ullah et al., 2022; Mu et al., 2017).

These control approaches optimize BESS frequency management and battery use. In addition, these optimized controllers improve frequency stability and battery longevity in renewable-heavy microgrids.

6.5 Flywheel energy storage systems for frequency control

FESS plays a vital role in applications requiring rapid response and high power over short durations. These systems store kinetic energy in a rotating mass, allowing them to instantly inject or absorb energy as needed. Modern advancements, such as the use of magnetic bearings and vacuum-sealed environments, have significantly improved their efficiency and durability. Applications of FESS include stabilizing renewable-dominated microgrids and mitigating frequency fluctuations. For example, in wind-integrated systems, FESS can smooth out short-term power variations, ensuring stable frequency control. Additionally, FESS is increasingly deployed in high-demand urban microgrids, where quick stabilization is essential to prevent power outages (Ji et al., 2024).

6.5.1 Introduction to flywheel energy storage technology

Kinetic energy storage allows FESS to instantly inject or absorb energy for frequency regulation. They spin a rotor to convert electrical energy to kinetic energy during low demand and release power when needed through an opposite process, making them efficient and durable for frequent, intense use. The connection of the FESS to the microgrid is depicted in Fig. 6.3. New flywheel technologies, including magnetic bearings and vacuum operation, have increased the efficiency and storage

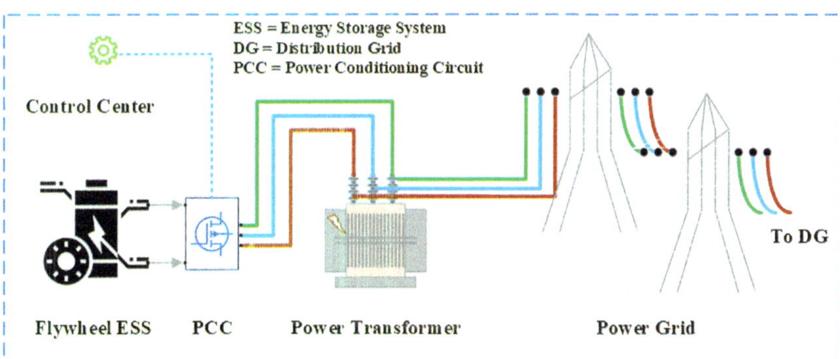

FIGURE 6.3 Schematic of the FESS's connection to the microgrid.

capacity, making them appropriate for microgrid frequency management. Microgrids, where frequent cycling is needed to regulate renewable source variations, benefit from FESS's longevity and capacity degradation resistance (Lei et al., 2023).

6.5.2 Advantages and limitations of FESS in frequency regulation

FESS can handle quick microgrid fluctuations and provide frequency control and voltage stabilization due to its instantaneous power supply. Their lower energy density restricts their usage for some applications, and expensive installation costs for specialized components may limit their use in smaller, budget-limited microgrids (Lei et al., 2023). The evaluation of the benefits and drawbacks of FESS is demonstrated in Table 6.8.

6.5.3 Applications and case studies of FESS in microgrid LFC

FESS offers a high-performance alternative to electrochemical batteries in applications where rapid response and high cycling frequency are required. Their unique ability to store energy mechanically as kinetic energy in a rotating mass makes them ideal for microgrids that demand short-duration,

TABLE 6.8 Comparison of advantages and limitations of flywheel energy storage systems (FESS).

Advantage	Limitation
Fast response time in milliseconds, suitable for high-frequency applications	Lower energy density compared to chemical batteries, limiting long-term storage capacity
High cycle life with minimal degradation over time	High initial capital cost due to precision components and materials
Reliable and robust for frequent cycling in variable load scenarios	Limited suitability for long-duration storage due to rapid discharge characteristics
Environmentally friendly with no harmful chemical disposal requirements	Requires vacuum-sealed enclosure and magnetic bearings, increasing system complexity
Low maintenance and high mechanical efficiency over the system's lifespan	Limited commercial adoption compared to mainstream battery solutions
Capable of operating in extreme environmental conditions (e.g., wide temperature ranges)	Energy losses due to friction and the standby power required to maintain rotational inertia

high-power interventions for frequency regulation. In practice, FESS has demonstrated effectiveness in a variety of microgrid setups. One well-documented case is the use of FESS in remote island microgrids powered primarily by solar PV. Here, sudden fluctuations in solar generation, such as from passing cloud cover, cause rapid frequency deviations. The flywheel's immediate power response smooths these variations, preventing generator tripping or blackouts. Data centers represent another promising application. Due to their strict power quality requirements, even slight deviations in voltage or frequency can lead to data loss or system crashes. FESS units, integrated at the microgrid level or as standalone uninterruptible power supplies (UPS), provide real-time buffering to protect sensitive equipment. Moreover, FESS has been successfully integrated into military and industrial microgrids where robustness and low maintenance are paramount. Because flywheels do not rely on chemical reactions, they experience minimal wear over time and require fewer lifecycle interventions. These systems are also environmentally advantageous, avoiding the use of toxic materials and offering full recyclability. Their long lifespan, combined with their near-instantaneous response and high round-trip efficiency, makes FESS a strategic asset for next-generation frequency regulation in resilient and sustainable microgrids (Xu et al., 2023).

6.6 Supercapacitors and other storage technologies

6.6.1 Overview of supercapacitor technology

Supercapacitors, or ultracapacitors, occupy a unique position in the energy storage landscape. They store energy via electrostatic rather than electrochemical means, allowing them to operate at much higher cycle speeds and with greater efficiency than traditional batteries.

The internal structure of a supercapacitor consists of two carbon-based electrodes separated by an electrolyte and dielectric medium. During operation, electrical energy is stored in the electric double layer formed at the electrode-electrolyte interface. This mechanism enables extremely rapid charging and discharging, often in fractions of a second.

While their energy density is significantly lower than that of lithium-ion batteries, their power density is unmatched. This means they can deliver high bursts of power on demand, making them particularly useful for frequency regulation, voltage sag mitigation, and load leveling in dynamic microgrid environments.

Supercapacitors are also known for their long service life. They can endure hundreds of thousands to millions of cycles without significant performance degradation. This durability, combined with their low internal resistance, results in exceptionally high round-trip efficiency, typically exceeding 95%.

Recent innovations have further improved their performance. New materials such as graphene and carbon nanotubes are being used to increase capacitance and reduce leakage currents, making modern supercapacitors more compact and effective.

Due to their rapid response and reliability, supercapacitors are increasingly being deployed alongside batteries in hybrid systems. In such configurations, they handle immediate power surges while the batteries take over sustained energy delivery, thereby reducing stress on the battery and enhancing overall system longevity (Shukla and Raju, 2024).

6.6.2 Role of supercapacitors in frequency regulation

In frequency regulation applications, supercapacitors excel in mitigating short-term power imbalances. Their hallmark feature (the ability to deliver or absorb power in microseconds) makes them ideal for scenarios where traditional energy sources or even batteries would respond too slowly.

Supercapacitors are especially effective in handling high-frequency transients, which can arise from sudden load changes, inrush currents, or renewable intermittency. For example, when a large motor starts up or when a cloud rapidly dims solar PV output, the frequency may deviate briefly. Supercapacitors can step in almost instantly, injecting or absorbing energy to neutralize these fluctuations and stabilize the frequency.

Another important application is in grid-forming inverters and virtual synchronous generators. When paired with intelligent control, supercapacitors can mimic the inertia of traditional spinning machines, enhancing the dynamic stability of low-inertia microgrids.

Moreover, because supercapacitors do not suffer from the degradation issues that plague many battery chemistries, they are especially useful in high-cycling environments. Their long lifespan and robust tolerance to charge/discharge cycles make them suitable for systems where frequent and rapid frequency adjustments are required.

Although their energy capacity is limited, their high efficiency and near-zero latency ensure that supercapacitors remain a powerful complementary technology for frequency regulation, particularly when integrated with other storage forms in a hybrid ESS configuration (Liu et al., 2023a,b,c).

6.6.3 Emerging energy storage technologies for microgrid LFC

Microgrid energy storage is being improved with solid-state batteries and hydrogen storage to improve efficiency, safety, and dependability. These devices handle energy and frequency longer than supercapacitors. Solid-state batteries offer higher energy density and improved safety due to the absence of liquid electrolytes. These are particularly suitable for long-duration applications in microgrids with variable renewable energy input. Hydrogen storage systems, on the other hand, provide a sustainable solution for large-scale, long-term energy storage, converting surplus electricity into hydrogen via electrolysis for later use.

6.6.3.1 Solid-state battery technology

In high-demand situations, solid-state batteries reduce leakage and overheating due to their solid electrolytes. Their increased energy density allows longer discharges, stabilizing micro-grids amid power fluctuations (Chowdhury et al., 2022). Advantages of solid-state batteries include.

- **High Energy Density**: Solid-state batteries store more energy than lithium-ion batteries, extending their lifespan.
- **Enhanced Safety**: These batteries are safer for energy backup applications due to lower thermal runaway danger.
- **Long Cycle Life**: Strong, solid-state batteries can withstand frequent charging and discharging without losing capacity.

Recent developments have shown that solid-state batteries are no longer confined to laboratory research. For example, Toyota has begun limited-scale deployment of solid-state battery technology in its next-generation electric vehicles, showcasing promising improvements in energy density and safety. Similarly, QuantumScape has initiated pilot production phases and shared performance data from independent testing that confirms enhanced cycle life and volumetric capacity.

6.6.3.2 Hydrogen storage systems

Hydrogen storage systems for long-term, large-scale microgrid energy storage convert surplus power into hydrogen via electrolysis, which may be utilized to create electricity (Ruban et al., 2023). Its benefits include.

- **Large-Scale Storage Capability**: Hydrogen can store a lot of energy, making it a good backup during low renewable generation.
- **Sustainability**: Hydrogen fuel cells create just water, making them environmentally friendly.
- **Flexibility**: The adaptability of hydrogen as an energy storage medium and fuel supports different microgrid needs.

Hydrogen storage is also finding real-world traction. Notably, projects such as the European Union's H2Haul initiative and the deployment of hydrogen-based energy buffering in the "Energy Observer" maritime vessel highlight the adaptability of hydrogen solutions across sectors. Additionally, hybrid microgrids incorporating electrolyzers and hydrogen fuel cells are being tested in rural and islanded regions, demonstrating feasible integration for long-duration energy storage.

6.6.4 Comparative applications in microgrids

Supercapacitors, solid-state batteries, and hydrogen storage satisfy diverse energy storage and response demands in microgrids.

- **Supercapacitors**: Their fast responsiveness and great power density make them ideal for high-frequency, short-duration assistance (Liu et al., 2023a,b,c).
- **Solid-State Batteries**: Medium-term energy storage with safety and efficiency for continuous load support (Yang et al., 2024b).
- **Hydrogen Storage**: Perfect for long-term, large-scale energy storage in microgrids with prolonged backup needs (Hossain et al., 2023).

These technologies make microgrids resilient and flexible, allowing them to operate smoothly even with varying energy demands and renewable energy input. Modern microgrids benefit from their robust load frequency control and energy management architecture, which complements classical BESS. Table 6.9, provides a comparative analysis of these technologies, highlighting their applications and benefits in microgrid systems.

6.7 Integration challenges of ESS and solutions

While energy storage systems offer significant potential for microgrid frequency regulation, their successful integration involves navigating a variety of technical, economic, and operational challenges.

TABLE 6.9 Comparison of supercapacitors and emerging energy storage technologies (Yang et al., 2024b; Liu et al., 2023a,b,c; Hossain et al., 2023; Sheelam et al., 2023).

Technology	Energy density (Wh/kg)	Response time	Typical application
Supercapacitors	5–10	Microseconds	High-frequency, short-duration support
Solid-state batteries	300–400	Milliseconds	Medium-term frequency regulation
Hydrogen storage	120–160	Seconds	Extended backup, large-scale storage
Redox flow batteries	20–50	Milliseconds to seconds	Scalable storage, energy arbitrage, load leveling
Hybrid capacitors	30–50	Milliseconds	Bridging supercapacitors and batteries, rapid-response buffer storage

One major hurdle is synchronization. ESS must operate in perfect coordination with distributed generators, variable loads, and other grid-forming elements. Any latency or control mismatch can destabilize the microgrid, leading to frequency oscillations or inverter conflicts.

Another challenge lies in control system design. The proliferation of different storage technologies, each with unique characteristics, necessitates complex multi-layered control strategies. These often involve combinations of local droop-based control, distributed cooperative algorithms, and centralized predictive optimization. Implementing such frameworks requires a high level of communication reliability, sensor precision, and computational capacity.

From an economic standpoint, high capital costs remain a barrier, particularly for advanced chemistries like lithium-sulfur or flow batteries. Additionally, lifecycle costs must account for degradation, replacement intervals, and efficiency losses.

There are also policy and regulatory gaps. Many jurisdictions lack clear standards for ESS interconnection, ownership models, and compensation schemes for grid services provided by storage assets.

Solutions to these challenges include the use of modular and scalable ESS architectures, which allow phased investment and simplify maintenance. Hierarchical control frameworks help manage system complexity while improving reliability. Economies of scale, along with supportive policy measures such as incentives and grid service compensation, are essential to lowering cost barriers.

Ultimately, integrating ESS into microgrids is not a one-size-fits-all endeavor. It requires tailored technical design, robust control systems, and enabling regulatory environments to fully realize the value of storage for frequency stability.

6.8 Integration of electric vehicles (EVs) into microgrids

EVs are evolving from mere transportation assets to dynamic energy resources, capable of playing a vital role in microgrid operations. With increasing battery capacities and the advent of Vehicle-to-Grid (V2G) technology, EVs now offer bidirectional energy flow, enabling them to act as both energy consumers and distributed storage providers.

The integration of EVs into microgrids opens up numerous possibilities for frequency control. During times of peak demand or sudden frequency dips, parked EVs with available capacity can be discharged back into the grid, supporting stability. Conversely, during periods of excess renewable generation, these vehicles can absorb surplus power, aiding in load balancing. Case studies in smart campuses and residential communities have demonstrated that coordinated EV fleets can provide synthetic inertia, peak shaving, and even black start capabilities. These systems use aggregator platforms to manage large numbers of vehicles, aligning their charging and discharging schedules with grid needs.

However, integrating EVs into microgrids is not without its challenges. Variability in user behavior makes forecasting aggregate energy availability difficult. Battery degradation concerns, infrastructure limitations, and the need for standardized communication protocols also present barriers to seamless integration.

To address these issues, smart charging strategies are being deployed that prioritize grid support while respecting user constraints and preserving battery life. Fast, bidirectional chargers with embedded communication modules are enabling real-time control and remote dispatch of EVs' power.

When well-managed, EVs represent a flexible and scalable solution for microgrid frequency regulation, transforming mobility assets into valuable contributors to grid resilience and energy sustainability (Ehsan et al., 2024). The connection of the EVs to the microgrid is depicted in Fig. 6.4, as illustrated in (Ghafoor et al., 2024; Jameel and Gulzar, 2023).

Advantages of Electric Vehicles (EVs).

- **Distributed Energy Storage**: Mobile energy storage from EVs helps stabilize microgrid load and frequency.
- **Peak Demand Support**: EVs alleviate grid pressure by discharging energy during peak demand.
- **Environmental Benefits**: EVs promote greener microgrid energy by lowering emissions and fossil fuel use.
- **Flexibility**: EVs can consume and provide energy, increasing microgrid flexibility.

However, EVs' incorporation into microgrids has drawbacks.

Drawbacks of Electric Vehicles (EVs).

- **Increased Load on Microgrid**: EVs charging strain microgrid resources, especially at peak hours.
- **Frequency Stability Challenges**: EVs' integration needs careful monitoring to avoid frequency swings from unexpected charging patterns.
- **Infrastructure Requirements**: Advanced chargers and grid management technologies are needed to integrate EVs.
- **Battery Degradation**: Frequent charging and consuming may shorten an EV's battery life.

6.9 Simulation studies and modeling approaches

Simulation studies are crucial in assessing the performance of ESS in microgrid scenarios characterized by high RES and EV integration. These

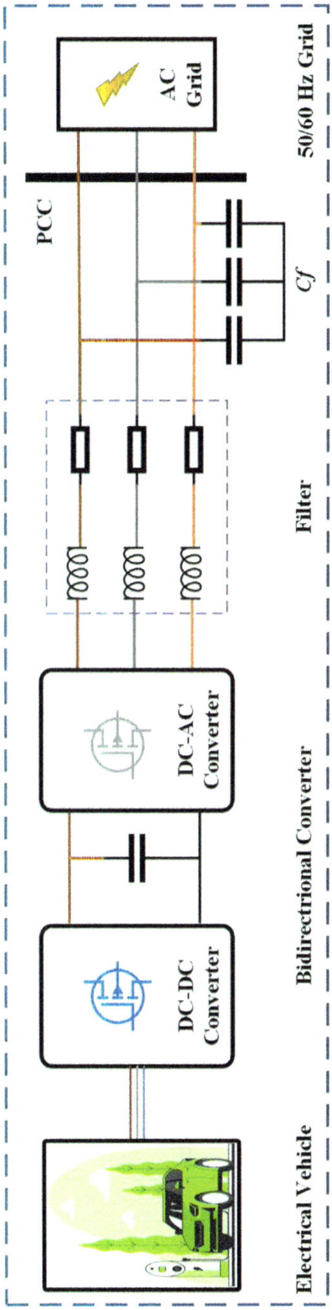

FIGURE 6.4 Schematic of the EV's connection to the microgrid.

simulations were designed to evaluate system stability, performance enhancements, and dynamic responses under various configurations of ESS. The primary objective of this study is to determine how ESS integration affects microgrid frequency regulation, particularly under varying load conditions. A three-area power system comprising photovoltaic (PV), wind, and thermal generation was analyzed using the PI–PD controller. The system layout is presented in Fig. 6.5, while the thermal, wind, PV, and EV system models are shown in Fig. 6.6.

The parameters of the controller are tuned using the Gray Wolf Optimization (GWO) algorithm. The cascaded PI–PD controller's layout is depicted in Fig. 6.7, while the flowchart of the optimization algorithm is presented in Fig. 6.8 (Gulzar et al., 2023; Nyamathulla and Dhanamjayulu, 2024).

The controller's gains are shown in Table 6.10, were tuned to assess the system's ability to manage frequency stability under a 15% load variation. The study investigates several scenarios, including the baseline system without any ESS, the integration of BESS, a combination of BESS and

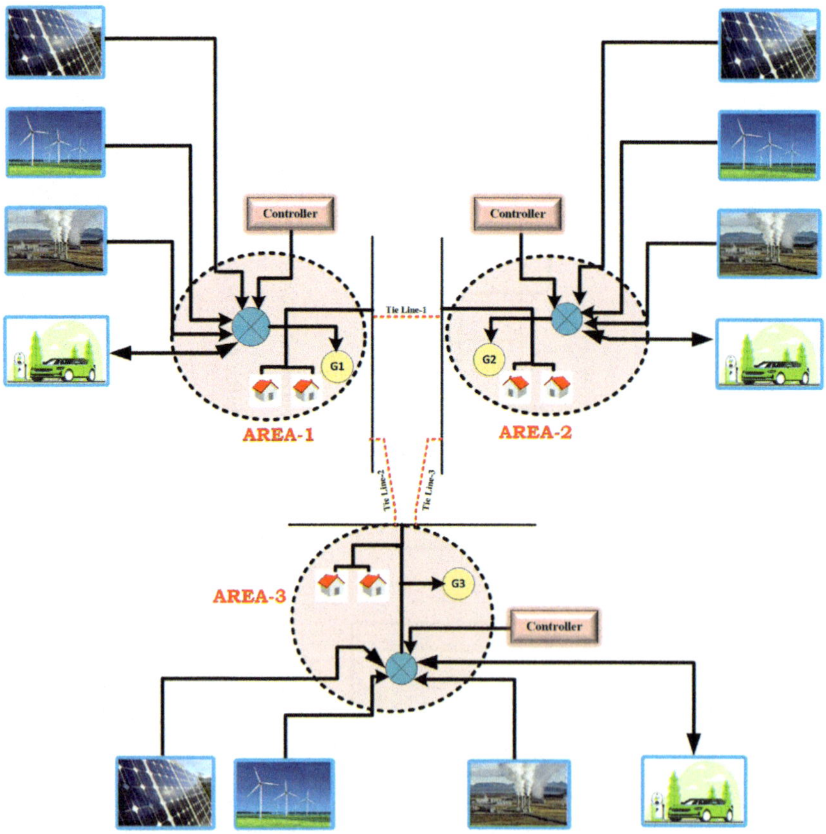

FIGURE 6.5 Model under investigation.

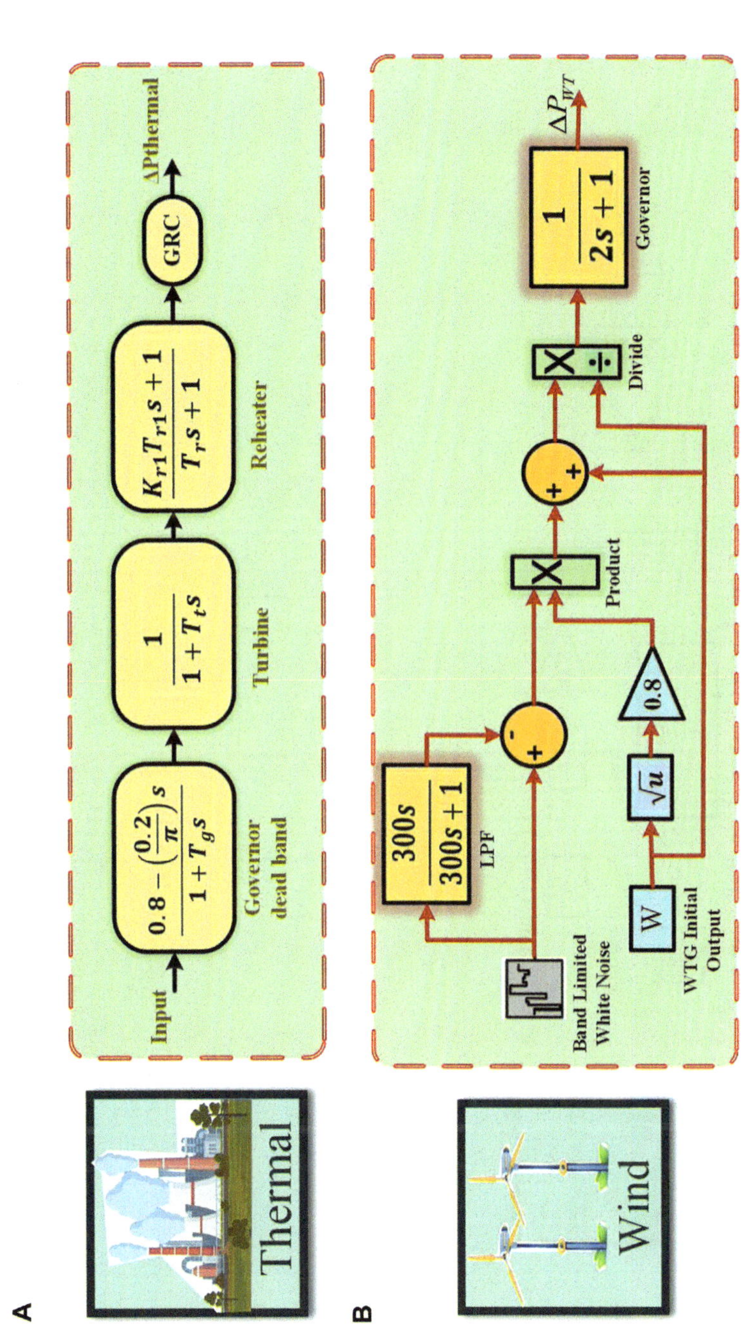

FIGURE 6.6 Power system modeling: (a) Thermal system model, (b) wind system model, (c) PV system model, and (d) EV system model.

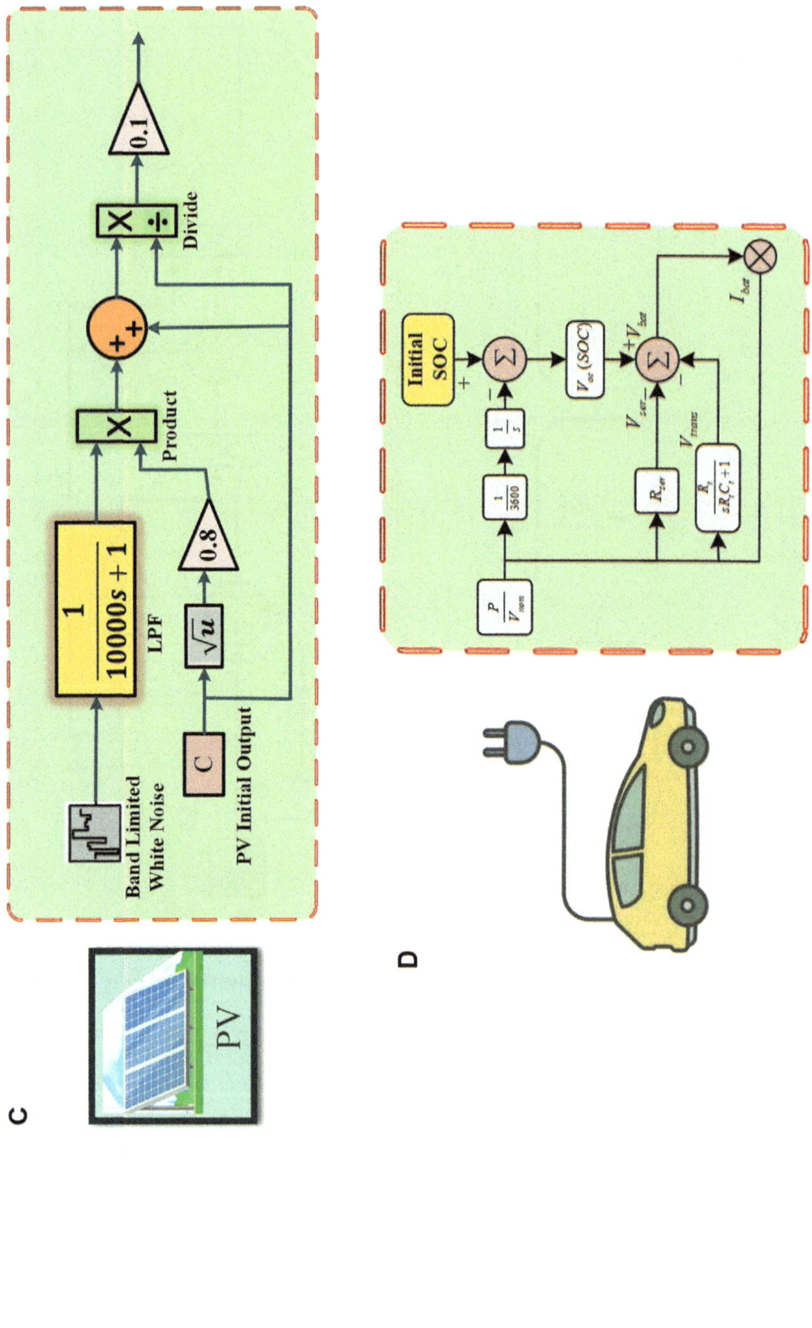

FIGURE 6.6 *Continued.*

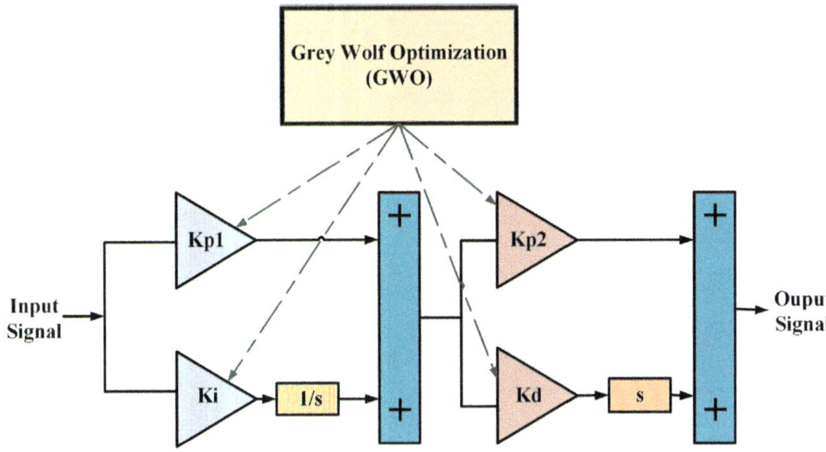

FIGURE 6.7 Cascaded PI-PD controller's structure.

FESS, and finally, the addition of EVs alongside BESS and FESS. The results reveal the progressive improvements in system stability as more advanced ESS configurations are incorporated.

The simulations were conducted using **MATLAB/Simulink**. The following modeling assumptions were made to ensure realism and tractability.

- Each area contains a conventional generator and a renewable generation unit.
- Load variations of 15% were simulated to reflect peak and off-peak scenarios.
- Ideal converters and standard system parameters were used based on industry norms.
- ESS units are assumed to operate with a known and controlled state-of-charge.
- Line losses, communication delays, and battery aging were neglected.

To validate the proposed control strategies and ESS configurations, the following methodology was employed.

- **Comparative Scenario Analysis:** The system was tested under four configurations: (i) no ESS, (ii) BESS only, (iii) BESS + FESS, and (iv) BESS + FESS + EVs.
- **Performance Metrics:** Overshoot, undershoot, and settling time were used to quantify dynamic performance.
- **Graphical Validation:** 12 presents the dynamic responses across scenarios, while Table 6.11 summarize numerical comparisons.

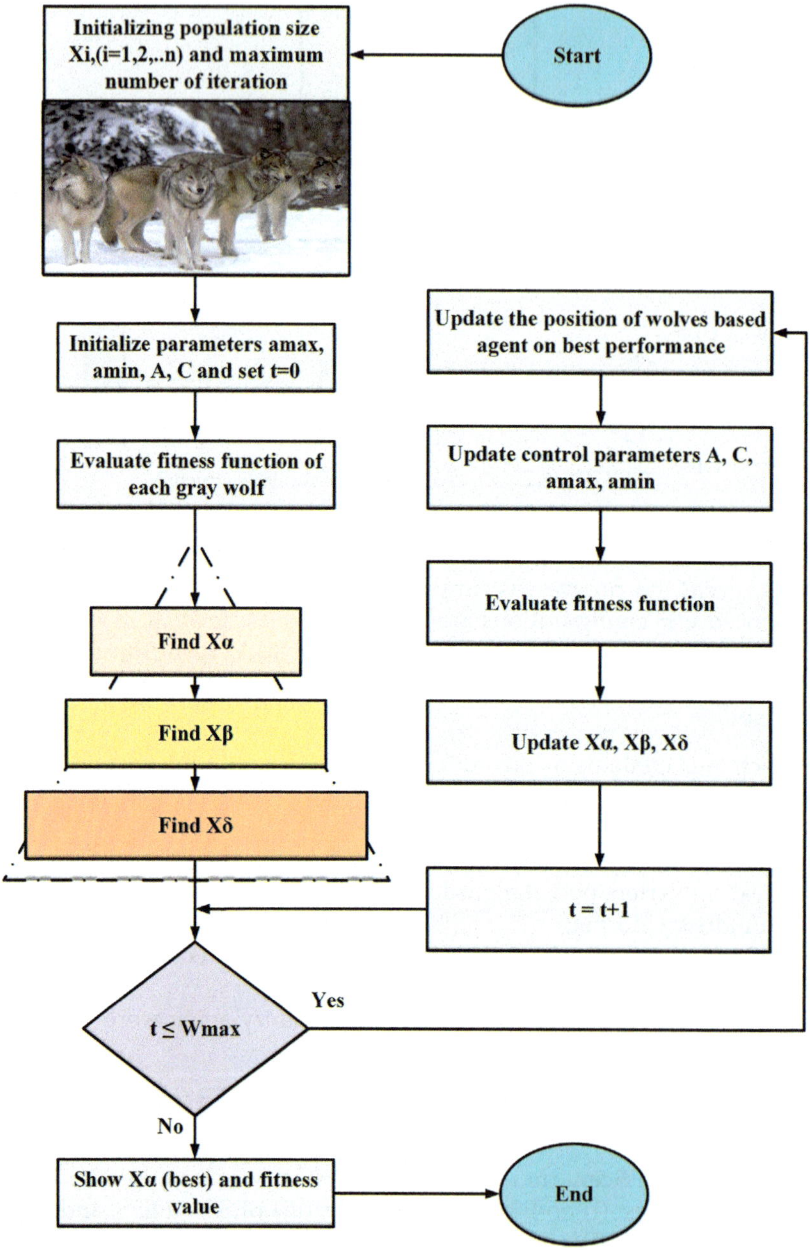

FIGURE 6.8 Flowchart of the Gray Wolf optimization (Goa) algorithm.

TABLE 6.10 Optimum settings of (GWO: PI-PD) controller.

Area No.	Kp_1	K_I	Kp_2	K_D
Area-1	−2.8771	−2.7874	−3.004	−2.0146
Area-2	−4.2809	−2.0014	−2.9084	−1.2979
Area-3	−4.3	−2	−2.9	−1.3

TABLE 6.11 System's performance across multiple ESS's penetration.

Specifications	Δf_1	Δf_2	Δf_3	ΔP_{tie1}	ΔP_{tie2}	ΔP_{tie3}
System without any ESS						
Overshoot (Hz)	0.37	0.38	0.38	7e-3	0.022	0.02
Undershoot (Hz)	0.05	0.053	0.053	0.044	3e-3	3e-3
Steady state (sec)	25.2	25.5	25.4	30	29.3	29.3
System with BESS						
Overshoot (Hz)	0.32	0.34	0.34	0.008	0.02	0.02
Undershoot (Hz)	0.045	0.05	0.05	0.04	3e-3	3e-3
Steady state (sec)	24.6	24.5	24.3	30	30.2	30.1
System with BESS & FESS						
Overshoot (Hz)	0.29	0.3	0.3	7e-3	0.019	0.019
Undershoot (Hz)	0.04	0.046	0.046	0.039	3e-3	3e-3
Steady state (sec)	27	25.7	25.9	30.7	29.7	29.8
System with BESS, FESS & EVs						
Overshoot (Hz)	1e-4	1e-4	1e-4	0	1e-5	1e-5
Undershoot (Hz)	0	0	0	2e-5	0	0
Steady state (sec)	18	17.3	17	21.7	21.9	22.3

6.9.1 System model under study

6.9.1.1 PV system

The PV array operates under standard test conditions (1000 W/m^2, 25°C) and uses MPPT control. The AC-to-DC voltage ratio is defined in Eq. (6.1).

$$X = \frac{V_{dc}}{V_{ac}} \tag{6.1}$$

The inverter gain is expressed in Eq. (6.2).

$$G_2(s) = \frac{s^2}{s^2 + W^2} \tag{6.2}$$

Assuming resistive loading, the instantaneous power is modeled by Eq. (6.3).

$$P(s) = \frac{V_m I_m}{2s}\left(\frac{s^2}{s^2 + (2W)^2}\right) \tag{6.3}$$

The ratio between average and instantaneous power is given by Eq. (6.4).

$$G_4(s) = \frac{s^2 + (4W)^2}{2\left(s^2 + (2W)^2\right)} \tag{6.4}$$

6.9.1.2 *Wind system*

The output mechanical power from the wind turbine is described in Eq. (6.5), which depends on air density, turbine area, power coefficient, and effective wind speed.

$$P_w = \frac{1}{2}\rho A_T C_p(\lambda, \beta) V^3{}_{\text{wake}} \tag{6.5}$$

Assuming lossless generator operation, the electrical output power and generator torque dynamics are captured in Eq. (6.6).

$$P_w = T_{\text{gen}}\omega_g, \frac{dT_{\text{gen}}}{dt} = -\frac{1}{\tau_{\text{gen}}}T_{\text{gen}} + \frac{1}{\tau_{\text{gen}}}T_{\text{control}} \tag{6.6}$$

6.9.1.3 *Thermal power plant*

In thermal systems, the governor response is based on power reference and frequency deviation as shown in Eq. (6.7).

$$\Delta P_g(s) = \Delta P_{\text{ref}}(s) - \frac{1}{R}\Delta f(s) \tag{6.7}$$

The overall response of the governor, turbine, and reheater is approximated using a general first-order system, as given in Eq. (6.8).

$$H(s) = \frac{G}{Ts + 1} \tag{6.8}$$

Area Control Error (ACE) is used to regulate frequency deviations across interconnected areas, as defined in Eq. (6.9).

$$\text{ACE} = B.\Delta f + \Delta P_{\text{tie}} \tag{6.9}$$

6.9.1.4 Electric vehicle model

EVs are modeled as aggregated battery units. The open-circuit voltage (OCV), which varies with the state of charge (SOC), is calculated using the Nernst equation shown in Eq. (6.10).

$$V_{OC}(\text{SOC}) = V_{\text{nomial}} + S\frac{RT}{F} ln\left(\frac{\text{SOC}}{C_{\text{nomial}} - \text{SOC}}\right) \tag{6.10}$$

This model supports accurate simulation of dynamic charging and discharging behavior in EVs.

6.9.2 Comparative analysis of system response under different ESS configurations

6.9.2.1 Scenario 1: Model without any ESS

In the first scenario, where the system operates without any ESS, a significantly ineffective response is observed across all areas. The frequency deviations are considerable, as shown in Fig. 6.9, with area one experiencing an overshoot of 0.37 Hz and an undershoot of 0.05 Hz. Similar trends are observed in areas two and 3, where the overshoot reaches 0.38 Hz, and the undershoot is slightly higher at 0.053 Hz. Tie-line power deviations also exhibit poor response, as ΔP_{tie1} shows an overshoot of 0.0075 and an undershoot of 0.044, while ΔP_{tie2} and ΔP_{tie3} both have overshoots of approximately 0.02 and undershoots of 0.0038, as shown in Table 6.11. Furthermore, the steady-state time for the system ranges from 25.2 to 30 s, reflecting a slow recovery from disturbances. These results demonstrate the inability of the microgrid to maintain frequency stability or effectively manage power fluctuations without ESS, highlighting the critical role of energy storage systems in addressing such challenges.

6.9.2.2 Scenario 2: Effect of adding BESS on the model performance

The integration of BESS in the second scenario significantly improves system performance. In area 1, the frequency's overshoot is reduced to 0.32 Hz, a 13.5% decrease compared to the no-ESS scenario, while the undershoot decreases to 0.045 Hz, as shown in Fig. 6.10. Areas two and three experience similar improvements, with overshoots of 0.34 Hz and undershoots of 0.05 Hz. These reductions in frequency excursions directly contribute to system stability, minimizing stress on generation units and

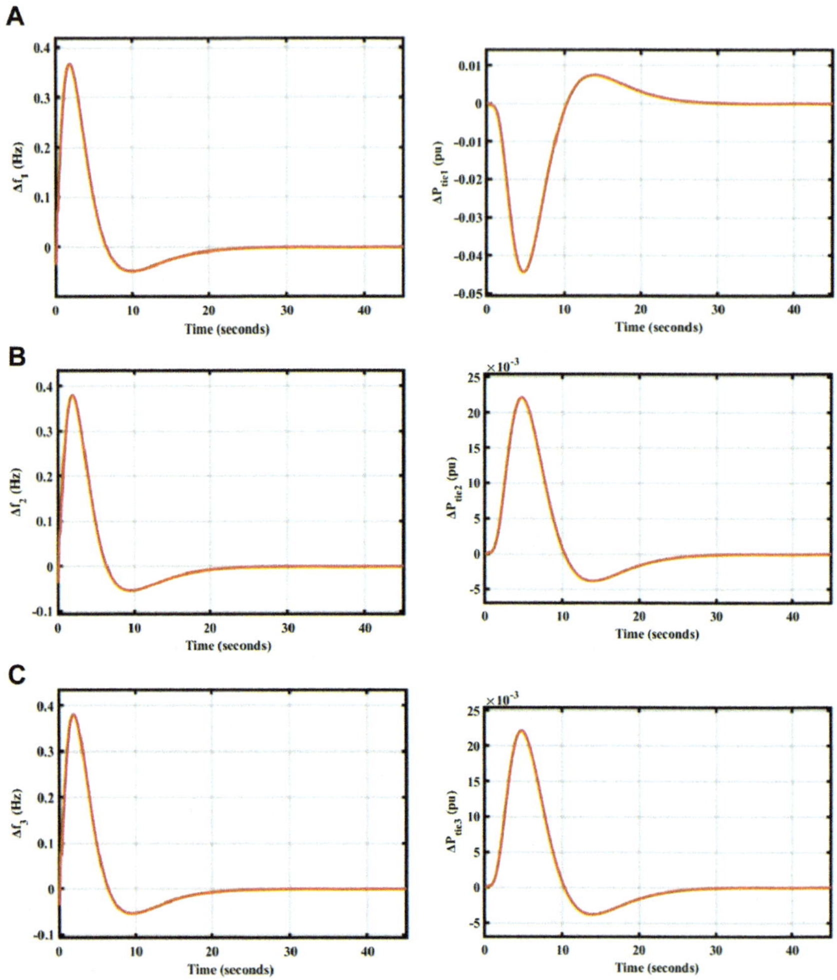

FIGURE 6.9 Frequency profiles and tie-line power deviations across all areas under 15% load perturbation without ESS. The figure illustrates substantial frequency overshoots (up to 0.38 Hz), undershoots, and long settling times (around 30 s), reflecting inadequate system response in the absence of energy storage systems.

load-following mechanisms. Although ΔP_{tie1} shows a slight increase in overshoot to 0.008, the undershoot improves to 0.04. For ΔP_{tie2} and ΔP_{tie3}, the overshoots remain at 0.02, with minor improvements in undershoots to 0.0039 and 0.0038, respectively. The steady-state time also improves slightly, ranging from 24.3 to 30.2 s, as detailed in Table 6.11. This faster settling time implies that the system reaches nominal conditions more quickly, reducing the duration of transient instability.

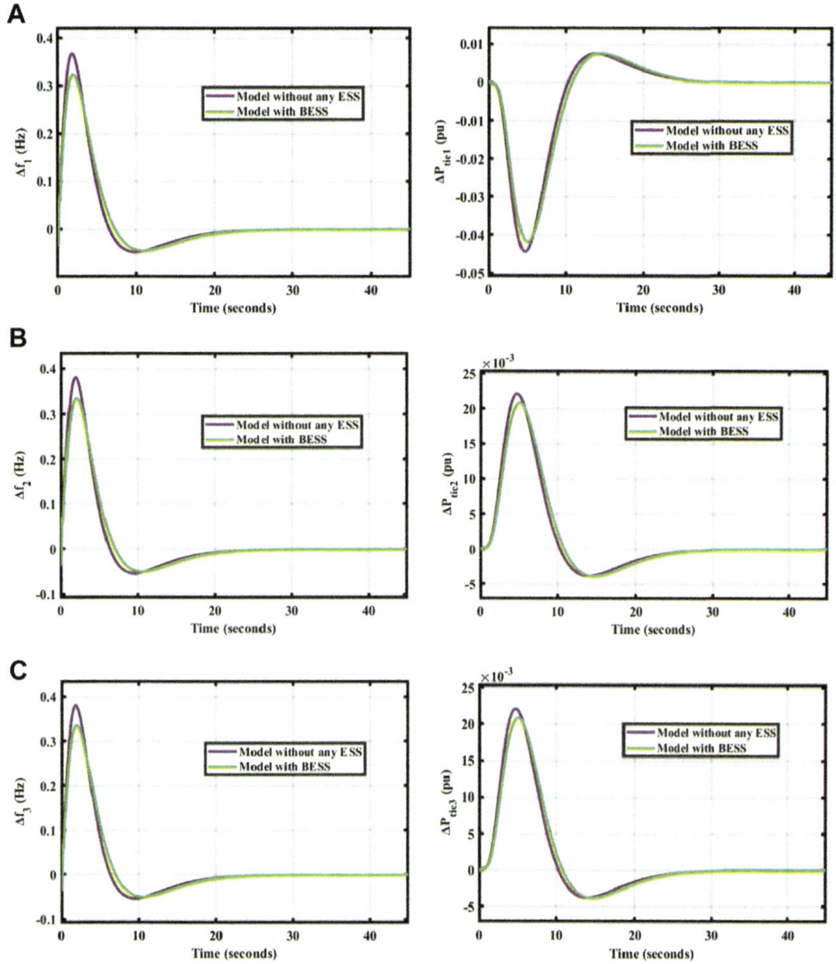

FIGURE 6.10 Impact of BESS integration on multi-area frequency regulation and tie-line dynamics under a 15% demand fluctuation. The integration of BESS reduces frequency overshoot by approximately 13%, shortens settling time to about 24.5 s, and smooths tie-line power profiles, enhancing transient performance.

These results indicate that BESS effectively mitigates frequency deviations and enhances dynamic response, particularly in Area 1, by absorbing surplus energy during generation surges and releasing it during demand peaks. The positive impact of BESS is evident in the smoother time-domain profiles and tighter frequency regulation. However, further optimization is needed to achieve faster stabilization and even smaller deviations.

6.9.2.3 Scenario 3: Effect of adding BESS and FESS on the model performance

The third scenario, which incorporates both BESS and FESS, demonstrates further enhancements in stability. In area 1, the frequency overshoot decreases to 0.29 Hz, while the undershoot drops to 0.04 Hz. Areas two and three exhibit similar trends, with overshoots of 0.3 Hz and undershoots of 0.046 Hz as depicted in Fig. 6.11.

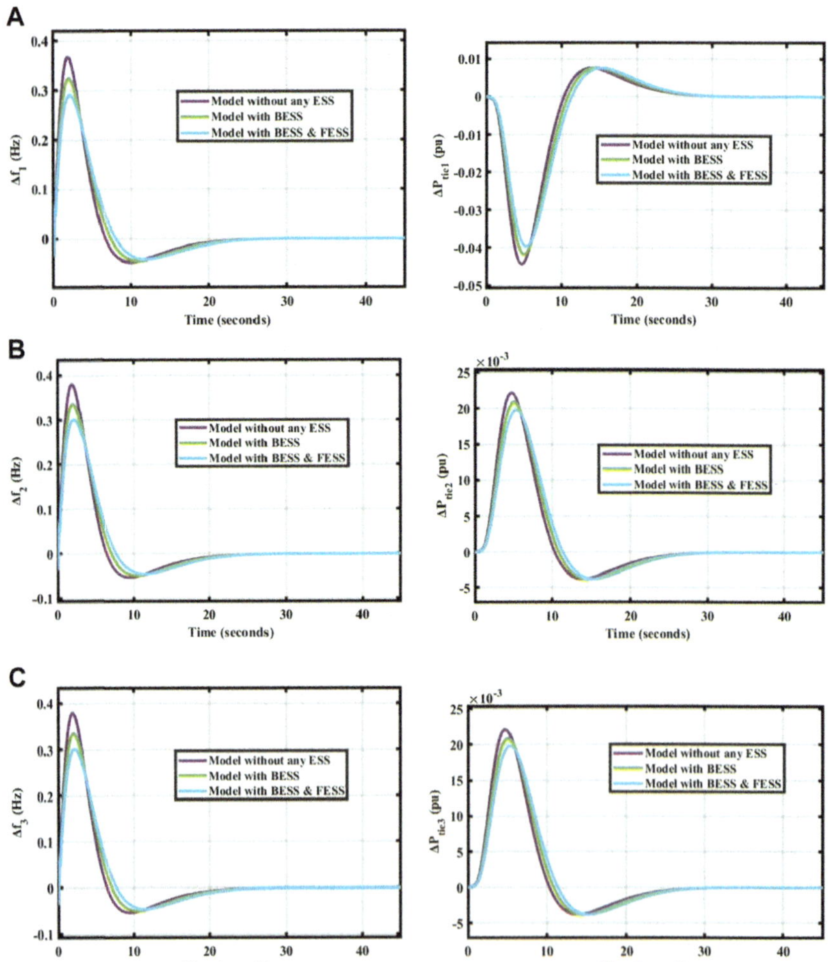

FIGURE 6.11 Frequency and power exchange behavior in a hybrid BESS–FESS setup during 15% load variation events. The combined use of BESS and FESS leads to further reduction in overshoot (as low as 0.29 Hz) and improved damping of oscillations, confirming the benefit of hybrid ESS in mitigating rapid and sustained disturbances.

Compared to the no-ESS and single-ESS cases, this scenario achieves a more rapid damping of oscillations and reduced transient swings. Tie-line power deviations also show improvement. For ΔP_{tie1}, the overshoot remains at 0.0075, but the undershoot decreases to 0.039. Similarly, ΔP_{tie2} and ΔP_{tie3} have overshoots of 0.0198 and undershoots of 0.0038. The steady-state time ranges from 25.7 to 30.7 s as presented in Table 6.11. Although marginally longer in some areas, the improved frequency and tie-line profiles reflect better energy balancing and robustness.

The combination of BESS and FESS leverages the complementary strengths of these technologies: BESS addresses sustained energy needs, while FESS effectively mitigates rapid fluctuations. The dual-ESS setup reduces frequency deviations and improves tie-line power regulation, demonstrating enhanced oscillation suppression and stability. The compounded benefit is most apparent in the smoother frequency curves and attenuated power swings, affirming the value of hybrid ESS integration.

6.9.2.4 Scenario 4: Effect of adding BESS, FESS and EVs on the model performance

In the final scenario, the integration of EVs alongside BESS and FESS delivers the most impressive results, achieving near-perfect frequency and tie-line power regulation as depicted in Fig. 6.12. According to Table 6.11, the frequency overshoot in all areas was reduced to 0.00,014 Hz, compared to 0.29–0.3 Hz with BESS and FESS, and 0.37–0.38 Hz in the base case without ESS. Likewise, the steady-state settling time improved significantly from 25.7 to 27 s (BESS + FESS) to just 17–18 s when EVs were included. Tie-line power deviations were also nearly eliminated: for instance, ΔP_{tie1} overshoot was reduced to 0, and ΔP_{tie2} and ΔP_{tie3} to as low as 1×10^{-5}. These results demonstrate a remarkable enhancement in dynamic stability and frequency response attributable to EVs integration. The improvement is especially notable considering that all other simulation parameters remained identical across scenarios, affirming the effectiveness of EVs as mobile storage resources that contribute significantly to load-frequency control and grid robustness.

To improve the clarity of Fig. 6.12, each subplot includes an inset magnification focusing on the initial 0–10 s is included, where key transient characteristics such as overshoot and settling behavior are most prominent. This allows for clearer visual differentiation among system configurations, especially highlighting the superior damping performance of the BESS + FESS + EVs setup. The figure also uses a consistent time axis and color-coded legend across all six subplots to support immediate comparison. Readers can track the progressive improvement in system dynamics from the violet curve (no ESS) to the maroon curve (full ESS), which demonstrates the smallest deviation range and quickest stabilization. The inclusion of EVs introduces additional flexibility to the

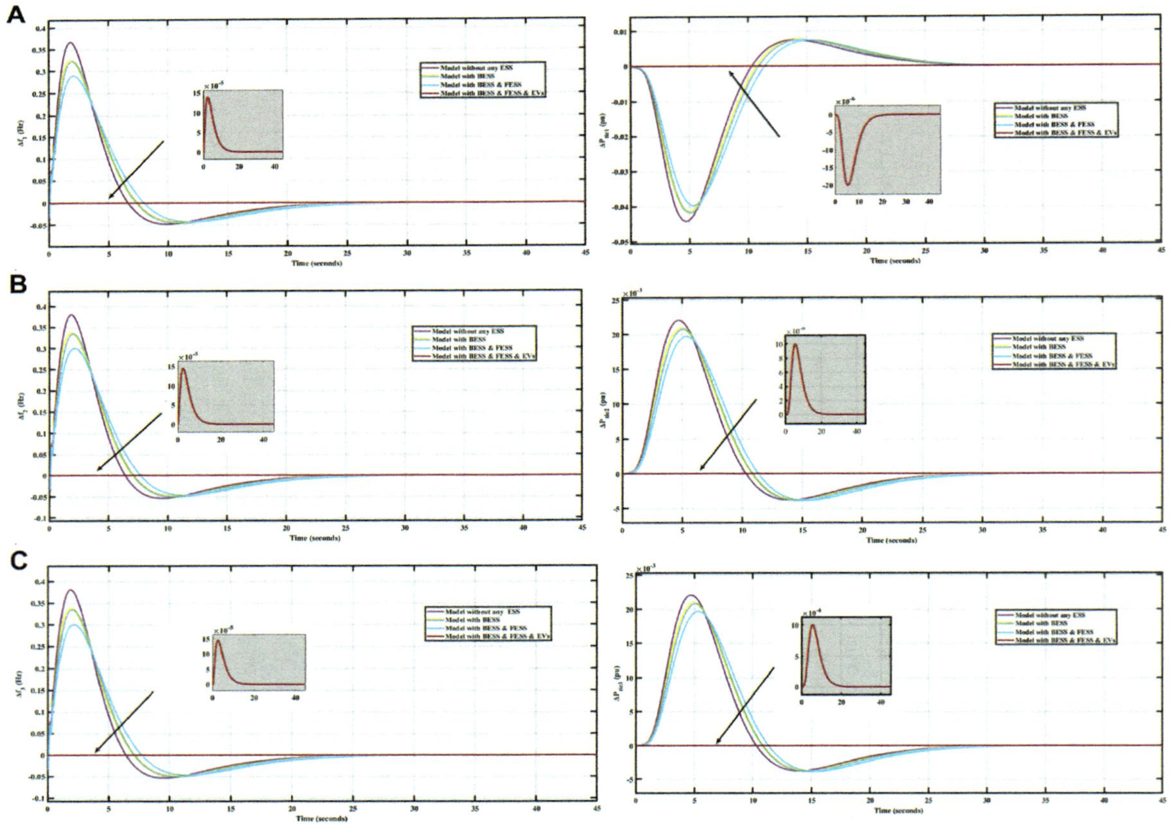

FIGURE 6.12 Enhanced system frequency response and tie-line power stabilization with BESS, FESS, and EVs support under a 15% load change. This scenario achieves near-perfect regulation with minimal overshoot (~0.00,014 Hz), fastest settling time (17–18 s), and negligible tie-line deviations, demonstrating the robustness and effectiveness of full hybrid ESS implementation.

system, as these mobile energy storage units contribute to balancing load and generation with remarkable efficiency. This configuration not only stabilizes the system but also demonstrates the transformative potential of EVs in achieving grid reliability and sustainability. In summary, the results clearly show a progressive enhancement in system stability as more advanced ESS configurations are integrated. The inclusion of BESS addresses basic stability needs, while the addition of FESS provides further improvements in dynamic response. The integration of EVs achieves optimal performance, effectively eliminating frequency and tie-line power deviations while significantly reducing stabilization time. These insights, together with the enhanced explanation of, ensure that the performance benefits of each ESS configuration are clearly communicated, reinforcing the importance of hybrid ESS integration in renewable-rich microgrids.

6.10 Future trends and emerging technologies

The future of ESS in microgrids is shaped by advancements such as autonomous EVs, wireless charging systems, and predictive control algorithms (Hadjipaschalis et al., 2009). Autonomous EVs can optimize their charging and discharging schedules based on real-time grid conditions, ensuring efficient energy utilization. Wireless charging technology further enhances the integration of EVs by enabling dynamic energy exchange without the need for physical connections. Predictive control systems are revolutionizing microgrid management by forecasting demand and renewable energy production. This allows ESS to adapt proactively, enhancing grid stability and efficiency. Furthermore, decentralized energy management systems are enabling microgrid components to operate independently while maintaining overall system cohesion and improving resilience during disturbances. Rapidly developing energy storage technologies might enhance microgrid frequency management, usefulness, and stability (Hadjipaschalis et al., 2009). Emerging trends in ESS for microgrid frequency control include.

- **Solid-State Batteries**: Solid-state batteries, with higher energy density, safety, and longevity than lithium-ion batteries, provide stable, long-term storage for high-demand microgrids, improving grid stability and efficiency.
- **Advanced Hydrogen Storage Systems**: Hydrogen storage converts surplus power into hydrogen for emission-free microgrid backup and load balancing.
- **Autonomous and Connected Electric Vehicles (EVs)**: Autonomous connected EVs optimize charging based on real-time needs to balance loads and boost energy availability.

- **Wireless Charging Technology for EVs**: Wireless charging lets EVs dynamically connect to the grid for smooth energy transfer, improving frequency regulation reaction times and facilitating renewable-heavy microgrids.
- **Predictive and Autonomous Control Systems**: Advanced control systems estimate demand and adapt ESS operations in real time to increase grid reliability and efficiency during renewable variations.
- **Decentralized Energy Management**: Decentralized energy management allows microgrid components to operate independently yet coordinate with the main grid, improving resilience and stability during disturbances.
- **Vehicle-to-Grid (V2G) Technology**: EVs can give power back to the grid via V2G to help manage frequency by boosting reserves during high demand and stabilizing load variations (Alfaverh et al., 2023).

ESS technology in renewable-heavy microgrids can improve energy efficiency, stability, and sustainability with these developments, as shown in Table 6.12.

TABLE 6.12 Emerging trends in ESS for microgrid frequency control.

Technology	Potential impact on frequency regulation
Solid-state batteries	Enhanced safety, longer life cycle, and higher energy density (Yang et al., 2024b)
Advanced hydrogen storage systems	Sustainable, large-scale, long-term energy storage, supporting backup and load balancing (Hossain et al., 2023)
Autonomous and connected EVs	Optimized charging and discharging scheduling, improved frequency support (Benhammou et al., 2023)
Wireless charging for EVs	Dynamic energy exchange, enabling flexible frequency regulation (Sagar et al., 2023)
Predictive and autonomous control systems	Real-time demand forecasting, optimized ESS operation, enhanced reliability (Chapaloglou et al., 2024)
Decentralized energy management	Independent operation of microgrid components, increased resilience and stability (Shirkhani et al., 2023)
Vehicle-to-grid (V2G) technology	Mobile energy storage, frequency support during peak demand and load fluctuations (Jie et al., 2024)

6.11 Conclusion

This chapter has comprehensively analyzed the role of ESS in microgrid frequency control, providing insights into their critical importance for addressing the challenges posed by renewable energy integration. Key findings demonstrate the effectiveness of ESS technologies, including BESS, FESS, and supercapacitors, in enhancing frequency stability and dynamic response under varying load and renewable energy conditions. The incorporation of Electric Vehicles (EVs) into microgrid systems further highlights their trans-formative potential as mobile energy storage units, enabling flexible and efficient frequency regulation. Emerging ESS technologies, such as solid-state batteries and hydrogen storage systems, present promising avenues for addressing long-term energy storage needs and improving grid reliability. However, challenges such as cost, scalability, and integration complexity remain significant barriers to widespread adoption. Although advanced control strategies including model predictive control and artificial intelligence-driven algorithms offer potential solutions for optimizing ESS operations in diverse microgrid environments, future research should focus on addressing these challenges by: exploring cost-effective ESS solutions that balance performance with affordability, investigating scalable and modular ESS designs suitable for various microgrid configurations, enhancing predictive and autonomous control systems to manage ESS operations more effectively under dynamic and uncertain conditions, and finally examining the long-term environmental and economic impacts of novel ESS technologies, such as hydrogen storage and solid-state batteries. By bridging these knowledge gaps, the field can move closer to realizing resilient, efficient, and sustainable microgrids, solidifying the role of ESS as a cornerstone of modern energy systems.

References

Abadlia, I., Bahi, T., Bouzeria, H., 2016. Energy management strategy based on fuzzy logic for compound RES/ESS used in stand-alone application. Int. J. Hydrogen Energy 41 (38), 16705–16717. https://doi.org/10.1016/j.ijhydene.2016.07.120.

Abouzeid, S.I., Guo, Y., Zhang, H.C., 2021. Cooperative control framework of the wind turbine generators and the compressed air energy storage system for efficient frequency regulation support. Int. J. Electr. Power Energy Syst. 130. https://doi.org/10.1016/j.ijepes.2021.106844.

Akram, U., Nadarajah, M., Shah, R., Milano, F., 2020. A review on rapid responsive energy storage technologies for frequency regulation in modern power systems. Renew. Sustain. Energy Rev. 120, 109626. https://doi.org/10.1016/j.rser.2019.109626.

Alfaverh, F., Denaï, M., Sun, Y., 2023. Optimal vehicle-to-grid control for supplementary frequency regulation using deep reinforcement learning. Elec. Power Syst. Res. 214, 108949. https://doi.org/10.1016/j.epsr.2022.108949.

Ali, M., Khalid, M., Gulzar, M.M., 2024. Transforming the Grid: AI, ML, Renewable, Storage, EVs, and Prosumers Intelligent Learning Approaches for Renewable and Sustainable Energy. Elsevier, Saudi Arabia, pp. 3–30. https://doi.org/10.1016/B978-0-443-15806-3.00001-2.

Alirezaei, R., Dashti, R., Mirhosseini, M., 2025. Analysis of energy supply scenarios using grid-connected hybrid systems: investigating time-of-use demand response strategy for improving operational performance. Energy 318, 134596. https://doi.org/10.1016/j.energy.2025.134596.

Antony, A.S.M., Sumithra, J., Rayudu, K., Dhanalakshmi, G., Agilesh Saravanan, R., 2024. Dynamic and model predictive controllers for frequency regulation of an isolated micro—grid with electrical vehicles and the ess inte- gration. Elec. Power Compon. Syst. 52 (3), 426–444.

Benhammou, A., Hartani, M.A., Tedjini, H., Rezk, H., Al-Dhaifallah, M., of Autonomy, I., 2023. Efficiency, and stress of fuel cell hybrid electric vehicle system using robust controller. Sustainability 15 (7), 5657. https://doi.org/10.3390/su15075657.

Calero, I., Canizares, C.A., Bhattacharya, K., 2019. Compressed air energy storage system modeling for power system studies. IEEE Trans. Power Syst. 34 (5), 3359–3371. https://doi.org/10.1109/tpwrs.2019.2901705.

Castillo, A., Gayme, D.F., 2014. Grid-scale energy storage applications in renewable energy integration: a survey. Energy Convers. Manag. 87, 885–894. https://doi.org/10.1016/j.enconman.2014.07.063.

Çelik, E., Öztürk, N., Houssein, E.H., 2023. Improved load frequency control of interconnected power systems using energy storage devices and a new cost function. Neural Computing and Applications 35 (1), 681–697. https://doi.org/10.1007/s00521-022-07813-1.

Chapaloglou, S., Alves, E.F., Trovato, V., Tedeschi, E., 2024. Optimal energy management in autonomous power systems with probabilistic security constraints and adaptive frequency control. IEEE Trans. Power Syst. 39 (1), 1543–1554. https://doi.org/10.1109/TPWRS.2023.3236378.

Chen, H., Cong, T.N., Yang, W., Tan, C., Li, Y., Ding, Y., 2009. Progress in electrical energy storage system: a critical review. Prog. Nat. Sci. 19 (3), 291–312. https://doi.org/10.1016/j.pnsc.2008.07.014.

Chidambaram, I.A., Paramasivam, B., 2013. Optimized load-frequency simulation in restructured power system with redox flow batteries and interline power flow controller. Int. J. Electr. Power Energy Syst. 50 (1), 9–24. https://doi.org/10.1016/j.ijepes.2013.02.004.

Chowdhury, M.R., Jobayer, A.M., Islam, M.S., Zhao, L., 2022. Potential Applications and Impacts of Solid-State Energy Storage in Power Grids Green Sustainable Process for Chemical and Environmental Engineering and Science: Solid-State Energy Storage - a Path to Environmental Sustainability. Elsevier, United States, pp. 319–342. https://doi.org/10.1016/B978-0-323-90635-7.00012-9.

Demiroren, A., Zeynelgil, H.L., Sengor, N.S., 2001. The application of ANN technique to load-frequency control for three-area power system. IEEE Porto Power Tech. Proc. 2, 465–469. https://doi.org/10.1109/PTC.2001.964793.

Dong, M., Liu, K., He, J., Li, X., Qiao, L., Zhao, X., 2022. Power grid load frequency control based on fractional order pid combined with pumped storage and battery energy storage. In: The 3rd International Conference on Power and Electrical Engineering, 9, pp. 1402–1411.

Ehsan, F., Habib, S., Gulzar, M.M., Guo, J., Muyeen, S., Kamwa, I., 2024. Assessing pol- icy influence on electric vehicle adoption in China: an in-depth study. Energy Strategy Rev. 54.

Elalfy, D.A., Gouda, E., Kotb, M.F., Bureš, V., Sedhom, B.E., 2024. Comprehensive review of energy storage systems technologies, objectives, challenges, and future trends. Energy Strategy Rev. 54. https://doi.org/10.1016/j.esr.2024.101482.

Gao, M., Han, Z., Zhao, B., Li, P., Wu, Di, Li, P., 2023. Optimal planning method of multi-energy storage systems based on the power response analysis in the integrated energy system. J. Energy Storage 73, 109015. https://doi.org/10.1016/j.est.2023.109015.

Ghafoor, R., Guanghua, L., Gulzar, M.M., Irfan, R., Alqahtani, M., Khalid, M., 2024. Load frequency control of multi-area power system incorporated renewable energy considering electrical vehicle effect using modified cascaded controller tuned by BESSO algorithm. Heliyon 10 (11). https://doi.org/10.1016/j.heliyon.2024.e31840.

Gulzar, M.M., Murawwat, S., Sibtain, D., Shahid, K., Javed, I., Gui, Y., 2022. Modified cascaded controller design constructed on fractional operator 'β' to mitigate frequency fluctuations for sustainable operation of power systems. Arabia Ener. 15 (20). https://doi.org/10.3390/en15207814.

Gulzar, M.M., Iqbal, A., Sibtain, D., Khalid, M., 2023a. An innovative converterless solar PV control strategy for a grid connected hybrid PV/Wind/Fuel-Cell system coupled with battery energy storage. Arabia IEEE Acc. 11, 23245–23259. https://doi.org/10.1109/ACCESS.2023.3252891.

Gulzar, M.M., Sibtain, D., Khalid, M., 2023b. Cascaded fractional model predictive controller for load frequency control in multiarea hybrid renewable energy system with uncertainties. Int. J. Energy Res. 2023. https://doi.org/10.1155/2023/5999997.

Gulzar, M.M., Umar, M.M., Al-Dhaifallah, M.M., 2023c. Arabia robust load frequency control of hybrid power system. In: International Conference on Control, Automation and Diagnosis, ICCAD 2023. https://doi.org/10.1109/ICCAD57653.2023.10152415.

Gulzar, M.M., Sibtain, D., Al-Dhaifallah, M., Alismail, F., Khalid, M., 2024. A new optimal 3° of freedom fractional order proportion integral derivative controller with model predictive controller for frequency regulation in high penetrated renewable based interconnected system. Comput. Electr. Eng. 119, 109651. https://doi.org/10.1016/j.compeleceng.2024.109651.

Hadjipaschalis, I., Poullikkas, A., Efthimiou, V., 2009. Overview of current and future energy storage technologies for electric power applications. Renew. Sustain. Energy Rev. 13 (6–7), 1513–1522. https://doi.org/10.1016/j.rser.2008.09.028.

Hartani, M.A., Rezk, H., Benhammou, A., Hamouda, M., Abdelkhalek, O., Mekhilef, S., Olabi, A.G., 2023. Proposed frequency decoupling-based fuzzy logic control for power allocation and state-of-charge recovery of hybrid energy storage systems adopting multi-level energy management for multi-DC-microgrids. Energy 278, 127703. https://doi.org/10.1016/j.energy.2023.127703.

Hossain, M.A., Islam, M.R., Hossain, M.A., Hossain, M.J., 2023. Control strategy review for hydrogen-renewable energy power system. J. Energy Storage 72. https://doi.org/10.1016/j.est.2023.108170.

Jameel, A., Gulzar, M.M., 2023. Load frequency regulation of interconnected muli-source multi-area power system with penetration of electric vehicles aggregator model. Electr. Eng. 105 (6), 3951–3968. https://doi.org/10.1007/s00202-023-01923-2.

Ji, W., Hong, F., Zhao, Y., Liang, L., Du, H., Hao, J., Fang, F., Liu, J., 2024. Applications of flywheel energy storage system on load frequency regulation combined with various power generations: a review. Renew. Energy 223. https://doi.org/10.1016/j.renene.2024.119975.

Jie, B., Baba, J., Kumada, A., 2024. Contribution to V2G system frequency regulation by charging/discharging control of aggregated EV group. IEEE Trans. Ind. Appl. 60 (1), 1129–1140. https://doi.org/10.1109/TIA.2023.3292814.

Khan, I.A., Mokhlis, H., Mansor, N.N., Illias, H.A., Jamilatul Awalin, L., Wang, L., 2023. New trends and future directions in load frequency control and flexible power system: a comprehensive review. Alex. Eng. J. 71, 263–308. https://doi.org/10.1016/j.aej.2023.03.040.

Khooban, M.H., Niknam, T., Blaabjerg, F., Davari, P., Dragicevic, T., 2016. A robust adaptive load frequency control for micro-grids. ISA Trans. 65, 220–229. https://doi.org/10.1016/j.isatra.2016.07.002.

Kousksou, T., Bruel, P., Jamil, A., El Rhafiki, T., Zeraouli, Y., 2014. Energy storage: applications and challenges. Sol. Energy Mater. Sol. Cell. 120, 59–80. https://doi.org/10.1016/j.solmat.2013.08.015.

Kumar Bhoi, S., Rath, S., Badatia, S., 2024. Lead-acid battery and supercapacitor based hybrid energy storage systems in microgrid for energy control system. Int. Res. J. Adv. Eng. Hub (IRJAEH) 2 (04), 1037–1044. https://doi.org/10.47392/irjaeh.2024.0144.

Lei, M., Meng, K., Feng, H., Bai, J., Jiang, H., Zhang, Z., 2023. Flywheel energy storage controlled by model predictive control to achieve smooth short-term high-frequency wind power. J. Energy Storage 63, 106949. https://doi.org/10.1016/j.est.2023.106949.

Li, S., Xu, Q., Huang, J., 2023. Research on the integrated application of battery energy storage systems in grid peak and frequency regulation. J. Energy Storage 59, 106459. https://doi.org/10.1016/j.est.2022.106459.

Liu, J., Hou, Y., Guo, J., Liu, X., Liu, J., 2023a. A cost-efficient virtual synchronous generator system based on coordinated photovoltaic and supercapacitor. IEEE Trans. Power Electron. 38 (12), 16219–16229. https://doi.org/10.1109/TPEL.2023.3317497.

Liu, J., Hu, H., Yu, S.S., Trinh, H., 2023b. Virtual power plant with renewable energy sources and energy storage systems for sustainable power grid-formation, control techniques and demand response. Energies 16 (9). https://doi.org/10.3390/en16093705.

Liu, J., Jia, Y., Han, X., Wang, P., 2023c. Power grid frequency regulation strategy of hybrid energy storage considering efficiency evaluation. J. Energy Storage 74, 109418. https://doi.org/10.1016/j.est.2023.109418.

Liu, X., Wang, C., Kong, X., Zhang, Y., Wang, W., Lee, K.Y., 2024. Tube-based distributed MPC for load frequency control of power system with high wind power penetration. IEEE Trans. Power Syst. 39 (2), 3118–3129. https://doi.org/10.1109/TPWRS.2023.3277997.

Makahleh, F.M., Amer, A., Manasrah, A.A., Attar, H., Solyman, A.A.A., Kamarposhti, M.A., Thounthong, P., 2023. Optimal management of energy storage systems for peak shaving in a smart grid. Comput. Mater. Continua 75 (2), 3317–3337. https://doi.org/10.32604/cmc.2023.035690.

Makola, C.S., Le Roux, P.F., Jordaan, J.A., 2023. Comparative analysis of lithium-ion and lead–acid as electrical energy storage systems in a grid-tied microgrid application. Appl. Sci. 13 (5), 3137. https://doi.org/10.3390/app13053137.

Mu, C., Tang, Y., He, H., 2017. Improved sliding mode design for load frequency control of power system integrated an adaptive learning strategy. IEEE Trans. Ind. Electron. 64 (8), 6742–6751. https://doi.org/10.1109/TIE.2017.2694396.

Naderipour, A., Abdul-Malek, Z., Davoodkhani, I.F., Kamyab, H., Ali, R.R., 2023. Load- frequency control in an islanded microgrid pv/wt/fc/ess using an optimal self-tuning fractional-order fuzzy controller. Environ. Sci. Pollut. Control Ser. 30.

Nyamathulla, S., Dhanamjayulu, C., 2024. A review of battery energy storage systems and advanced battery management system for different applications: challenges and recommendations. J. Energy Storage 86, 111179. https://doi.org/10.1016/j.est.2024.111179.

Oshnoei, A., Kheradmandi, M., Muyeen, S.M., 2020. Robust control scheme for distributed battery energy storage systems in load frequency control. IEEE Trans. Power Syst. 35 (6), 4781–4791. https://doi.org/10.1109/TPWRS.2020.2997950.

Oshnoei, S., Oshnoei, A., Mosallanejad, A., Haghjoo, F., 2021. Novel load frequency control scheme for an interconnected two-area power system including wind turbine generation and redox flow battery. Int. J. Electr. Power Energy Syst. 130, 107033. https://doi.org/10.1016/j.ijepes.2021.107033.

Ram Babu, N., Bhagat, S.K., Saikia, L.C., Chiranjeevi, T., Devarapalli, R., García Márquez, F. P., 2023. A comprehensive review of recent strategies on automatic generation control/load frequency control in power systems. Arch. Comput. Methods Eng. 30 (1), 543–572. https://doi.org/10.1007/s11831-022-09810-y.

Ribeiro, P.F., Johnson, B.K., Crow, M.L., Arsoy, A., Liu, Y., 2001. Energy storage systems for advanced power applications. Proc. IEEE 89 (12), 1744–1756. https://doi.org/10.1109/5.975900.

Rimsha, Murawwat, S., Gulzar, M.M., Alzahrani, A., Hafeez, G., Khan, F.A., Abed, A.M., 2023. State of charge estimation and error analysis of lithium-ion batteries for electric vehicles using Kalman filter and deep neural network. J. Energy Storage 72. https://doi.org/10.1016/j.est.2023.108039.

Ruban, N., Rudnik, V., Askarov, A., Maliuta, B., 2023. Frequency control by the PV station in electric power systems with hydrogen energy storage. Feder. Int. J. Hydro. Energy 48 (73), 28262–28276. https://doi.org/10.1016/j.ijhydene.2023.04.048.

Safari, A., Babaei, F., Farrokhifar, M., 2021. A load frequency control using a PSO-based ANN for micro-grids in the presence of electric vehicles. Int. J. Ambient Energy 42 (6), 688–700. https://doi.org/10.1080/01430750.2018.1563811.

Sagar, A., Kashyap, A., Nasab, M.A., Padmanaban, S., Bertoluzzo, M., Kumar, A., Blaabjerg, F., 2023. A comprehensive review of the recent development of wireless power transfer technologies for electric vehicle charging systems. IEEE Acc. 11, 83703–83751. https://doi.org/10.1109/ACCESS.2023.3300475.

Sheelam, A., McLeod, W.T., Badam, R., King, M., Bell, J.G., 2023. Comparison between Supercapacitors and Other Energy Storing Electrochemical Devices. Elsevier BV, pp. 673–712. https://doi.org/10.1016/b978-0-323-90530-5.00022-8.

Shim, J.W., Verbic, G., Kim, H., Hur, K., 2019. On droop control of energy-constrained battery energy storage systems for grid frequency regulation. IEEE Acc. 7, 166353–166364. https://doi.org/10.1109/ACCESS.2019.2953479.

Shirkhani, M., Tavoosi, J., Danyali, S., Sarvenoee, A.K., Abdali, A., Mohammadzadeh, A., Zhang, C., 2023. A review on microgrid decentralized energy/voltage control structures and methods. Energy Rep. 10, 368–380. https://doi.org/10.1016/j.egyr.2023.06.022.

Shukla, H., Raju, M., 2024. Combined frequency and voltage regulation in an interconnected power system using fractional order cascade controller considering renewable energy sources, electric vehicles and ultra capacitor. J. Energy Storage 84, 110875. https://doi.org/10.1016/j.est.2024.110875.

Siti, M., Mbungu, N., Tungadio, D., Banza, B., 2022. Ngoma, Application of load fre- quency control method to a multi-microgrid with energy storage system. J. Energy Storage 52.

Song, H., Gu, M., Liu, C., Amani, A.M., Jalili, M., Meegahapola, L., Yu, X., Dickeson, G., 2023. Multi-objective battery energy storage optimization for virtual power plant applications. Appl. Energy 352. https://doi.org/10.1016/j.apenergy.2023.121860.

Terkes, M., Arikan, O., Gokalp, E., 2024. The effect of electric vehicle charging demand variability on optimal hybrid power systems with second-life lithium-ion or fresh Na–S batteries considering power quality. Energy 288, 129760. https://doi.org/10.1016/j.energy.2023.129760.

Tolba, M.S., Fanni, M., Nasser, G.A., Umezu, S., Fath El-Bab, A.M.R., 2022. IEEE computer society Egypt design, fabrication, and control of micro-heater based on joule effect for low-cost medical device. In: IECON Proceedings (Industrial Electronics Conference), 2022. https://doi.org/10.1109/IECON49645.2022.9969048.

Tolba, M.S., Gulzar, M.M., Arishi, A., Soliman, M., Murtaza, A.F., 2025. A novel MPC-based cascaded control for multi-area smart grids: tackling renewable energy and EV integration challenges. ISA Trans. https://doi.org/10.1016/j.isatra.2025.06.024.

Tolba, M.S., Al-Wajih, Y.A., Shafiullah, M., Al-Dhaifallah, M.M., 2025. Arabia identifying the magnetic levitation system (33-210) through neural Networks. Transp. Res. Procedia 84, 59–64. https://doi.org/10.1016/j.trpro.2025.03.045.

Ullah, S., Khan, L., Sami, I., Ro, J.S., 2022. Voltage/frequency regulation with optimal load dispatch in microgrids using SMC based distributed cooperative control. IEEE Acc. 10, 64873–64889. https://doi.org/10.1109/ACCESS.2022.3183635.

Vladimir, P., Iurii, P., 2023. Supercapacitor energy storages in hybrid power supplies for frequency-controlled electric drives: review of topologies and automatic control systems. Energies 16 (7), 3287. https://doi.org/10.3390/en16073287.

Wali, S.B., Hannan, M.A., Jern Ker, P., Rahman, S.A., Le, K.N., Begum, R.A., Tiong, S.K., Indra Mahlia, T.M., 2024. Grid-connected lithium-ion battery energy storage system towards sustainable energy: a patent landscape analysis and technology updates. J. Energy Storage 77, 109986. https://doi.org/10.1016/j.est.2023.109986.

Wang, S., Li, F., Zhang, G., Yin, C., 2023. Analysis of energy storage demand for peak shaving and frequency regulation of power systems with high penetration of renewable energy. Energy 267, 126586. https://doi.org/10.1016/j.energy.2022.126586.

Xu, Y., Li, C., Wang, Z., Zhang, N., Peng, B., 2018. Load frequency control of a novel renewable energy integrated micro-grid containing pumped hydropower energy storage. IEEE Acc. 6, 29067–29077. https://doi.org/10.1109/ACCESS.2018.2826015.

Xu, K., Guo, Y., Lei, G., Zhu, J., 2023. A review of flywheel energy storage system technologies. Energies 16 (18), 6462. https://doi.org/10.3390/en16186462.

Yang, H., Chen, Q., Liu, Y., Ma, Y., Zhang, D., 2024a. Demand response strategy of user-side energy storage system and its application to reliability improvement. J. Energy Storage 92, 112150. https://doi.org/10.1016/j.est.2024.112150.

Yang, R., Xie, Y., Li, K., Tran, M.K., Fowler, M., Panchal, S., Deng, Z., Zhang, Y., 2024b. Comparative study on the thermal characteristics of solid-state lithium-ion batteries. IEEE Trans. Transp. Electrif. 10 (1), 1541–1557. https://doi.org/10.1109/TTE.2023.3289997.

Zhang, S., Mishra, Y., Shahidehpour, M., 2016. Fuzzy-Logic based frequency controller for wind farms augmented with energy storage systems. IEEE Trans. Power Syst. 31 (2), 1595–1603. https://doi.org/10.1109/TPWRS.2015.2432113.

Zhang, H., Liu, J., Xu, S., 2020. H-infinity load frequency control of networked power systems via an event-triggered scheme. IEEE Trans. Ind. Electron. 67 (8), 7104–7113. https://doi.org/10.1109/TIE.2019.2939994.

Zhao, C., Andersen, P.B., Træholt, C., Hashemi, S., 2023. Grid-connected battery energy storage system: a review on application and integration. Renew. Sustain. Energy Rev. 182. https://doi.org/10.1016/j.rser.2023.113400.

CHAPTER

7

Design and implementation of optimized cascaded controllers for frequency stabilization in modern microgrid systems

K. Peddakapu[1,2], *Dalila Mat Said*[1,2,5], *P. Srinivasarao*[3], *and D. J Krishna Kishore*[4]

[1] Centre of Electrical Energy Systems (CEES), Institute of Future Energy (IFE), Universiti Teknologi Malaysia (UTM), Johor Bahru, Johor, Malaysia; [2] Faculty of Electrical Engineering, Universiti Teknologi Malaysia (UTM), Johor Bahru, Johor, Malaysia; [3] Faculty of Electrical & Electronics Engineering Tech., Universiti Malaysia Pahang Al-Sultan Abdullah, Pekan, Malaysia; [4] Department of Electrical & Electronics Engineering, Sri Vasavi Engineering College, Pedatadepalli, Andhra Pradesh, India; [5] Department of Engineering, Faculty of Advanced Technology and Multidiscipline, Universitas Airlangga, Surabaya, East Java, Indonesia

7.1 Introduction

Continuous increments in load demand, population, and industrial expansions accelerate traditional energy sources worldwide for the generation of electricity. These conventional sources prominently affect climate and human health. Moreover, fossil fuels are responsible for approximately 75% of greenhouse gas emissions worldwide. Hence, the electrical industry needs changes, and renewable energy technologies should be required for the transition to a clean environment. Many countries in recent years have increased the utilization of renewable energy sources (RESs) for electricity (Beyza and Yusta, 2021; Pérez-Denicia et al., 2017). However, some countries are still in the early stages of RE

Frequency Control in Microgrids
https://doi.org/10.1016/B978-0-443-36476-1.00009-2

integration. Although this globe has plenty of RESs, still 13% of people are living in remote areas without any power supply. illustrates worldwide electricity access for the global population (Peddakapu et al., 2024a). Microgrids are intended to be an alternative power generation unit for overcoming these challenges and enhancing electricity accessibility in rural areas Fig. 7.1.

Various local sources, namely, solar, biomass, wind, microturbines, and other distributed energy resources, are prevalently available, and those are used to integrate into the microgrid for efficient power generation. Further, technological advancements and cost-effectiveness have accelerated the popularity of microgrids. The benefits of the microgrids are to contribute to the uninterrupted power supply without violating any electrical parameters. Considering these privileges, exceptional communication and advanced control techniques have transformed the conventional isolated microgrid structure into a multi-microgrid (MMG) system with the integration of tie-lines (Uddin et al., 2023; Mbungu et al., 2023). However, the challenging task in the microgrid system is to equalize the power production and varying load without impacting the frequency due to sudden variations in atmospheric conditions.

The stability and longevity of the microgrid system primarily depend on the specified frequency and voltage values. Hence, many researchers over the years have shown increasing fascination with automatic generation control (AGC) to alleviate the variations in frequency. The governor can control and sustain the scheduled frequency through the interaction of AGC (Verma et al., 2024; Ullah et al., 2023). However, due to sudden changes in atmospheric conditions, AGC cannot sustain the prescribed frequency in the RES-integrated microgrid system. Concerning this, an additional controller is necessary to reduce the variations in frequency and tie-line power. Over the past 10 years, different controllers, namely,

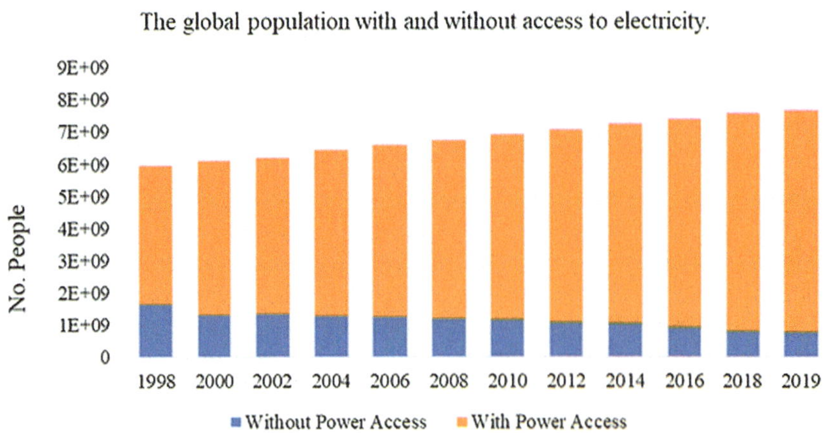

FIGURE 7.1 **Electricity access across the world for the population** World population.

proportional integral derivative (PID), PI, and integral double derivative (IDD), were implemented by several researchers (Khare et al., 2024; Tumari et al., 2024). Nevertheless, these conventional controllers can't show effective performance when integrating one microgrid with another microgrid. To overcome these difficulties, some researchers have used different control approaches such as fuzzy logic controllers (FLC) Nanda and Sakkaram, (2003), decentralized controllers (Gbadega and Sun, 2024), Sliding mode controllers (Dahiya et al., 2016), Model predictive control (Monteiro and Igreja, 2015), optimal controllers (Arya, 2018a), and intelligent methods (Arya, 2018b). Moreover, type-2 fuzzy controller (Sahu et al., 2018), fuzzy adaptive model predictive control (Kayalvizhi and Vinod Kumar, 2017), a robust mixed H_2/H_∞ controller based on static output feedback (Khan et al., 2021), linear quadratic gaussian (LQG)-based virtual inertial control (Mohan et al., 2023), modified tilt integral derivative (TID) (Ahmed et al., 2022), tilt integrator derivative and acceleration operators (TIDA)+1 Error Signal Regulator (Shayeghi et al., 2025), and 2° of two-degree-of-freedom tilt-integral-derivative controller with fractional derivative (2DOF-TID$^{\mu}$) (Moschos and Parisses, 2023). They are utilized recently by researchers to alleviate the fluctuations in microgrid systems during perturbations. However, many of these techniques rely on obtaining the system's state information, and implementing them in real-time applications is challenging.

Researchers are fascinated by fractional-order (FO) controllers due to their flexible structure and facilitation of many tuning factors. Many parameters enable finer adjustments, which can enhance the system's outcomes and stability. Hence, FO-based controllers are also used enormously in the electrical industry to mitigate the changes under disturbance. FO-based controllers such as FOI, FO-based fuzzy PID (FOFPID), FO-based nonlinear PID (FONPID), and Tilt-FOID were employed by researchers to solve AGC challenges in interconnected power systems (Arya, 2019; Daraz et al., 2021). Besides, some researchers designed cascade-based FO controllers like CF-FOIDN, FFOPI-FOPID, CFOID-FOPIDN, 2DOF PID-FOPDN, hybrid FOFPID-LQG, FOTID-PD, FOPI-FOPID, fuzzy-FOPID, and TID-FOPTID are used for improving the frequency response and system stability (Mishra et al., 2020; Ahmad et al., 2024). Noman et al. (2024) cascaded the one plus tilt derivate (1 + TD) and FOTID with filer (FOTIDF) was developed to sustain reliable power supply between generation and load demand. El-Sousy et al. (2023) proposed a 1 + PII2D with FOPD control technique in the load frequency control (LFC) system to improve the frequency and voltage profiles. A comprehensive review of the literature reveals that the application of FO-based cascade controllers has been minimally explored. Therefore, it becomes crucial to examine AGC utilizing a cascaded FO controller, which forms the core motivation for this research. This study introduces an innovative cascade configuration of FOPD and FOPIDN (CFOPD-FOPIDN) to alleviate the frequency variations and stabilize tie-line power flows.

Due to the incorporation of RESs and continuously varying load demand, the fixed gain parameters could not achieve optimal performance. Hence, optimization methods are needed to adjust the gain parameters to sustain the system stability in the event of nonlinearities. The bacterial foraging algorithm (BFA) is used in the AGC system to improve the neural network (Hakimuddin et al., 2022). Other optimization methods such as particle swarm optimization (PSO) (Falehi, 2017), bat algorithm (BAT) (Peddakapu et al., 2020), mine blast algorithm (MBA) (Fathy et al., 2020), multi-objective optimization (MBO) (Nallolla et al., 2023), Jaya algorithm (JA) (Singh et al., 2017), Archimedes optimization algorithm (AOA) (Peddakapu et al., 2024), Whale optimization algorithm (WOA) (Satapathy et al., 2018), and jellyfish search optimization algorithm (JSOA) Karnavas and Nivolianiti (2023) employed by researchers to obtain the finest gains. Besides, various optimized controllers for mitigating frequency variations in different micro grid systems, types of generators, and Energy storage systems are summarized in Table 7.1. According to the findings of previous studies, achieving performance improvement with the designed CFOPD-FOPIDN controller alone is quite challenging. As a result, the secondary objective of this work is to implement an innovative optimization technique, such as Moth Flame Optimization (MFO), to tackle the challenges in two-area microgrid systems. The proposed MFO-based CFOPD-FOPIDN technique's potency is also assessed in comparison to other methods, including PSO, BAT, and WOA, under various uncertain conditions.

The novel contributions of this research are outlined as follows:

1. Designed an AGC framework for a two-area microgrid system integrating diverse power sources, including wind turbine generators (WTG), microturbines, PV, diesel generators (DG), fuel cells (FC), and BESS.
2. Developed a new cascaded FOPD and FOPIDN (CFOPD-FOPIDN) for improving the responses of frequency and tie-line power under various uncertain circumstances.
3. A newly designed MFO algorithm is used to attain the finest gains from the CFOPD-FOPIDN controller. Its proficiency is compared against other optimizations like PSO, BAT, and WOA.
4. The proposed MFO-based CFOPD-FOPIDN strategy is investigated under three conditions namely, changes in load, wind power variations, and changes in solar power.

Part of your document could not be imported as it contained an unsupported element. For more information about unsupported elements, please visit the Support Center.

TABLE 7.1 Previous studies with different optimized controllers in microgrid systems.

S. No.	Microgrid configuration	Controller type	Optimization method	Type of generators	Energy storage systems	References
1	Two-area	Fuzzy PIDF	Marine predator algorithm (MPA)	WTG, PV, DG	BESS	Ahmad et al. (2024)
2	Two-area	C-PDF (1+ PI)	Imperialist competitive algorithm (ICA)	DG, MT	None	Nayak et al. (2021)
3	Single area	FPD-TID	Chaotic crow search algorithm (CCSA)	PV,DG	BESS	Khokhar et al. (2021)
4	Single area	FPDN-FPTID	Dandelion optimization	PV,WTG, DG	BESS	Khalil et al. (2024)
5	Two-area	$PI^{\lambda}D^{\mu}$-PD^{μ}	Coot algorithm	WTG,FC, DG	BESS	Shukla and Raju (2024)
6	Two-area	FOPI-FOPD	Dragonfly search algorithm (DSA)	PV,DG	None	Çelik, (2021)
7	Two-area	FOPID/ (1 + PI)	Equilibrium optimizer (EO) algorithm	PV,WTG, DG	BESS	Das et al. (2024)
8	Two-area	HFFOC PDF (1 + PI)	Hybrid modified sine cosine algorithm-pattern search algorithm	PV,WTG, MT	BESS	Patel et al. (2023)
9	Two-area	1 + PII2D/ FOPID	Artificial rabbits optimizers	DG,SMES, PV	SMES	El-Sousy et al. (2023)

7.1.1 Design of a two-area microgrid system

This study considered two microgrids and connected them via a tie-line to investigate the impacts of renewable and conventional sources. The control structure of the suggested microgrid system is presented in Fig. 7.2.

Power sources such as WTG, microturbines, PV, DG, FC, and BESS are incorporated into each microgrid. Further, the designed controller

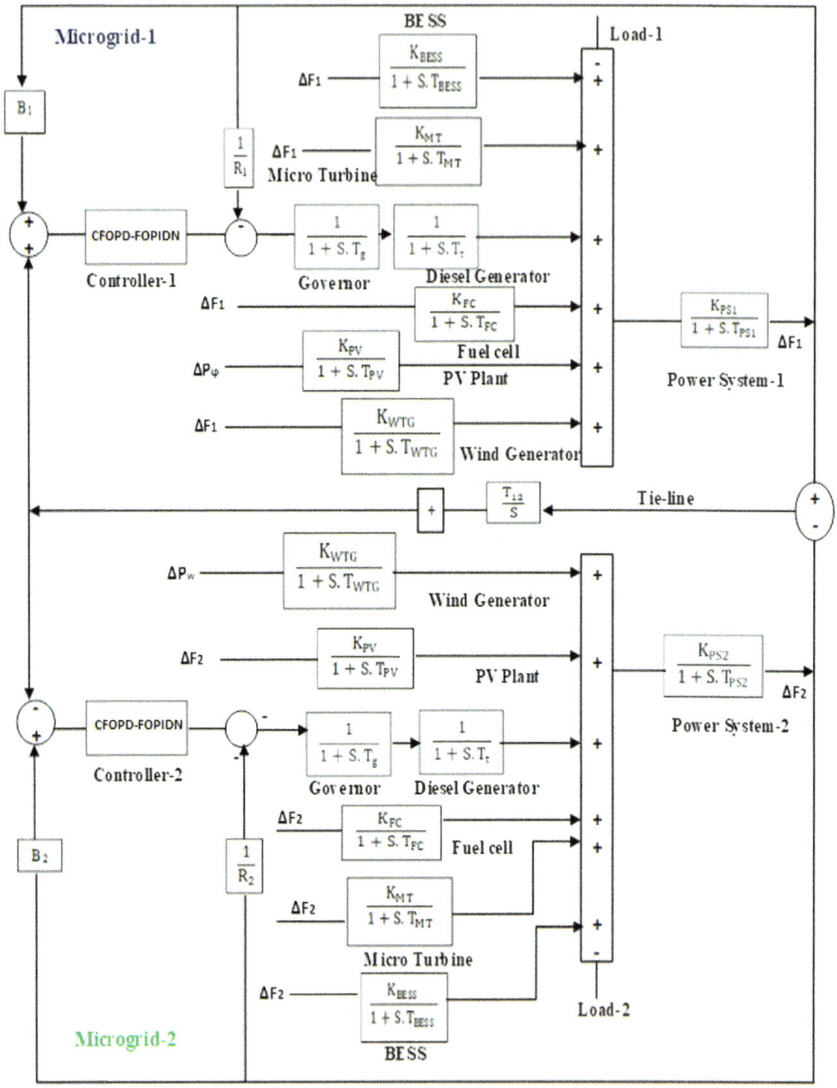

FIGURE 7.2 **Proposed two-area microgrid system** Two-area microgrid system.

(CFOPD-FOPIDN) is presented in each microgrid system as a secondary loop controller, which can contribute the control signals to the diesel engine. To attain the finest gains from the proposed controller, a new metaheuristic algorithm called MFO is developed in this work. The suggested two-area microgrid system specifications are noted in Table 7.2. The following sections describe the model of each source.

TABLE 7.2 Designed system specifications.

Specification	Value
Two-area microgrid	f = 60 Hz, Area1 = Area2 = 1200 MW, base power = 1200 MW
PS	K_{PS} = 120Hz/puMW
	T_{PS} = 6 s
PV	K_{PV} = 1
	T_{PV} = 1.8 s
DG	K_{DG} = 1
	T_{DG} = 1 s
MT	K_{MT} = 1
	T_{MT} = 2 s
WTG	K_{WTG} = 1
	T_{WTG} = 1.5 s
FC	K_{FC} = −1/5
	T_{FC} = 1 s
BESS	K_{BE} = −1/300
	T_{BE} = 0.1 s
Other configurations	R_1=R_2= 2.4Hz/puMW
	$2\pi T_{12}$ = 0.545pu MW/Hz
	B_1=B_2= 0.425pu MW/Hz

7.1.2 Diesel generator (DG)

The DG performs a key role in a microgrid system because of the continuous, unpredictable changes in renewable energy sources (WTG & PV). It can provide consistent and reliable power to end users. The DG model can be shown as (Peddakapu et al., 2024a, 2024b):

$$G_{DEG}(s) = \frac{K_{DEG}}{1 + S.T_{DEG}} \tag{7.1}$$

7.1.3 Wind turbine generator (WTG)

Significant variations in wind turbine mechanical power can cause continuous changes in wind. It is described as (Peddakapu et al., 2024a, 2024b):

$$P_{Wind} = \frac{1}{2}\rho A_r C_P V_W^3 \tag{7.2}$$

Here, ρ and A_r indicates the air density (1.25 kg/m^3) and blades swept area (1735 m^2), respectively. V_W shows the wind speed, which is measured in m/s. C_P indicates the power coefficient and it is expressed as (Peddakapu et al., 2024a, 2024b):

$$C_P = (0.44 - 0.0167\beta).\sin\left[\frac{\lambda - 3}{15 - 0.3\beta}\right] - 0.0184(\lambda - 3)\beta \qquad (7.3)$$

Where β shows the pitch angle of the blade (0.1745). Similarly, λ represents the tip speed ratio $\left(R_{blade}\omega_{blade}/V_W\right)$. ω_{blade}&R_{blade} indicate the length and rotation of turbine blades, respectively. The transfer function of the WTG (Peddakapu et al., 2024a, 2024b):

$$G_{WTG}(s) = \frac{P_{WTG}}{P_{Wind}} = \frac{K_{WTG}}{1 + S.T_{WTG}} \qquad (7.4)$$

7.1.4 PV system

The electric power from the PV can be expressed as (Peddakapu et al., 2024a, 2024b):

$$P_{pv} = \eta \varnothing_{Ir} S[1 - 0.005(T_a + 25)] \qquad (7.5)$$

Where η & $\varnothing_{Ir}$ indicates as efficiency and irradiance (W/m^2) of PV, respectively. Similarly, S & T_a shows the area (m^2) of the PV array and ambient temperature (25°C), respectively. Besides, distinct types of power electronic components like converters are employed for AC microgrid synchronization. The PV transfer function (Peddakapu et al., 2024a, 2024b) is:

$$G_{PV}(s) = \frac{\Delta P_{PV}}{\varnothing_{Irr}} = \frac{K_{PV}}{1 + S.T_{PV}} \qquad (7.6)$$

7.1.5 Fuel cell (FC)

In a microgrid system, the FC does key performance for ameliorating the system stability in the event of uncertainties. Since the integration of PV and WGT in the proposed system, the FC can deliver the required power to the system due to sudden changes in irradiance and wind speed. The transfer function of FC (Peddakapu et al., 2024a, 2024b) is:

$$G_{FC}(s) = \frac{K_{FC}}{1 + S.T_{FC}} \qquad (7.7)$$

7.1.6 Battery energy storage systems (BESS)

Energy storage systems are essential to deliver a reliable power supply to the end users as PV and wind plants are integrated into the microgrid systems. They can have charging and discharge modes for transferring and getting power from other sources. The transfer function of the BESS (Peddakapu et al., 2024a, 2024b) is:

$$G_{BESS}(s) = \frac{K_{BESS}}{1 + S.T_{BESS}} \tag{7.8}$$

Here, T_{BESS} & K_{BESS} are time and gain constants of the BESS, and their values are 0.1 and $-1/300$, respectively.

7.1.7 Microturbines

They typically operate on fuels such as natural gas or diesel and are known for their compact size, efficiency, and ability to provide both electricity and heat simultaneously. Microturbines also feature air or oil-free bearings to reduce friction and improve operational efficiency. Their flexibility in fuel usage and relatively low emissions make them suitable for a range of commercial, industrial, and residential applications. The microturbine transfer function is represented as (Peddakapu et al., 2024a, 2024b):

$$G_{MT}(s) = \frac{K_{MT}}{1 + S.T_{MT}} \tag{7.9}$$

7.2 Cascade controller designing

In this paper, two Fractional order (FO)- based controllers are used such as Fractional ordered proportion plus differential (FOPD) and FOPIDN are integrated as CFOPD-FOPIDN to achieve dynamic output under situations of uncertainty. Different controllers are connected one by one manner in cascade connection to obtain the essential output. In this cascade connection, each controller output is connected next controller input to obtain the desired response. It consists of two loops inner and outer in which the inner loop mitigates the specific process variable while the outer loop makes a desired set point. Moreover, the inner loop shows faster performance in contrast to the outer loop. The advantage this hybrid controller exhibits greater response under nonlinearities moreover, reduces the disturbance and increases the system accuracy. In this system, each controller's main objective has been focused on nonlinear behavior. Both integral and derivative controllers have been encountering non-linear behavior and improving the overall system performance. The cascaded CFOPD-FOPIDN Controller has been illustrated in Fig. 7.3.

From above Figures, D(S) and R(S) are about signals of disturbance and reference, respectively. The TF of controllers is represented as G_{C1}, G_{C2} respectively. The G_{P1} and G_{P2} are TFs of inner and outer loops, respectively. At last, the outcome is indicated by Y(S).

The proposed FOPD and FOPIDN TFs are illustrated as:

$$G_{C1} = K_P + K_D.S^{\mu} \tag{7.10}$$

$$G_{C2} = K_P + \frac{K_{I_1}}{S^{\lambda_1}} + K_{D_1}.S^{\mu_1}\left(\frac{N}{N+S^{\mu_1}}\right) \tag{7.11}$$

In this hybrid controller, FOPD-FOPIDN consists of 10 optimized parameters such as $K_P/K_I/K_{P_1}/K_D/K_{D_1}/\lambda_1/\mu/\mu_1/N$. Since this controller has multiple fractional order parameters, the tuning process will be computationally intensive. Moreover, this controller needs high-precision numerical approximation of fractional operators. Nevertheless, the hybrid FOPD and FOPIDN controller shows greater performance with respect to settling time, oscillations, and undershoots. By using frictional

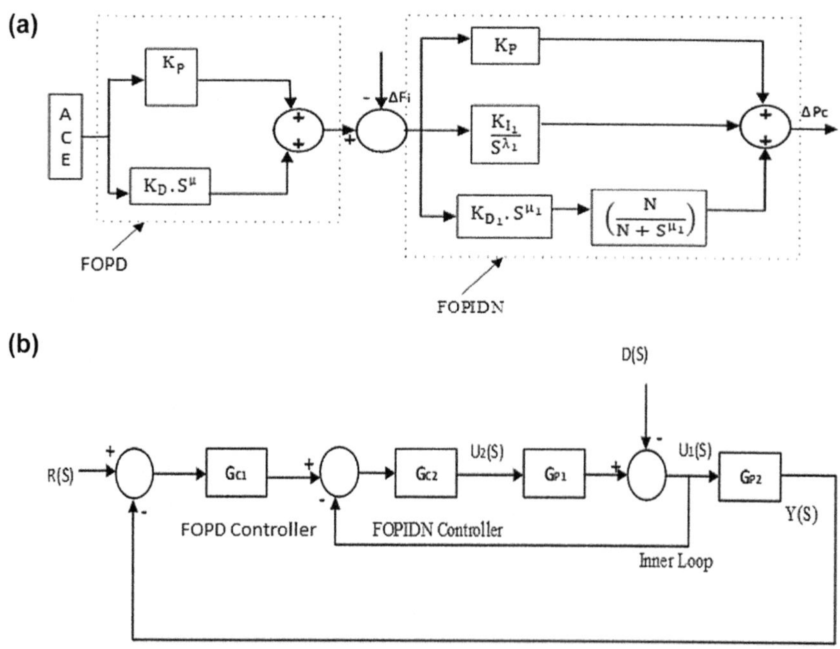

FIGURE 7.3 **Schematic diagram of CFOPD-FOPIDN** cascaded CFOPD-FOPIDN controller.

calculus, the FOPD, and FOPIDN have been developed to extend the traditional controller of PD and PIDN. To calculate real-time activities of the frictional order transfer function, the integer order transfer function (TF) has been used. To assess the performance, Oustaloup's CRONE approximation has employed, as it is also more appropriate for recursive distribution transfer functions, as demonstrated:

$$S^{\alpha} = K_f \prod_{K=-N}^{N} \frac{S + \omega_K^t}{S + \omega_K} \tag{7.12}$$

The poles are denoted as ω_K, denoted as:

$$\omega_K = \omega_b \left(\frac{\omega_h}{\omega_b}\right)^{\left(K+N+\left(\frac{1}{2}\right)(1+\alpha)/2N+1\right)} \tag{7.13}$$

Similarly, ω_K^t shows the zeros and indicated as:

$$\omega_K^t = \omega_b \left(\frac{\omega_h}{\omega_b}\right)^{\left(K+N+\left(\frac{1}{2}\right)(1-\alpha)/2N+1\right)}, K_f = \omega_\alpha^h \tag{7.14}$$

The filter gain and integration-differentiation coefficient are indicated as α and K_f. In addition, predefined frequency and analog filter order bands are $[\omega_b, \omega_h]$ and 2N+ 1, respectively. The predefined frequency span from 10^{-2} to 10^2 rad/s and the N value is 6.

The output of the designed CFOPD-FOPIDN is the input (ΔP_C) and illustrated as:

$$\Delta P_C = \left[ACE\left(\frac{K_P}{S^{\lambda}} + K_D.S^{\mu}\right) - \Delta F\right] * \left[K_P + \frac{K_{I_1}}{S^{\lambda_1}} + K_{D_1}.S^{\mu_1}\left(\frac{N}{N + S^{\mu_1}}\right)\right] \tag{7.15}$$

The hybrid controller of CFOPD-FOPIDN has been used in this paper to improve the system performance in response to changes in load, various irradiance, and wind speed.

7.2.1 Objective function

Integral Time Absolute Error (ITAE) is considered in this work as an objective function as it shows greater and more successful performance.

$$J = ITAE = \int_0^{t_s} \{|\Delta F_1| + |\Delta F_2| + |\Delta P_{tie}|\} t \cdot dt \tag{7.16}$$

Minimize J subject to:

$$
\begin{aligned}
&K_{P_i}^{min} \le K_{P_i} \le K_{P_i}^{max} \\
&K_{I_i}^{min} \le K_{I_i} \le K_{I_i}^{max} \\
&K_{I_{1i}}^{min} \le K_{I_{1i}} \le K_{I_{1i}}^{max} \\
&K_{D_i}^{min} \le K_{D_i} \le K_{D_i}^{max} \\
&K_{D_{1i}}^{min} \le K_{D_{1i}} \le K_{D_{1i}}^{max} \\
&\lambda_{1i}^{min} \le \lambda_{1i} \le \lambda_{1i}^{max} \\
&\mu_i^{min} \le \mu_i \le \mu_i^{max} \\
&\mu_{1i}^{min} \le \mu_{1i} \le \mu_{1i}^{max} \\
&N_i^{min} \le N_i \le N_i^{max}
\end{aligned}
\tag{7.17}
$$

From Eq. (7.17), minimum (min.) and maximum (max.) values of the designed CFOPD-FOPID are mentioned. The min. and max. bands of

$$
K_{P_i}^{min} \text{is} [02], [02] \text{for} K_{D_i}^{min} \text{and} K_{D_{1i}}^{min}, [04] \text{for} K_{I_i}^{min} \text{and} K_{I_{1i}}^{min},
$$

$$
[02] \text{for} \mu_i^{min} \text{and} \mu_{1i}^{min}, [01] \text{for} \lambda_i^{min} \text{and} \lambda_{1i}^{min}, \text{and} [0500] \text{for} N_i^{min} \tag{7.18}
$$

7.2.2 Control design steps of a microgrid controller

7.2.2.1 *Model the microgrid system*

Develop a detailed dynamic model of the microgrid, incorporating distributed energy resources (DERs), energy storage systems, inverters, and load dynamics.

7.2.2.2 *Define control objectives*

Establish clear control objectives, such as minimizing frequency deviations, stabilizing tie-line power flows, improving transient performance (settling time, overshoot), and maintaining system stability under various disturbances.

7.2.2.3 *Select controller structure*

Choose an appropriate controller structure (e.g., PI, PID, fractional-order PID (FOPID), adaptive, or intelligent controllers), depending on system requirements and performance goals.

7.2.2.4 Formulate optimization problem

Define an objective function that captures performance metrics (such as Integral Time Absolute Error - ITAE), and identify system constraints for the controller tuning process.

7.2.2.5 Optimize controller parameters

Apply an optimization algorithm (e.g., Moth-Flame Optimization (MFO), Particle Swarm Optimization (PSO), Whale Optimization Algorithm (WOA), etc.) to fine-tune the controller parameters for optimal performance.

7.2.2.6 Conduct stability analysis

Perform stability checks, such as eigenvalue analysis or frequency-domain analysis, to ensure the optimized controller guarantees stable operation under various conditions.

7.2.2.7 Validate controller performance

Test the controller under different scenarios (such as load variations, renewable generation fluctuations) through simulation studies, and, where possible, through MATLAB.

7.3 Moth flame optimization

7.3.1 Inspiration

The insects of moths are similar to the family of butterflies. In nature, there are more than 160,000 various species that have existed in nature. The larva and adult stages are the main milestones in their life. The larva turns into a moth by cocoon. A metaheuristic-based nature-inspired moth flame algorithm (MFO) has been introduced by Mirjalili et al. (Moschos and Parisses, 2023). The moth has a unique nature of navigation during the nighttime time and the nature of the navigation mechanism can be called transverse orientation. The moth flies at a fixed angle to the moon. Furthermore, the moths travel long distances in a straight line toward the moon with a spiral manner around the light. A schematic model of moth flame behavior as shown in Fig. 7.4.

This method is very effective for long traveling distances in a straight line however, may struck due to artificial lights around the moon. However, the MFO technique involves focusing on the direction of flame while developing their behavior in an optimized manner. The main advantage of this method has few control parameters for global optimization problems. Moreover, due to the nature of approximation, there is an opportunity for effective searching. The flowchart for MFO is shown in Fig. 7.5.

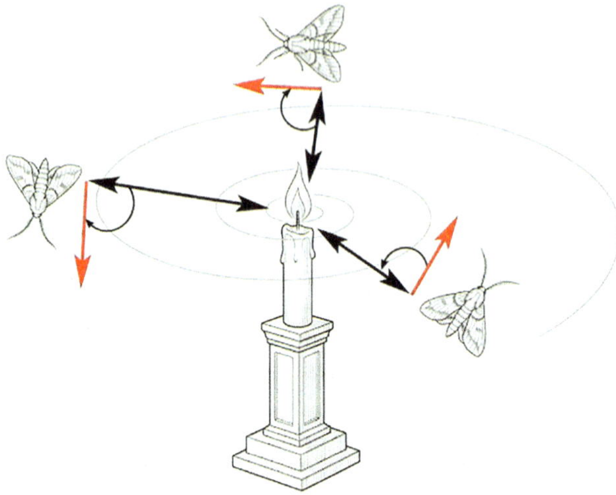

FIGURE 7.4 **Schematic diagram of the moth flame** Moth flame algorithm.

7.3.2 MFO algorithm

The moth's position in a search space can be represented as a problem variable whereas moths themselves create a solution. Through changing the location of vector moth fly in three-dimensional ways. The moth's group can be represented in matrix form as shown below (Moschos and Parisses, 2023):

$$M = \begin{bmatrix} m_{1,1} m_{1,2} m_{1,3} \ldots\ldots m_{1,d} \\ m_{2,1} m_{2,2} m_{2,3} \ldots\ldots m_{2,d} \\ .. \\ .. \\ .. \\ m_{n,1} m_{n,2} m_{n,3} \ldots\ldots m_{n,d} \end{bmatrix} \tag{7.19}$$

Where n and d are the months and variables.

Fitness evaluations are based on moth population data, accumulated and represented in a matrix format.

$$OM = \begin{bmatrix} OM_1 \\ OM_2 \\ OM_3 \\ . \\ . \\ OM_n \end{bmatrix} \tag{7.20}$$

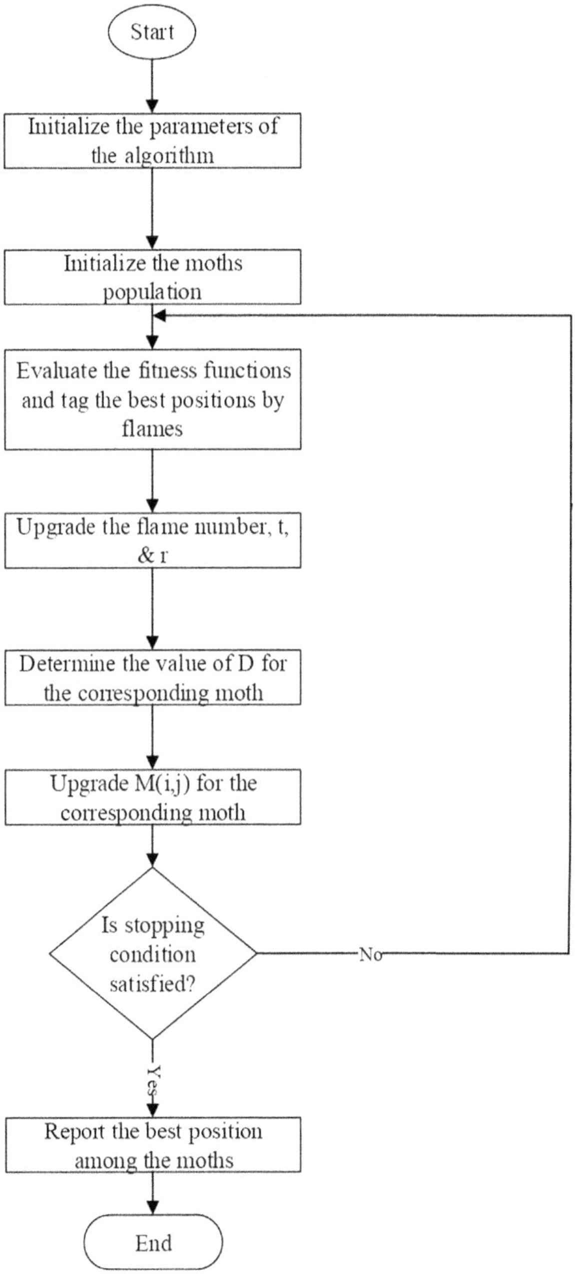

FIGURE 7.5 **MFO Algorithm** Flowchart of MFO.

Here moths is represented by n.

Similarly, the following matrix represents as number of flames:

$$OF = \begin{bmatrix} f_{1,1}f_{1,2}f_{1,3}\ldots\ldots f_{1,d} \\ f_{2,1}f_{2,2}f_{2,3}\ldots\ldots f_{2,d} \\ .. \\ .. \\ .. \\ f_{n,1}f_{n,2}f_{n,3}\ldots\ldots f_{n,d} \end{bmatrix} \quad (7.21)$$

Where f, d shows the number of flames and variables.

For all the flames, the fitness values can be stored in the following matrix:

$$OF = \begin{bmatrix} OF_1 \\ OF_2 \\ OF_3 \\ . \\ . \\ OF_n \end{bmatrix} \quad (7.22)$$

Where n indicates the number of flames.

The moths and flames are the potential solutions, however; they can be distinguished by their mechanism and treatment of every iteration. Moreover, the moths are the real search agents navigating through the entire search area where whereas flame denotes the optimum solutions that can be accomplished by the moths.

In this work, the Moth-Flame Optimization (MFO) algorithm is employed to optimize the parameters of controllers designed for power-frequency stabilization in microgrids. The process begins with modeling the dynamic behavior of the microgrid, considering inverter-based distributed energy resources (DERs), loads, and storage systems. A suitable controller structure, such as PI, PID, or fractional-order PID (FOPID), is selected, and an objective function is formulated to minimize key performance metrics, including frequency deviation, control effort, and transient response characteristics. An initial population of candidate controller parameter sets is generated and evaluated through simulation. The MFO algorithm then iteratively refines these candidates by balancing global exploration and local exploitation, progressively improving the population based on the performance of the best solutions ("flame"). In addition to performance metrics, stability of the optimized controllers is ensured through appropriate stability checks, such as eigenvalue analysis or frequency-domain criteria. Upon convergence, the optimized and stability-verified controller parameters are extracted and validated through detailed simulation studies ensuring robust and stable microgrid performance across a range of operating conditions.

Moths can improve their position about the flame as illustrated in Eq. (7.23).

$$M_{i,j} = S\left(M_i, F_j\right) \tag{7.23}$$

The notations M_i, F_j, and S denote the *i*th moth, the *j*th flame, and the spiral function, respectively.

The MFO has a spiral function and is illustrated as:

$$S\left(M_i, F_j\right) = D_i.e^{bt}.\cos(2\pi t) + F_j \tag{7.24}$$

D_i is the distance between i^{th} moth and j^{th} flame and yielded below:

$$D_i = |F_j - M_i| \tag{7.25}$$

b is the constant of the shape of a spiral, and t represents a random value within [1, −1].

The flame can be upgraded using the following Eq. (7.26).

$$flameno. = round\left(N - l * \frac{N-1}{T}\right) \tag{7.26}$$

Here, l indicates the current iteration, N represents the number of flames and T signifies the total number of iterations.

7.4 Results and discussion

This study focuses in all three scenarios analyzed—load variations, wind fluctuations, and solar irradiance changes—the CFOPD-FOPIDN controller optimized via the MFO algorithm consistently demonstrated superior dynamic performance over other methods (WOA, BAT, and PSO). The improved performance is evident in the faster settling times, lower overshoots, and minimal undershoots, indicating that the proposed strategy responds swiftly and stably to disturbances are mentioned. The three conditions are mentioned below.

Case 1. Variation in load

Case 2. Fluctuations in wind power

Case 3. Differences in irradiance power

7.4.1 Case 1: Variations in load only

The MFO-based controller rapidly compensates for sudden demand shifts, stabilizing system frequency and tie-line power more effectively than the alternatives. This implies better load-sharing and control signal accuracy, which are crucial in practical grid operations. Area-1 microgrid has considered various load patterns to investigate the effectiveness of the MFO-based CFOPD-FOPIDN strategy. Moreover, the changes in wind speed ($\Delta P\varphi = 0$) and sun irradiation ($\Delta Pw = 0$) are taken as zero. In, Fig. 7.6 indicates the frequency and tie-line power of the two-area microgrid system with different optimizations-based CFOPD-FOPIDN. As per the observations from, the settling times (STs) of the area 1 frequency response are 3.1 s for MFO-based CFOPD-FOPIDN, 4.2 s for WOA-based CFOPD-FOPIDN, 4.8 s for BAT-based CFOPD-FOPIDN and 6.2 s for PSO-based CFOPD-FOPIDN. In addition to area-2, the STs are 3.7 s for MFO-based CFOPD-FOPIDN, 4.9 s for WOA-based CFOPD-FOPIDN, 5.3 s for BAT-based CFOPD-FOPIDN, and 7.1 s for PSO-based CFOPD-FOPIDN. Concerning tie-line power, the STs are 3.3 s for MFO-based CFOPD-FOPIDN, 4.7 s for WOA-based CFOPD-FOPIDN, 4.9 s for BAT-CFOPD-FOPIDN and 6.5 s for PSO-based CFOPD-FOPIDN. The numerical values of the observations from are noted in Table 7.3. As per the in-depth findings, the MFO-based CFOPD-FOPIDN shows greater performance in reducing the oscillations and improving the system stability during load variations. The gains of the CFOPD-FOPIDN controller are noted in Table 7.4.

7.4.2 Case 2: Fluctuations in wind speed

The proposed strategy demonstrates robustness by dampening the high variability associated with wind power. Unlike PSO and BAT, which showed more oscillatory behavior, the MFO-tuned controller maintains smoother responses, highlighting its adaptability to real-time renewable uncertainties.

In this case, varying wind power values are considered and depicted in Fig. 7.7. These values are applied to Area-1 (Microgrid-1) to evaluate the system's frequency and tie-line power responses using different algorithms. Fig. 7.7 presents the frequency responses of Area-1 and Area-2, along with the tie-line power, utilizing various optimization-based CFOPD-FOPIDN approaches. The simulation outcomes from Fig. 7.8, the proposed MFO-based CFOPD-FOPIDN strategy yields effective performance with regards to ST, overshoot (OS), and undershoot (US) as compared to other strategies like WOA, BAT, and PSO. The ST, OS, and US of the frequency and tie-line power responses are indicated in Table 7.5.

7.4.3 Scenario 3: Solar irradiance variations

The results showcase the controller's ability to manage intermittent solar input effectively. The MFO approach ensures minimal deviations

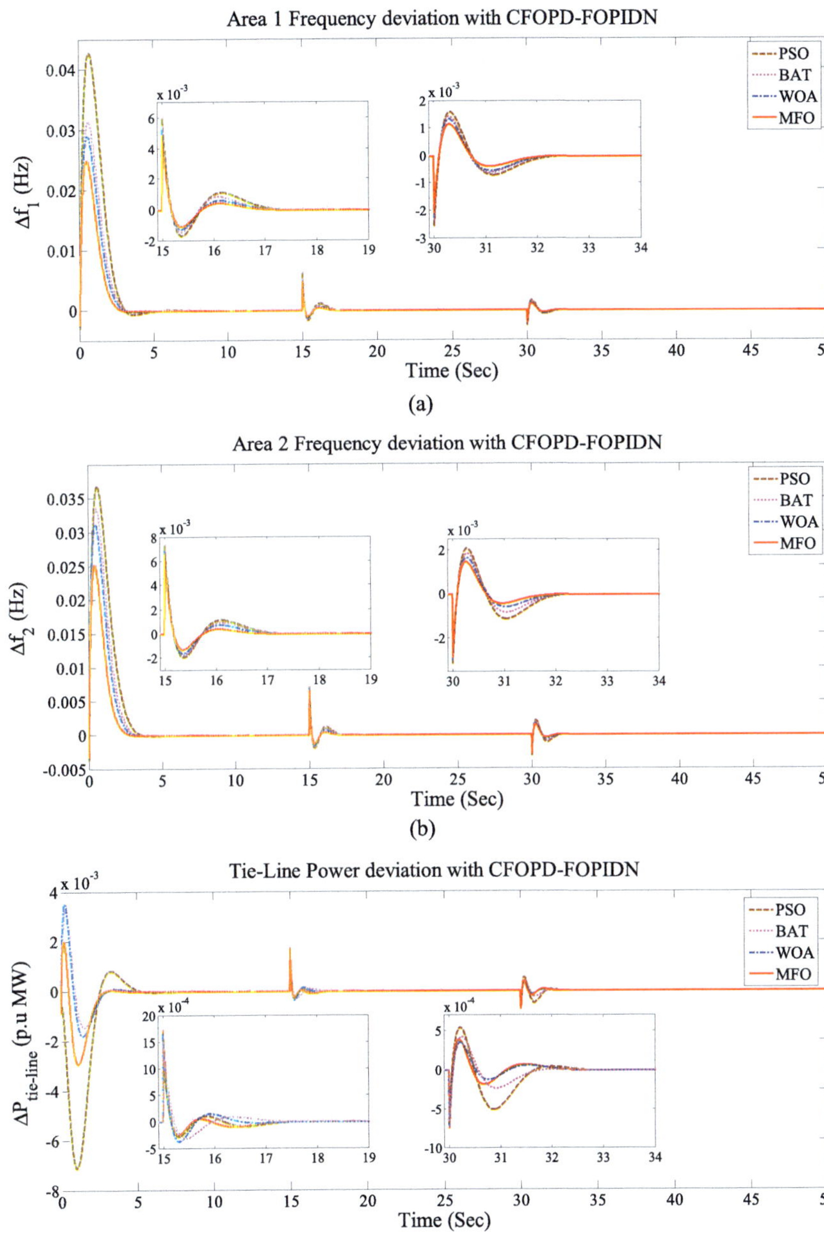

(c)

FIGURE 7.6 **Frequency and tie-line power responses of the two-area microgrid system.**

TABLE 7.3 Numerical values of carried-out control strategies at load fluctuations.

Controller	Response	Parameter	PSO	BAT	WOA	MFO
CFOPD-FOPIDN	ΔF_1	OS	0.045	0.032	0.028	0.025
		US	0	0	0	0
		ST	6.2	4.8	4.2	3.1
	ΔF_2	OS	0.036	0.033	0.031	0.025
		US	0	0	0	0
		ST	7.1	5.3	4.9	3.7
	$\Delta P_{tie-line}$	OS	0	0.0038	0.0036	0.0025
		US	−0.007	−0.0017	−0.0019	−0.0027
		ST	6.5	4.9	4.7	3.3
ITAE			0.0044	0.0041	0.0038	0.0032

TABLE 7.4 Gains of the designed CFOPD-FOPIDN.

Controller	Profile	CFOPD-FOPIDN gains	MFO	PSO	BAT	WOA
CFOPD-FOPIDN	**Area 1**	K_P	1.932	1.531	1.642	1.721
		K_D	1.583	1.231	1.342	1.471
		μ	0.821	0.622	0.711	0.781
		K_{P1}	1.826	1.563	1.642	1.711
		K_{I_1}	2.893	2.103	2.312	2.605
		K_{D_1}	1.592	1.172	1.351	1.492
		N	97.35	74.65	81.22	88.53
		λ_1	0.752	0.552	0.633	0.698
	Area 2	K_P	1.645	1.421	1.421	1.421
		K_D	0.389	0.318	0.318	0.318
		μ	0.710	0.635	0.635	0.635
		K_{P1}	1.325	1.124	1.124	1.124
		K_{I_1}	1.621	1.412	1.412	1.412
		K_{D_1}	0.272	0.201	0.201	0.201
		N	82.61	68.33	68.33	68.33
		λ_1	0.572	0.491	0.581	0.581
		μ_1	0.668	0.581	0.581	0.581

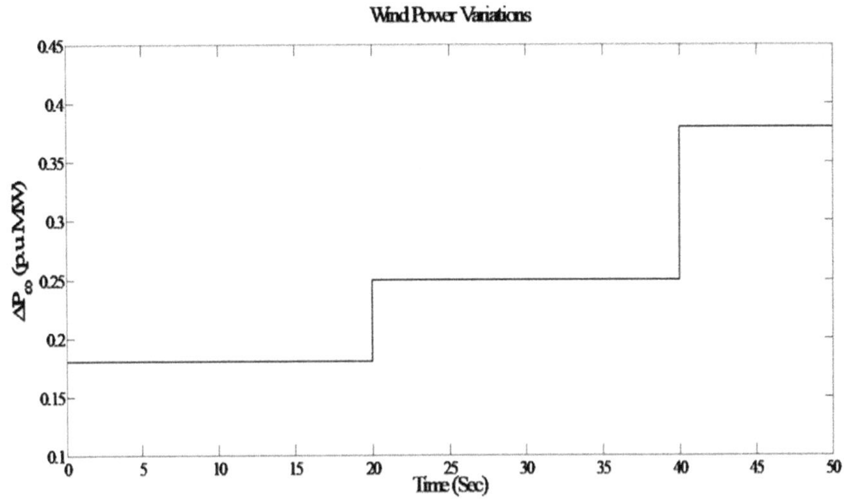

FIGURE 7.7 **Fluctuations in wind power** Wind power fluctuations.

and quicker stabilization even when irradiance varies sharply, reflecting its suitability for solar-integrated grids. Diverse irradiance levels are considered in this work and applied to Microgrid-1 to investigate the effectiveness of the proposed strategy. The different irradiance levels are presented in Fig. 7.9. The frequency responses of area-1 & 2 and tie-line power at irradiance power fluctuations are shown in Fig. 7.10. These are worth mentioning in Fig. 7.10, the STs of the area-1 frequency responses are 2.9 s for MFO-based CFOPD-FOPIDN, 3.3 s for WOA-based CFOPD-FOPIDN, 3.8 s for BAT-based CFOPD-FOPIDN, and 4.5 s for PSO-CFOPD-FOPIDN. Conversely, the STs of the area-2 frequency response are 2.7 s for MFO-tuned CFOPD-FOPIDN, 3.1 s for WOA-based CFOPD-FOPIDN, 3.5 s for BAT-optimized CFOPD-FOPIDN, and 4.7 s for PSO-CFOPD-FOPIDN. For tie-line power, the STs of the proposed control strategies are 3.2 s for MFO-based CFOPD-FOPIDN, 3.7 s for WOA-CFOPD-FOPIDN, 4.2 s for BAT-optimized CFOPD-FOPIDN, and 4.8 s for PSO-CFOPD-FOPIDN. The ST, OS, and US of the frequency and tie-line power responses are indicated in Table 7.6. The simulation outcomes reveal that the MFO-based CFOPD-FOPIDN strategy performs better in mitigating the oscillations and improving system stability during changes in irradiance when compared to other approaches.

Across all scenarios, the trends suggest that the hybrid cascaded FO controller structure offers better tuning flexibility, and the MFO algorithm's search mechanism achieves global optima more reliably. This synergistic combination results in improved system resilience and reliability, particularly in multi-source, multi-area microgrids.

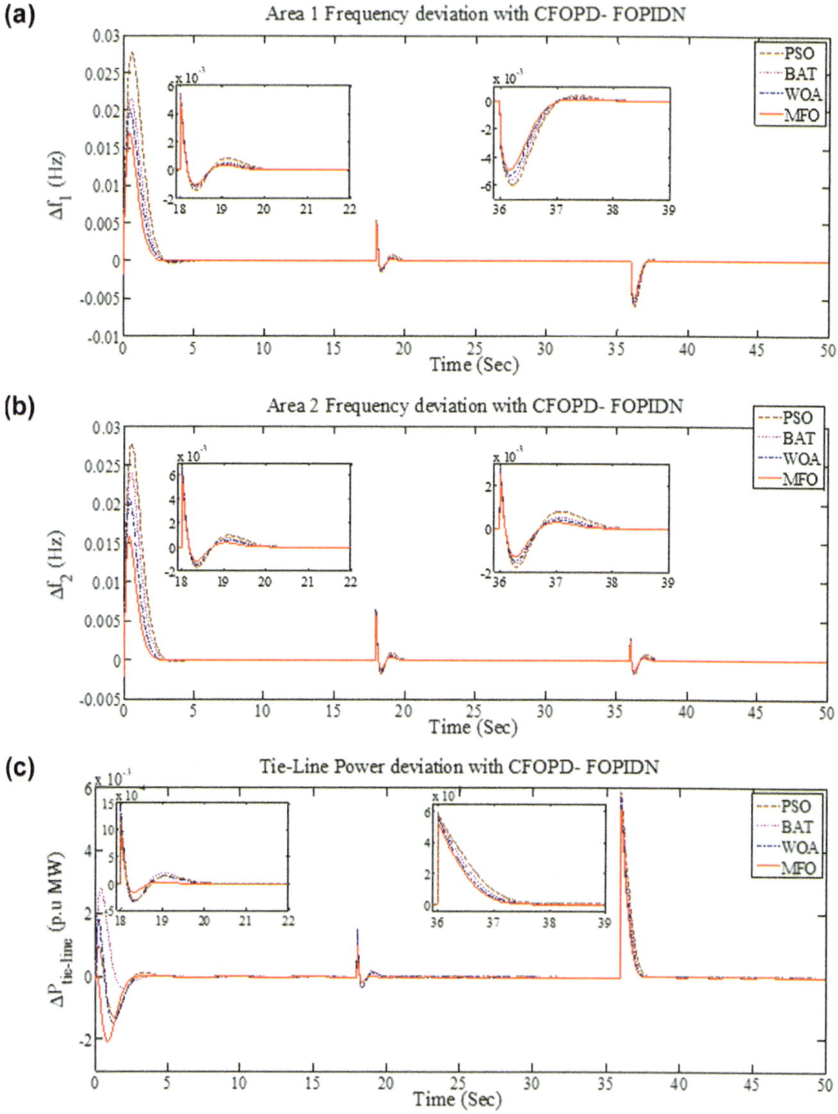

FIGURE 7.8 **Two-area microgrid systems responses to wind power variations** Frequency and tie-line power responses during wind power variations.

TABLE 7.5 ST, OS, and US of Frequency and tie-line power responses under wind power variations.

Controller	Response	Parameter	PSO	BAT	WOA	MFO
CFOPD-FOPIDN	ΔF_1	OS	0.027	0.023	0.019	0.017
		US	0	0	0	0
		ST	5.2	4.7	4.3	3.3
	ΔF_2	OS	0.028	0.024	0.022	0.015
		US	0	0	0	0
		ST	4.8	4.5	3.8	2.9
	$\Delta P_{tie-line}$	OS	0.0019	0.003	0.008	0.004
		US	−0.009	−0.007	−0.004	−0.0031
		ST	4.9	4.6	4.1	2.8
ITAE			0.0039	0.0034	0.0029	0.0026

7.5 Conclusion

The current study designed a two-area microgrid system by integrating the PV, WGT, microturbines, BESS, and FC to investigate the AGC performance under various uncertain conditions. A newly developed cascade combination of FO-based controllers like CFOPD-FOPIDN is used as a centralized controller to sustain the prescribed frequency range. In order to determine the finest gains from CFOPD-FOPIDN, the MFO algorithm is developed. Furthermore, the proposed two-area system was investigated using the MFO-based CFOPD-FOPIDN strategy under three scenarios: changes in load, fluctuations in wind power, and variations in solar irradiance power. The effectiveness and reliability of the MFO-based CFOPD-FOPIDN are tested and compared with other strategies such as WOA-based CFOPD-FOPIDN, BAT-based CFOPD-FOPIDN, and PSO-based CFOPD-FOPIDN. As per the simulation findings across three conditions, the MFO-based CFOPD-FOPIDN has shown a reduction in ST, OS, and US and an enhancement in system stability. This work has examined three conditions, such as wind power variations, load fluctuations, and changes in solar irradiance, excluding other perturbations like communication delays, component failures, and cyber-physical attacks. Further research can be carried out using hybrid algorithms and machine learning models to improve the system's dynamic performance under various uncertain conditions.

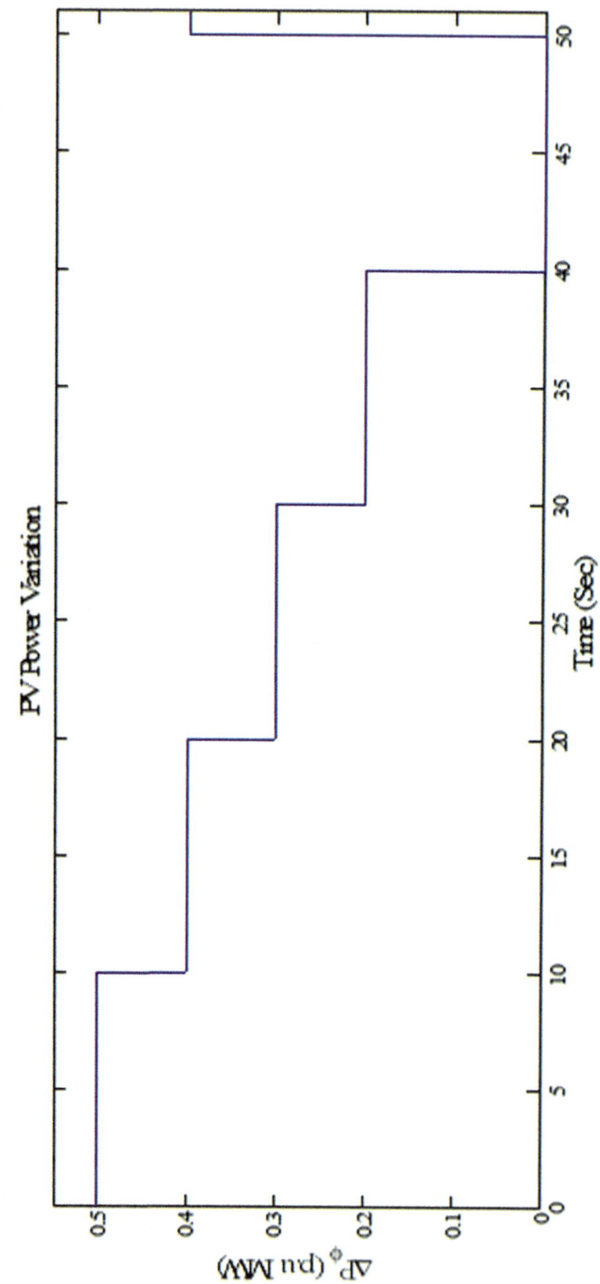

FIGURE 7.9 Variations in solar irradiance power Solar irradiance power fluctuations.

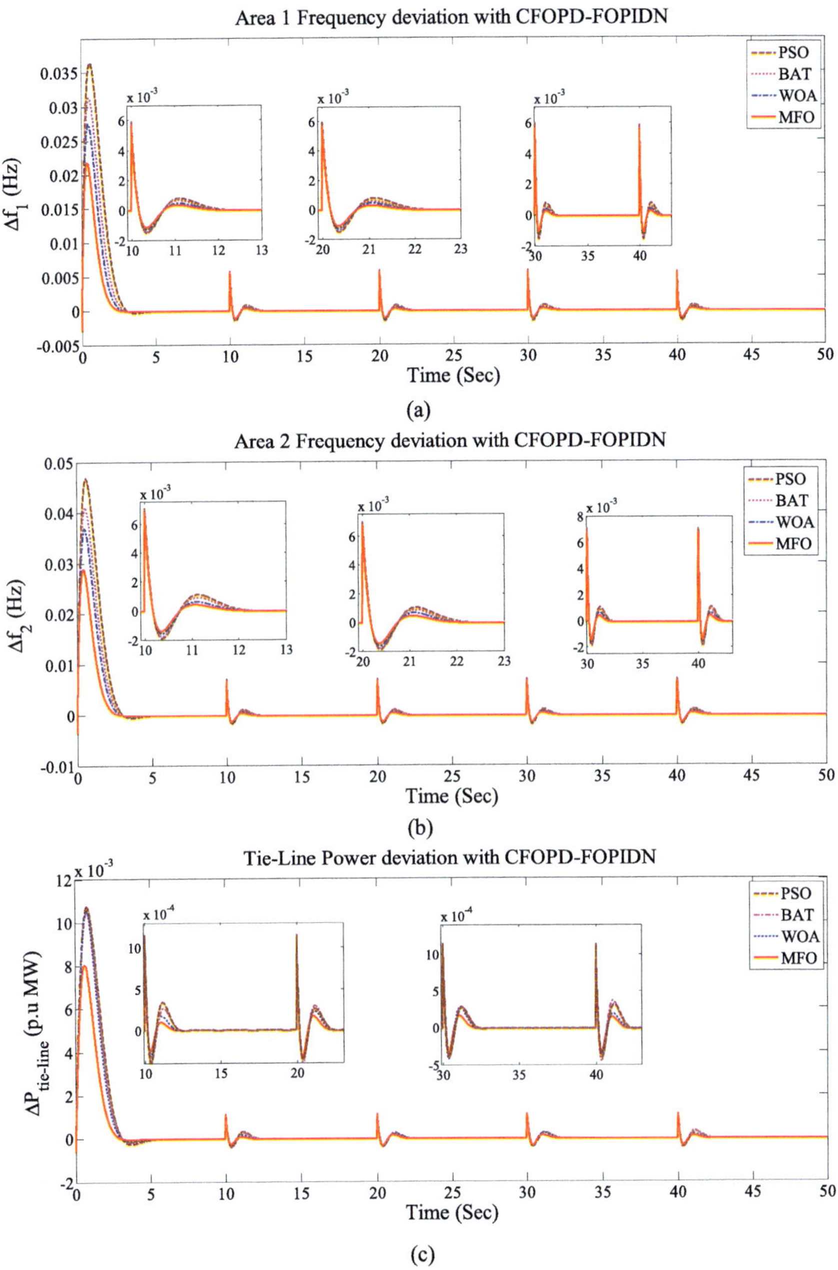

FIGURE 7.10 Frequency and tie-line power at changes in irradiance power.

TABLE 7.6 Numerical values for the systems responses under irradiance power variations.

Controller	Response	Parameter	PSO	BAT	WOA	MFO
CFOPD-FOPIDN	ΔF_1	OS	0.037	0.032	0.027	0.021
		US	0	0	0	0
		ST	4.5	3.8	3.3	2.9
	ΔF_2	OS	0.048	0.042	0.038	0.033
		US	0	0	0	0
		ST	4.7	3.5	3.1	2.7
	$\Delta P_{tie-line}$	OS	0.012	0.011	0.010	0.009
		US	0	0	0	0
		ST	4.8	4.2	3.7	3.2
ITAE			0.0052	0.0045	0.0039	0.0034

Abbreviations

FO Fractional Order
FOPD Fractional Order Proportional-Derivative
FOPIDN Fractional Order Proportional-Integral-Derivative with Filter
FOPID Fractional Order -based fuzzy Proportional-Integral-Derivative
FONPID Fractional Order -based nonlinear Proportional-Integral-Derivative
MFO Moth Flame Optimization
RES Renewable Energy Sources
PV Photovoltaic
AGC Automatic Generation Control
PID Proportional Integral Derivative
IDD Integral Double Derivative
TID Tilt Integral Derivative
1+TD One plus tilt derivate
BFA Bacterial Foraging Algorithm
PSO Particle Swarm Optimization
BAT Bat Algorithm
MBA Mine Blast Algorithm
MBO Multi-Objective Optimization
JA Jaya Algorithm
AOA Archimedes Optimization Algorithm
WOA Whale Optimization Algorithm
JSOA Jellyfish Search Optimization Algorithm

Acknowledgments

This research was carried out at the Faculty of Electrical Engineering and Centre of Electrical Energy Systems (CEES), Universiti Teknologi Malaysia (UTM). Additionally, the work was supported by UTM through Hi-Tech research grants Q.J130000.4609.00Q42 and R.J130000.7113.07E67.

References

Ahmad, R., Arya, Y., Ahmer, M.F., Nasiruddin, I., 2024. Enhanced frequency/voltage control in multi-source power system using CES and SSA-optimized cascade TID-FOPTID controller. IETE J. Res. 1–16.

Ahmed, M., Magdy, G., Mohamed, K., Kamel, S., 2022. Modified TID controller for load frequency control of a two-area interconnected diverse-unit power system. Int. J. Electr. Power Energy Syst. 135. https://doi.org/10.1016/j.ijepes.2021.107528.

Arya, Y., 2019. A new optimized fuzzy FOPI-FOPD controller for automatic generation control of electric power systems. J. Franklin Inst. 356 (11), 5611–5629. https://doi.org/10.1016/j.jfranklin.2019.02.034.

Arya, Y., 2018a. Automatic generation control of two-area electrical power systems via optimal fuzzy classical controller. J. Franklin Inst. 355 (5), 2662–2688. https://doi.org/10.1016/j.jfranklin.2018.02.004.

Arya, Y., 2018b. Improvement in automatic generation control of two-area electric power systems via a new fuzzy aided optimal PIDN-FOI controller. ISA (Instrum. Soc. Am.) Trans. 80, 475–490. https://doi.org/10.1016/j.isatra.2018.07.028.

Beyza, J., Yusta, J.M., 2021. The effects of the high penetration of renewable energies on the reliability and vulnerability of interconnected electric power systems. Reliab. Eng. Syst. Saf. 215. https://doi.org/10.1016/j.ress.2021.107881.

Çelik, E., 2021. Design of new fractional order PI–fractional order PD cascade controller through dragonfly search algorithm for advanced load frequency control of power systems. Soft Comput. 25 (2), 1193–1217.

Dahiya, P., Sharma, V., Naresh, R., 2016. Automatic generation control using disrupted oppositional based gravitational search algorithm optimised sliding mode controller under deregulated environment. IET Gener. Transm. Distrib. 10 (16), 3995–4005. https://doi.org/10.1049/iet-gtd.2016.0175. www.ietdl.org/IET-GTD.

Daraz, A., Malik, S.A., Waseem, A., Azar, A.T., Haq, I.U., Ullah, Z., Aslam, S., 2021. Automatic generation control of multi-source interconnected power system using FOI-TD controller. Energies 14 (18), 5867. https://doi.org/10.3390/en14185867.

Das, S., Nayak, P.C., Prusty, R.C., Panda, S., 2024. Design of fractional order multistage controller for frequency control improvement of a multi-microgrid system using equilibrium optimizer. Multiscale Multidiscip. Model. Exp. Des. 7 (2), 1357–1373.

El-Sousy, F.F.M., Aly, M., Alqahtani, M.H., Aljumah, A.S., Almutairi, S.Z., Mohamed, E.A., 2023. New cascaded 1+ PII2D/FOPID load frequency controller for modern power grids including superconducting magnetic energy storage and renewable energy. Fractal Fract. 7 (9), 672.

Falehi, A.D., 2017. Optimal design of fuzzy-AGC based on PSO & RCGA to improve dynamic stability of interconnected multi area power systems. Int. J. Autom. Comput. 1–11.

Fathy, A., Kassem, A.M., Abdelaziz, A.Y., 2020. Optimal design of fuzzy PID controller for deregulated LFC of multi-area power system via mine blast algorithm. Neural Comput. Appl. 32 (9), 4531–4551. https://doi.org/10.1007/s00521-018-3720-x.

Gbadega, P.A., Sun, Y.X., 2024. Iterative LMI-tuned PID controller for robust decentralized multi-area automatic generation control in deregulated electricity markets. Int. J. Eng. Res. Afr. 71, 109–140. https://doi.org/10.4028/p-lv4h7u.

Hakimuddin, N., Khosla, A., Garg, J.K., 2022. Comparative performance investigation of genetic algorithms (GAs), particle swarm optimization (PSO) and bacteria foraging algorithm (BFA) based automatic generation control (AGC) with multi source power plants (MSPPs). Elec. Power Compon. Syst. 49 (20), 1513–1524. https://doi.org/10.1080/15325008.2022.2134510.

Karnavas, Y.L., Nivolianiti, E., 2023. Optimal load frequency control of a hybrid electric shipboard microgrid using jellyfish search optimization algorithm. Appl. Sci. (Switzerland) 13 (10). https://doi.org/10.3390/app13106128.

Kayalvizhi, S., Vinod Kumar, D.M., 2017. Load frequency control of an isolated micro grid using fuzzy adaptive model predictive control. IEEE Access 5, 16241–16251. https://doi.org/10.1109/ACCESS.2017.2735545.

Khalil, A.E., Boghdady, T.A., Alham, M.H., Ibrahim, D.K., 2024. A novel cascade-loop controller for load frequency control of isolated microgrid via dandelion optimizer. Ain Shams Eng. J. 15 (3), 102526.

Khan, M., Sun, H., Xiang, Y., Shi, Di, 2021. Electric vehicles participation in load frequency control based on mixed H2/H. Int. J. Electr. Power Energy Syst. 125. https://doi.org/10.1016/j.ijepes.2020.106420.

Khare, P., Raju, M., Gupta, S.C., Shukla, H., 2024. Liberalized automatic generation control of interconnected thermal-hydro-gas system using firefly algorithm optimized PID controller. Int. J. Numer. Model. Electron. Network. Dev. Field. 37 (1). https://doi.org/10.1002/jnm.3136.

Khokhar, B., Dahiya, S., Parmar, K.P.S., 2021. A novel hybrid fuzzy PD-TID controller for load frequency control of a standalone microgrid. Arab. J. Sci. Eng. 46 (2), 1053–1065.

Mbungu, N.T., Ismail, A.A., AlShabi, M., Bansal, R.C., Elnady, A., Hamid, A.K., 2023. Control and estimation techniques applied to smart microgrids: a review. Renew. Sustain. Energy Rev. 179. https://doi.org/10.1016/j.rser.2023.113251.

Mishra, S., Patel, S., Prusty, R.C., Panda, S., 2020. MVO optimized hybrid FOFPID-LQG controller for load frequency control of an AC micro-grid system. World J. Eng. 17 (5), 675–686. https://doi.org/10.1108/WJE-05-2019-0142.

Mohan, A.M., Meskin, N., Mehrjerdi, H., 2023. LQG-based virtual inertial control of islanded microgrid load frequency control and DoS attack vulnerability analysis. IEEE Access 11, 42160–42179. https://doi.org/10.1109/ACCESS.2023.3271012.

Monteiro, L.M., Igreja, J.M., 2015. On performance of distributed model predictive control in power system frequency regulation. Lect. Notes Electr. Eng. 321, 71–80. https://doi.org/10.1007/978-3-319-10380-8_8.

Moschos, I., Parisses, C., 2023. Combined frequency and voltage control of two-area multi-source interconnected microgrids via the 2DOF-TIDμ controlle. e-Prime - Adv. Electr. Eng. Electron. Energy 5. https://doi.org/10.1016/j.prime.2023.100268.

Nallolla, C.A., Vijayapriya, P., Chittathuru, D., Padmanaban, S., 2023. Multi-objective optimization algorithms for a hybrid AC/DC microgrid using RES: a comprehensive review. Electronics 12 (4), 1062. https://doi.org/10.3390/electronics12041062.

Nanda, J., Sakkaram, J.S., 2003. Automatic generation control with fuzzy logic controller considering generation rate constraint. In: Sixth International Conference on Advances in Power System Control, Operation and Management - Proceedings, vol. 2. Institute of Electrical Engineers, India, pp. 770–775. https://doi.org/10.1049/cp:20030683.

Nayak, P.C., Prusty, U.C., Prusty, R.C., Panda, S., 2021. Imperialist competitive algorithm optimized cascade controller for load frequency control of multi-microgrid system. Energ. Source. Part A 1–23.

Noman, A.M., Aly, M., Alqahtani, M.H., Almutairi, S.Z., Aljumah, A.S., Ebeed, M., Mohamed, E.A., 2024. Optimum fractional tilt based cascaded frequency stabilization with MLC algorithm for multi-microgrid assimilating electric vehicles. Fractal Fract. 8 (3). https://doi.org/10.3390/fractalfract8030132.

Patel, S., Ghosh, A., Ray, P.K., 2023. Improved power flow management with proposed fuzzy integrated hybrid optimized fractional order cascaded proportional derivative filter (1+ proportional integral) controller in hybrid microgrid systems. ISA Trans. 136, 742–754.

Peddakapu, K., Mohamed, M.R., Srinivasarao, P., Arya, Y., 2024a. Assessment of energy storage and renewable energy sources-based two-area microgrid system using optimized fractional order controllers. J. Energy Storage 86. https://doi.org/10.1016/j.est.2024.111191.

Peddakapu, K., Mohamed, M.R., Srinivasarao, P., Licari, J., 2024b. Optimized controllers for stabilizing the frequency changes in hybrid wind-photovoltaic-wave energy-based maritime microgrid systems. Appl. Energy 361. https://doi.org/10.1016/j.apenergy.2024.122875.

Peddakapu, K., Mohamed, M.R., Sulaiman, M.H., Srinivasarao, P., Veerendra, A.S., Leung, P. K., 2020. Performance analysis of distributed power flow controller with ultra-capacitor for regulating the frequency deviations in restructured power system. J. Energy Storage 31. https://doi.org/10.1016/j.est.2020.101676.

Pérez-Denicia, E., Fernández-Luqueño, F., Vilariño-Ayala, D., Manuel Montaño-Zetina, L., Alfonso Maldonado-López, L., 2017. Renewable energy sources for electricity generation in Mexico: a review. Renew. Sustain. Energy Rev. 78, 597–613. https://doi.org/10.1016/j.rser.2017.05.009.

Sahu, P.C., Mishra, S., Prusty, R.C., Panda, S., 2018. Improved -salp swarm optimized type-II fuzzy controller in load frequency control of multi area islanded AC microgrid. Sustain. Energy Grids Netw. 16, 380–392. https://doi.org/10.1016/j.segan.2018.10.003.

Satapathy, P., Mishra, S.P., Sahu, B.K., Debnath, M.K., Mohanty, P.K., 2018. Design and implementation of Whale optimization algorithm based PIDF controller for AGC problem in unified system. In: Communications in Computer and Information Science, 837, pp. 837–846. https://doi.org/10.1007/978-981-13-1936-5_85.

Shayeghi, H., Rahnama, A., Bizon, N., Szumny, A., 2025. Interconnected microgrids load-frequency control using stage-by-stage optimized TIDA+1 error signal regulator. Eng. Rep. 7 (1). https://doi.org/10.1002/eng2.13095.

Shukla, H., Raju, M., 2024. Combined frequency and voltage regulation in an interconnected power system using fractional order cascade controller considering renewable energy sources, electric vehicles and ultra capacitor. J. Energy Storage 84, 110875.

Singh, S.P., Prakash, T., Singh, V.P., Babu, M.G., 2017. Analytic hierarchy process based automatic generation control of multi-area interconnected power system using Jaya algorithm. Eng. Appl. Artif. Intell. 60, 35–44. https://doi.org/10.1016/j.engappai.2017.01.008.

Tumari, M.Z.M., Ahmad, M.A., Ghazali, M.R., Suid, M.H., 2024. Optimizing PID controller parameters for robust automatic voltage regulator system through indirect design approach-2. Glob. Energy Interconnect. 7 (5), 682–696. https://doi.org/10.1016/j.gloei.2024.10.009.

Uddin, M., Mo, H., Dong, D., Elsawah, S., Zhu, J., Guerrero, J.M., 2023. Microgrids: a review, outstanding issues and future trends. Energy Strategy Rev. 49. https://doi.org/10.1016/j.esr.2023.101127.

Ullah, K., Ullah, Z., Aslam, S., Salam, M.S., Salahuddin, M.A., Umer, M.F., Humayon, M., Shaheer, H., 2023. Wind farms and flexible loads contribution in automatic generation control: an extensive review and simulation. Energies 16 (14), 5498. https://doi.org/10.3390/en16145498.

Verma, R., Gawre, S.K., Patidar, N.P., Nandanwar, S., 2024. A state of art review on the opportunities in automatic generation control of hybrid power system. Elec. Power Syst. Res. 226. https://doi.org/10.1016/j.epsr.2023.109945.

CHAPTER

8

Frequency control in multi-microgrid systems

Amil Daraz[1], *Muhammad Majid Gulzar*[1,2,3], *and Suheel Abdullah Malik*[4]

[1] Center for Sustainable Energy Systems, King Fahd University of Petroleum & Minerals, Dhahran, Saudi Arabia; [2] Department of Control & Instrumentation Engineering, King Fahd University of Petroleum & Minerals, Dhahran, Saudi Arabia; [3] Center for Smart Mobility and Logistics, King Fahd University of Petroleum & Minerals, Dhahran, Saudi Arabia; [4] Department of Electrical and Computer Engineering, Faculty of Engineering and Technology, International Islamic University, Islamabad, Pakistan

8.1 Introduction to multi-microgrid (MMG) load frequency control

Hybrid power systems, typically refers to a localized setup which combine diverse energy sources such as wind, solar, biomass and diesel generators (Zohuri, 2017). A multi-microgrid (MMG) system represents an advanced evaluation of power networks where several microgrids, each potentially consisting of hybrid energy sources and storage units, are interconnected to operate in a coordinated and resilient manner. A multi-microgrid becomes more complex by virtue of its size, disturbances, and external uncertainties (Mehrizi-Sani and Iravani, 2010). The introduction of non-conventional energy sources like solar, wind, etc., and fluctuating loads cause instability of the system frequency, which in turn leads to unstable system operation. In order to maintain a nominal load frequency, automatic load frequency control plays the most important role (Siti et al., 2022). A balance between the total amount of power produced and the amount of power that is consumed is through load frequency control. In view of the intermittent nature of renewable energy sources, there has

Frequency Control in Microgrids
https://doi.org/10.1016/B978-0-443-36476-1.00012-2

been an increase in the tie line and load frequency deviations to a high level of concern (Prusty et al., 2022). A variety of automatic load frequency control methods have been proposed in earlier studies in order to maintain the trade-off between system operation, efficiency, security and reliability (Daraz, 2023).

The integration of renewable energy sources (RESs), energy storage systems (ESS,s), and other distributed generations (DGs) units can occur in several configurations, including islanded microgrids, grid-connected microgrids, or interconnected microgrid systems. In the islanded mode, each microgrids operate independently to supply powers to the local loads. A multi-microgrid setups, on the other hands, enable coordination among several microgrids each with its own RESs and ESSs through a network architecture that support both DC and AC power sources like wind turbines, photovoltaic systems, and energy storage systems (Singh and Arya, 2023; Smith et al., 2015). Multi microgrid systems, due to their interconnected nature and cooperative operation, enhance grid reliability, increase flexibility, and provide greater power availability through shared management strategies. Nevertheless, both islanded and multi-microgrid systems faces frequency stability and regulation challenges primarily due to the intermittency of RESs and the dynamic behaviors of the electric vehicles. The fluctuations in power generation caused by wind turbines and photovoltaics can create imbalances between supply and demand within microgrids (Fathollahi et al., 2022; Abubakr et al., 2022). Thus, extreme RES diffusion in microgrids can lessen inertia response and frequency stabilization, where even slight frequency variations can cause interruptions (Freitas et al., 2022). For the effective operation of a microgrid, frequency stabilization must be prudently managed among a number of microgrids. MMG networks require sophisticated control tactics that are capable of adapting to fluctuations in load and generation.

In traditional power systems, proportional integral derivative controllers are usually inadequate because of their instability to handle nonlinearity, as well as uncertainties that are common to microgrid systems (Mohamed et al., 2024; Naderi et al., 2023). A variety of advanced control strategies are being proposed in order to improve the stability and responsiveness of the system, such as model predictive controllers (MPC) (Gulzar et al., 2022; Gulzar, 2023) and cascaded base controllers (Choudhary et al., 2022; Elkasem et al., 2022). The fractional order controller (Zhang et al., 2023) provides enhanced tuning flexibility, whereas fuzzy and adaptive-based controllers (Arya et al., 2023; Sharma et al., 2020; Sabahi et al., 2021) are highly effective at handling renewable energy fluctuations and system uncertainties. These controllers are adjusted to attain optimal performance employing natured inspired metaheuristic optimization algorithms, which play an essential role in accomplishing optimal performance. Several approaches have validated substantial improvements in reducing frequency aberrations and enhancing stability margins over the past few decades, including particle swarm optimization (Mishra et al., 2020), pattern search optimization with hybridization of Dragon fly algorithm (Mishra et al., 2023), and imperialist competitive algorithm (Nayak et al., 2025). Several other optimization methods, such as gray wolf optimization (Can et al., 2022), jellyfish search optimization (Daraz et al., 2023), marine predator algorithm (Yakout et al., 2021), and world cup optimization algorithm (V. S. R. Pavan et al., 2023), have shown superior convergence properties and robustness in determining controller parameter values. To overwhelmed the restrictions of conventional algorithms, researchers have developed innovative hybrid algorithms, such as the fitness-dependent optimization method hybridized with sine cosine algorithm (Daraz et al., 2022).

Despite these progression in controller designing for conventional and standalone microgrid systems, the rise of multi-microgrid systems has introduced additional challenges. MMG systems encompass the incorporation of multiple microgrids that works in both islanded and grid connected modes, often connected diverse energy sources and energy storage units (Guan et al., 2022). This distributed and decentralized structure strengthens the need of more sophisticated control techniques to manage inter-microgrid coordination, ensure frequency and volage stability and deal with uncertainties from intermittent resources (Si et al., 2024). Hence, adapting and extending controller designs and metaheuristic optimization strategies for MMG systems has become a critical area of research and development. Therefore, a properly optimized controller can manage frequency stabilization very effectively by minimizing overshoot, reducing settling time, and increasing system resilience under varied load conditions and renewable energy penetration. To maintain power balance and system stability, load frequency control is an important aspect of MMG systems. Advanced strategies ensure minimal frequency variations, quick response times, and reliable performance across a broad range of applications. Adaptive, predictive, and optimization-based control approaches are becoming increasingly important in MMG networks since they enhance reliability, efficiency, and robustness (Siti et al., 2022; Gulzar et al., 2022). Table 8.1 demonstrates the comparative literature review of LFC systems.

8.1.1 Need of multi-microgrid systems

The microgrid is a collection of distributed generators escorted by energy storage devices and resident loads that act as a power distribution system. Renewable energy sources such as fuel cells, photovoltaics, micro turbines, and wind turbines are being used to generate electricity. The environmental conditions may also cause these sources to deviate from

TABLE 8.1 Comparative literature review of LFC systems.

Ref.	Techniques	Key findings	Power systems
	Classical & PID-based controllers		
Mishra et al. (2020)	PSO-optimized PIDF controller	Superior dynamic response with faster settling times	Multi-islanded AC microgrids
Can et al. (2022)	PI–(1 + DD) controller with gray wolf optimization	Improved load frequency control	Non-reheat multi-area power system with solar and wind
V. S. R. Pavan et al. (2023)	GA-optimized PID controller	Improved stability in a non-linear dynamic system	Microgrid systems
Fractional order controllers (FOPID, FOI-TD, etc.)			
Prusty et al. (2022)	Fractional order type-2 fuzzy based PID	Improved frequency regulation using moth swarm optimization	Hybrid renewable microgrid
Mohamed et al. (2024)	FO-PI with virtual inertia emulator	Robust frequency control using EVs	MMG with distributed energy
Choudhary et al. (2022)	Cascade FOPI-FOPTID controller	Significant reduction in frequency deviations	Interconnected single and multi-area systems
Zhang et al. (2023)	FO PI-PDF with driver-training optimization	Improved load frequency stabilization	Diverse hybrid systems
Daraz et al. (2023)	Modified FOPID with jellyfish optimization	Frequency regulation of hybrid systems	Hybrid conventional/ renewable
Daraz et al. (2022)	FO integral-tilt derivative controller	Effective control under high renewable penetration	Interconnected hybrid systems
Intelligent/Fuzzy/AI-based controllers			
Elkasem et al. (2022)	Cascaded TD-TI with Chaos Game optimizer	Superior frequency response under renewables	MMG with wind and solar

TABLE 8.1 Comparative literature review of LFC systems.—cont'd

Ref.	Techniques	Key findings	Power systems
Classical & PID-based controllers			
Arya et al. (2023)	AI/Data-based control	Improved predictive modeling and control	Smart grid power systems
Sharma et al. (2020)	Layered RANN with robust control	Better performance under varying load	DFIG systems
Sabahi et al. (2021)	Adaptive type-2 fuzzy PID	Efficient under intermittent resources	AC microgrid with DG
Mishra et al. (2023)	Pattern search + dragonfly algorithm	Optimized adaptive fuzzy PID for LFC	MMG systems
Yakout et al. (2021)	Fuzzy PID with filter + marine predator algorithm	Robustness under load, irradiance, and wind fluctuations	MMG with renewables & ESS
Adaptive/Model predictive controllers			
Siti et al. (2022)	Adaptive heuristic-based MPC	Enhanced frequency stability with ESS	Renewable-integrated MMG
Gulzar et al. (2022)	MPC	Improved robustness against uncertainties	Hybrid renewable systems
Other/Hybrid approaches			
Daraz (2023)	Cascaded PI-PD with Sewing Training optimization	LFC for renewables and ESS	Marine microgrids
Nayak et al. (2025)	ICA-based cascade PD-(1+PI) controller	Better trade-off between stability and response	MMG with renewables

their nominal operation. In addition to supply and demand mismatches, there is a potential problem of load frequency and tie-line power aberrations (Siti et al., 2022). Energy storage systems with rapid response capabilities are needed to support the frequency regulation strategy in order to meet rapidly changing demands. In essence, a properly designed controller that detects disturbances effectively is necessary for effective load frequency control of the power system. Moreover, the microgrid can also be connected through tie lines, so in case of disruption, power will be switched (Abubakr et al., 2022). As a result of this, the microgrid will become more reliable, robust, responsive, durable, and sustainable.

8.1.2 Challenges of frequency control in multi-microgrid systems

It is particularly challenging to manage frequency in multi-microgrid systems due to the inherent complexity and variability of these systems. MMG's systems include power storage systems along with wind turbines and solar panels. Renewable energy sources such as wind and solar have intermittent characteristics that pose a variety of challenges. MMGs experience a number of problems as a result of their decentralized structure, which leads to a number of issues. To guarantee global microgrid steadiness, they must work organized in order to manage power acquaintances in an appropriate style. Microgrid distractions can cause momentous frequency instabilities if not addressed suitably. Power output variations can cause considerable interruptions in system frequencies if they are recurrent and erratic (Saha et al., 2023). There is a key contest in sustaining frequency steadiness in these states due to swift variations in generation and load circumstances across many unified microgrids. Furthermore, diverse microgrids may have unlike control goals and restrictions, making co-ordinated frequency supervision challenging. In order to dazed these effective problems, MMGs must pay intricate control tactics that can familiarize to the fundamental nonlinearities and reservations of the system (V. S. R. Pavan et al., 2023).

8.1.3 Significance of integrated LFC strategies

The microgrid system is a small-scale power system that has considerable benefits due to its capability to proficiently contribute nearby distributed energy resources and efficiently control distribution demands. This has caused in the extensive deployment of microgrids in diverse places, which offer a solution to the energy needs of rural settlements that face intrinsic complications in connecting to the primary electric grid (Mohamed et al., 2024). As a result of this, and the potential environmental impact of the conventional fossil fuel issue, the usage of renewable energy sources as a replacement for traditional fossil fuels has grown in popularity in microgrids. Microgrids have issues due to the intermittent nature of renewable energy resources and the low inertia of distributed generation sources, resulting in a rapid rate of frequency change. These characteristics have a substantial impact on the microgrid's reliability and security. The microgrid structure is critical in creating a self-sufficient and stable electrical power supply. Furthermore, these microgrids are interconnected, allowing for increased total power capacity and the establishment of a broader network (Naderi et al., 2023).

8.2 Coordination methods for multi-microgrid power systems

A well-coordinated method is required to attain the maximum efficiency of power grid systems by utilizing the most prevalent variable renewable energies, such as wind power and photovoltaic sources. The low cost of installation and maintenance of renewable energy technologies makes them less competitive than traditional energy sources due to their intermittent nature and high installation and maintenance costs (Li and Zhou, 2024). In contrast to conventional energy sources, renewable energy sources offer a number of compensations, including the ability to lessen greenhouse gas emissions and a reduction in reliance on fossil fuels, both of which can help reduce running costs. It is also important to consider the location and coordination of renewable energy sources. To achieve a better microgrid coordination, the optimization process must be a better strategy to apply for an excellent dispatch model (She et al., 2023). Although optimization is important, it does not necessarily mean that the microgrid operation will be efficient as well. To attain this, it is necessary to understand dynamic behavior in order to pretend the system's operation based on an optimal control approach for an ideal solution to the problem.

8.2.1 Decentralized versus centralized coordination

A comparison of centralized against decentralized coordination is provided in Table 8.2 (Li and Zhou, 2024; She et al., 2023; Casavola et al., 2022).

8.2.2 Multi-microgrid (MMG) communication protocols

In order for multi-microgrid systems to work properly, it is necessary to have a reliable communication protocol. Microgrid systems require efficient communication links in order to maintain system stability, ensure optimal power distribution, and actively participate in the energy market. The exchange of data between microgrids is crucial, especially for real-time tasks such as load balancing, frequency control, and fault management, which require synchronization between microgrids. It is essential for multi-microgrid systems to have high reliability, scalability and low

TABLE 8.2 A Comparison of centralized versus decentralized coordination.

Feature	Centralized coordination	Decentralized coordination
Control structure	A single controlling entity manages all microgrids	Control is distributed among individual microgrids
Decision-making	Centralized decision-making ensures optimal system-wide control	Local controllers make real-time decisions based on current conditions
Response time	Response may be slower due to communication and processing delays	Faster response times as decisions are made locally
Resilience to failures	Susceptible to single points of failure	More resilient; individual microgrids can operate independently
Data management	Centralized data analysis allows for comprehensive insights	Local data analysis can result in differing operational objectives
System synchronization	Easier to achieve system-wide synchronization	More challenging to maintain synchronization; may require advanced algorithms
Communication Overhead	Higher communication demands between microgrids and the central controller	Reduced communication overhead; local decisions require less information exchange
Adaptability	Slower to adapt to localized issues	Faster adaptation to local disturbances
Operational efficiency	Can optimize overall performance but may overlook local specifics	Potential inefficiencies due to conflicting local priorities

latency, as well as robust security measures. The system would be vulnerable to instability or disruption if these components were absent, particularly in the coordination of supplementary services, which are vital to system stability. A number of communication protocols have addressed these criteria, including the IEC 61,850 standard, which promotes interoperability among devices from different manufacturers (Yoo et al., 2011). Furthermore, DNP3 is a crucial component of SCADA systems in multi-microgrids, since it works effectively under low-bandwidth, remote, or resource-constrained conditions. However, despite the availability of these protocols, multi-microgrid systems encounter difficulties despite the availability of these protocols (Boyanapalli, 2018). A number of legacy systems and different protocols might have difficulty integrating due to interoperability issues. It may be challenging to achieve low latency and adequate bandwidth for real-time control in remote areas with insufficient communication infrastructures due to a lack of adequate communication infrastructure. There is still a serious problem of cybersecurity in multi-microgrids, as increased connectivity and reliance on internet-based communications creates the possibility of an intrusion. In order to maintain system integrity and avoid harmful disruptions, keeping communication channels protected from unauthorized access is of utmost importance.

8.3 Control architecture for multi-microgrid systems

A capable control assembly is obligatory for energy supervision, management, and system dependability in multi-microgrid power systems. This kind of arrangement comprises of several microgrids that are consistent and run autonomously of each other but are measured by a centralized control system that permits them to slog together (Guan et al., 2022). In this way, reliability and flexibility are improved, local energy supplies are adjusted, and network permanency is guaranteed through a decentralized structure. A multi-microgrid power systems is portrayed schematically in Fig. 8.1.

8.3.1 Integrated control structure

A co-ordinated control process between numerous microgrids is conceivable through unified control erections which permit them to work concurrently, within the context of an incorporated control structure (Casavola et al., 2022). By executing these skills, real-time statistics distribution and verdict making are made conceivable, which leads to a better power distribution, a stable load demand, and the manipulation of local energy resources. A control system integrated with centralized and decentralized control approaches is generally referred to as a centralized control system (She et al., 2023). In addition to overseeing the network, the centralized controller also focuses on market engagement, grid stability, and economic dispatch on the network. In addition, microgrids that have decentralized controls can be automatically adjusted to changes in load or generation without requiring guidance from centralized centers.

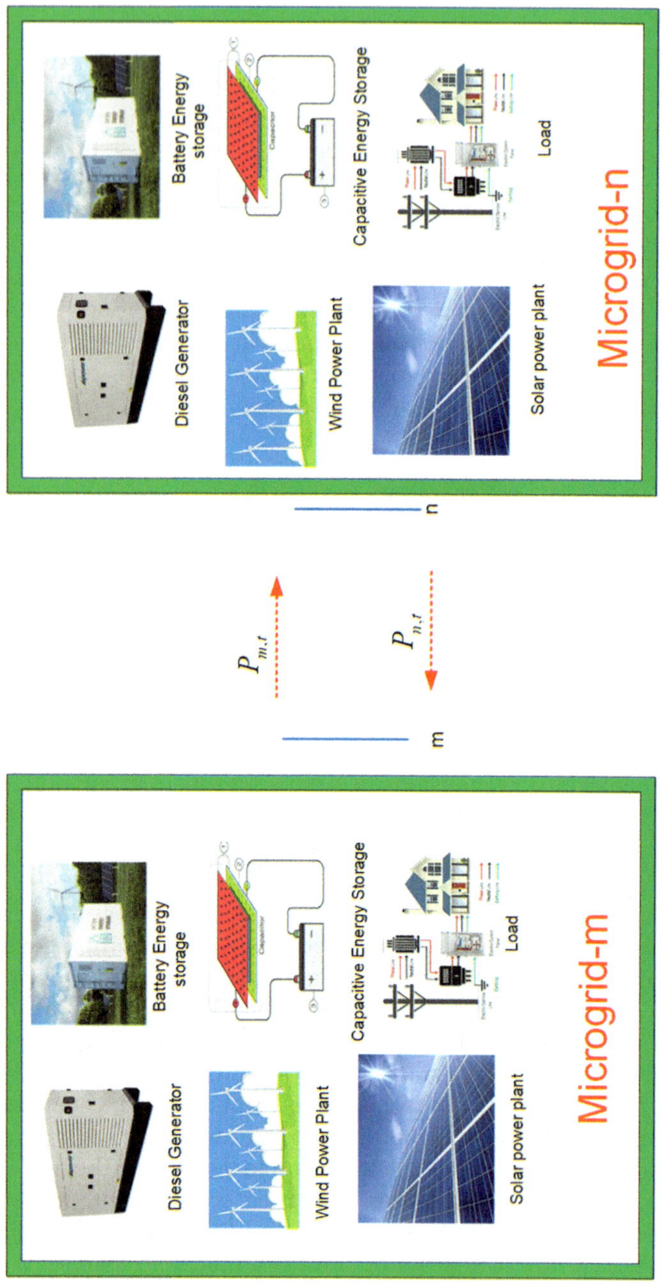

FIGURE 8.1 A schematic illustration of multi-microgrid power systems.

8.3.2 Hierarchical control structure

The implementation of hierarchical control structures is very common in multi-microgrid systems, which are divided into three levels: the primary level, the secondary level, and the tertiary level (Guan et al., 2022). Initially, voltage and frequency are managed by the primary control level in each microgrid, which is fully decentralized and operates as a single power plant. As a result, local disturbances within each microgrid can be quickly addressed to ensure stability within the system. The secondary control system works in conjunction with the primary control system in order to restore normal operating conditions in all microgrids following the primary response to a disturbance. With a combination of decentralized and centralized control mechanisms, it manages to synchronize key parameters throughout the entire power systems. Tertiary Control level optimizes power exchanges between microgrids and external grids, supporting energy trade, demand management, and integration with the external energy market. The tertiary control aims to enhance energy efficiency and economic performance across the entire system, guided by centralized decision-making.

8.3.3 Hybrid approaches for improved performance

In order to improve the performance of multiple microgrids, a hybrid control strategy is increasingly being adopted in order to balance local autonomy with system-wide coordination and, in result, to improve their performance. A decentralized, distributed, and centralized approach is combined to maximize their strengths. The goal of hybrid systems is to increase efficiency, reliability, and adaptability of the system. There are two types of microgrid controllers: one that manages the critical functions, such as the economic dispatch and the system optimization, while there is one that addresses the real-time operations in each microgrid. As a result, local responses can be provided quickly without compromising broader objectives such as reducing energy consumption and saving costs. By integrating renewable energy sources and storage systems, hybrid control offers flexibility in energy management (Siti et al., 2022). These architectures maximize renewable energy utilization and maintain system reliability by blending centralized optimization with localized decision making, which makes them ideal for optimizing renewable energy utilization.

8.4 Multi-microgrid load frequency control challenges and solutions

In load frequency control, coordination of multiple interconnected microgrids with distinct operational dynamics and control requirements that arise from the design and operation of multiple microgrids with different operational dynamics poses a challenge. A microgrid typically contains a variety of energy storage systems along with a variety of energy sources and loads that fluctuate because of the variability of renewable energy sources. The challenges associated with load frequency control of multi-microgrid systems and their solutions are given in Table 8.3.

TABLE 8.3 Challenges and solutions for LFC in multi-microgrid systems (Siti et al., 2022; Smith et al., 2015; Gulzar et al., 2022; Gulzar, 2023; Sharma et al., 2020; Mishra et al., 2020; Si et al., 2024; Saha et al., 2023).

Challenges in load frequency control for multi-microgrid systems	Proposed solutions
Diverse operational dynamics and control requirements	Implementation of coordinated control strategies tailored for different microgrids with varying energy sources, storage systems, and load profiles
Frequency instability due to fluctuating renewable energy sources	Advanced control strategies like model predictive control (MPC), adaptive controllers, and fractional-order PID controllers to dynamically adjust to frequency deviations
Rapid load changes during synchronized control	Real-time load balancing using robust local controllers and hierarchical control structures for better synchronization
Controller lag and poor communication infrastructure	Use of faster communication protocols, improved controller tuning, and predictive control techniques
Interoperability issues between legacy systems and new control strategies	Adoption of standard communication protocols and development of compatible interfaces for seamless integration
Cybersecurity vulnerabilities in distributed and decentralized systems	Implementing secure communication networks, encryption techniques, and intrusion detection systems
Managing large volumes of data for real-time decision-making	Development of sophisticated algorithms and machine learning models to process data efficiently and enable predictive decision-making
Need for advanced forecasting and adaptive protocols	Use of predictive algorithms for load and generation forecasting and adaptive communication frameworks for real-time responsiveness

8.4.1 Resilience and robustness Considerations in load frequency control

In order to ensure the reliable operation of a multi-microgrid power system, it is necessary to have a load frequency control system that is robust and resilient. Despite disturbances such as generator problems, changes in load demand, and grid interruptions, a resilient system can maintain operational stability. For multi-microgrid systems to be resilient in the event of power outages, redundant communication pathways must be implemented. Therefore, the total breakdowns of communication are reduced, which would otherwise lead to issues with control if they continued. The term robustness refers to the ability of a system to function effectively over a long period of time, regardless of fluctuations in load and renewable energy generation. The development of fault-tolerant control algorithms can improve robustness by identifying and compensating for component failures without undermining overall system stability. As a result of these algorithms, even if parts of the infrastructure are compromised, the system remains operational. Further strengthening system robustness can be achieved by utilizing advanced control strategies. Machine learning and advanced model-based controls enable these strategies to dynamically respond to real-time changes, thereby improving overall performance in unpredictable situations (Gulzar et al., 2022; Zhang et al., 2023; Sharma et al., 2020).

8.5 Case study on multi-microgrid LFC

As shown in Fig. 8.2, the interconnected microgrid power systems under investigation include two typical microgrids. Each microgrid consists of reheat thermal power, wind power, solar power, capacitive energy storage, battery energy storage system and load. The implementation of GRC with a reheat turbine in each control section greatly increases nonlinearity constraints in this network model. Table 8.4 shows the nominal parameters of the model. The proposed FOI-TD controller is compared to several classic PI, PID, and FOPID controllers. A mother optimization approach in MATLAB is used to fine-tune the controller coefficients.

8.5.1 Controller structure

LFC systems often use fractional-order or non-integer controllers because of their versatility, disturbance rejection capabilities, and design methodologies. The proposed Fractional Order Integrator-Tilt Derivative (FOI-TD) controller is the cascaded form of the controller that integrates

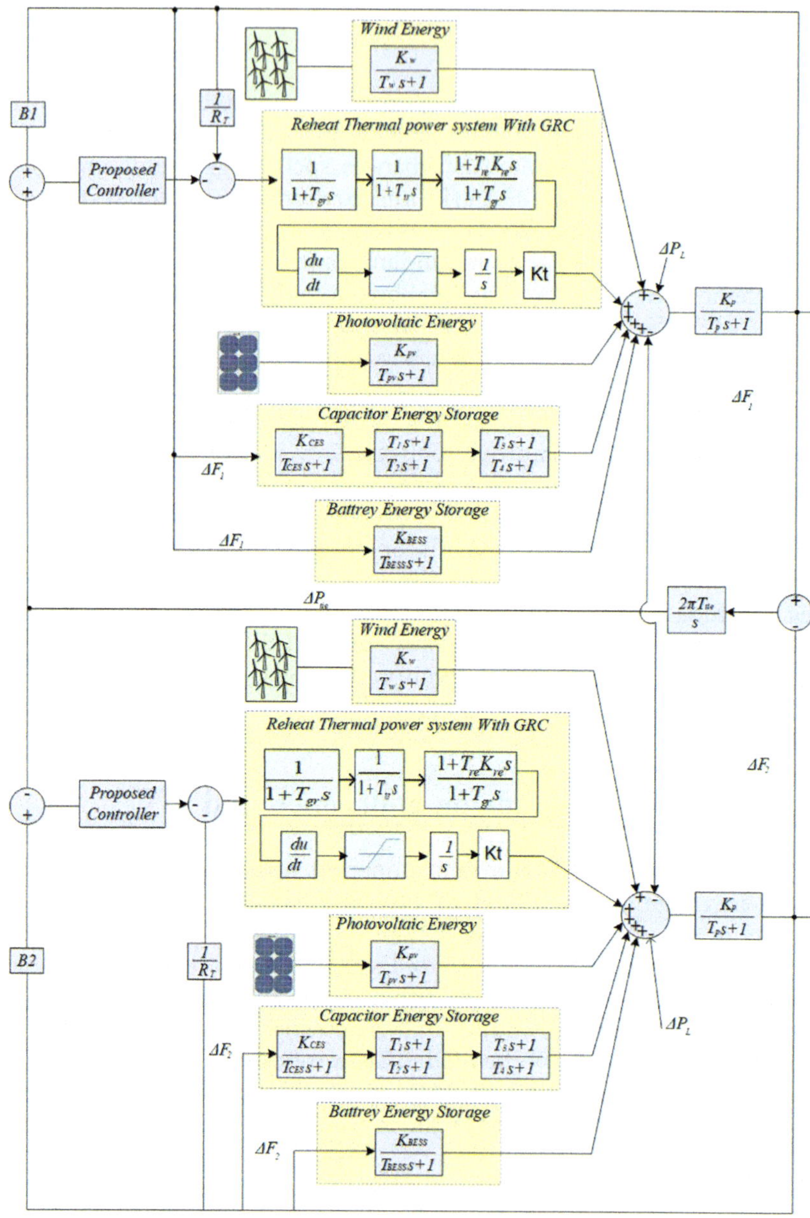

FIGURE 8.2 Mathematical modeling of two-area microgrid system.

fractional order integrator with tilt derivative. Fractional order integrators (FOI) introduce an extra degree of freedom through its fractional term (λ) which allows the fine tuning of the systems transient and steady state

TABLE 8.4 Nominal parameters of two area microgrid systems (Daraz, 2023; Choudhary et al., 2022; Daraz et al., 2022).

T_p	11.49 s	K_p	68.97Hz/puMW
R_T	2.4 Hz/puMW	T_{tie}	0.0866 puMW/rad
Kt	0.60	B1	0.4312 Hz/puMW
R_T	2.4 Hz/puMW	B2	0.4312 Hz/puMW
K_{BESS}	0.003	T_{BESS}	0.10 s
T_{gr}	0.08 s	T_{tr}	0.3sec
T_{re}	10 s	K_{re}	0.3
K_w	1	K_{CES}	0.3
K_{PV}	1	T_{CES}	0.046 s
T_{pv}	0.3 s	T_1	0.280 s
T_W	1.5 s	T_2	0.025 s
T_3	0.0411 s	T_4	0.39 s
Kt	0.60	ΔP_L	0.01puMW

behavior. On the other hand, the tilt derivative (TD) controller is the modification of traditional derivate with the addition of tilt gain structure that enhance the controller's ability to respond to rapid changes while reducing sensitivity to noise. By combining these two controllers in a cascade configuration can enhance the ability of the controller. The transfer function for FOI and TD are given below.

$$G_{FOI}(s) = \frac{K_i}{s^{\lambda}} \tag{8.1}$$

$$G_{TD}(s) = \frac{K_T}{\frac{1}{s^n}} + K_d s \tag{8.2}$$

The structure of the proposed FOI-TD controller is shown in Fig. 8.3.

8.5.2 Mother optimization algorithm (MOA)

The Mother Optimization Algorithm (MOA), is a metaheuristic algorithm which has been used for the tuning of the proposed FOI-TD controller to control the load frequency multi-microgrid systems. The fundamental inspiration of the mother optimization algorithm is to stimulated the mother care of children in three phases: (1) Instruction; (2) Counseling; and (3) Coaching. This technique is based on population-based optimization and improves potential solutions for optimization problems iteratively.

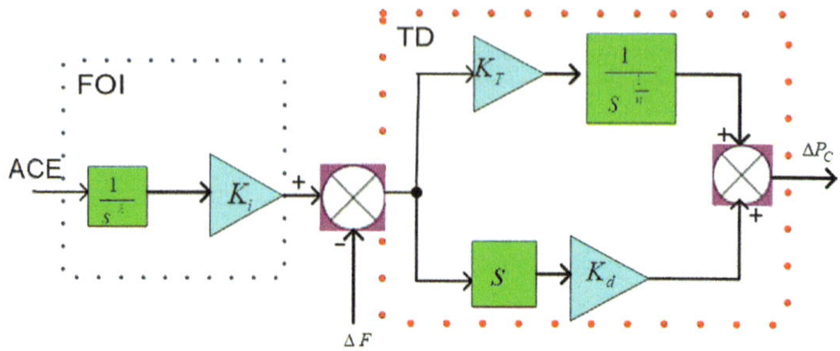

FIGURE 8.3 Proposed FOI-TD controller structure.

Mathematically, the population of candidate solutions can be represented in Eq. (8.3) as a matrix (Matoušová et al., 2023). The optimization process begins with the generation of an initial population using Eq. (8.4) (Matoušová et al., 2023).

$$X = \begin{bmatrix} X_1 \\ \vdots \\ X_i \\ \vdots \\ X_N \end{bmatrix}_{N\times d} = \begin{bmatrix} x_{1,1} & \cdots & x_{1,i} & \cdots & x_{1,d} \\ \vdots & \ddots & \vdots & \ddots & \vdots \\ x_{i,1} & \cdots & x_{i,j} & \cdots & x_{i,d} \\ \vdots & \ddots & \vdots & \ddots & \vdots \\ x_{N,1} & \cdots & x_{N,j} & \cdots & x_{N,d} \end{bmatrix}_{N\times d} \tag{8.3}$$

$$x_{i,j} = lb_j + \text{r.}\left(ub_j - lb_j\right), i = 1, 2, 3\ldots\ldots N, J = 1, 2, \ldots, d \tag{8.4}$$

The population matrix in the MOA is represented as X, where N is the population size and m is the number of decision variables. $X_i = (x_{i,1}, \ldots, x_{i,j}, \ldots, x_{i,m})$ is an applicant solution that consists of m variables, where $x_{i,j}$ indicates the jth variable of the ith solution. The symbols lb_j and ub_j represent the lower and upper bounds of the jth decision variable, respectively.

MOA considers each population member to be a potential solution to the problem under optimization, and the objective function can be computed as below (Matoušová et al., 2023):

$$F = \begin{bmatrix} F_1 \\ \vdots \\ F_i \\ \vdots \\ F_N \end{bmatrix}_{N\times 1} = \begin{bmatrix} F(X_1) \\ \vdots \\ F(X_i) \\ \vdots \\ F(X_N) \end{bmatrix}_{N\times 1} \tag{8.5}$$

MOA's algorithm population is updated in three phases based on mathematical modeling of mother-child interaction, which is described below.

Stage 1: Education (Exploration) Phase:

In this stage, each solution is updated using Eq. (8.6) (Matoušová et al., 2023). The position of the corresponding member is accepted if the objective function value improves in the new position, as represented in Eq. (8.7) (Matoušová et al., 2023).

$$x_{i,j}^{P1} = x_{i,j} + rand(0,1).\left(M_j - rand(2).x_{i,j}\right) \tag{8.6}$$

$$X_i = \begin{cases} X_i^{P1}, F_i^{P1} < F_i; \\ X_i, Otherwise \end{cases} \tag{8.7}$$

Where M_j represents the mother position with jth dimensions and F_i^{P1} signifies its cost function value. X_i^{P1} represents the new position and $x_{i,j}^{P1}$ shows the new position with jth dimension.

Stage 2: Advice (Exploration) Phase:

In MOA design, the position of other members of the population with greater values of the objective function than the member's own is considered deviant behavior. Based on the objective function value, Eq. (8.8) (Matoušová et al., 2023) determines the bad behavior (*BBi*) of each member. To simulate preventing the child from engaging in bad behavior, each member must create a new position using Eq. (8.9) (Matoušová et al., 2023). Consequently, by Eq. (8.10) (Matoušová et al., 2023), if the objective function improves, this new position replaces the previous one.

$$BB_i = \{X_k, F_k > F_i \wedge k\varepsilon\{1, 2, \ldots\ldots, N\}\}, where i = 1, 2, 3, \ldots\ldots N \tag{8.8}$$

$$x_{i,j}^{P2} = x_{i,j} + rand(0,1).\left(x_{i,j} - rand(2).SBB_{i,j}\right) \tag{8.9}$$

$$X_i = \begin{cases} X_i^{P2}, F_i^{P2} < F_i; \\ X_i, Otherwise \end{cases} \tag{8.10}$$

Stage 3: Upbringing (Exploitation) Phase:

The aim of this phase is to enhance the algorithm's capability to carry out local searches and exploit the search space. Using Eq. (8.11) (Matoušová et al., 2023), a new position is created for each member of the population to simulate the upbringing phase. Similarly, Eq. (8.12) (Matoušová et al., 2023), specifies that the previous position of the corresponding member is

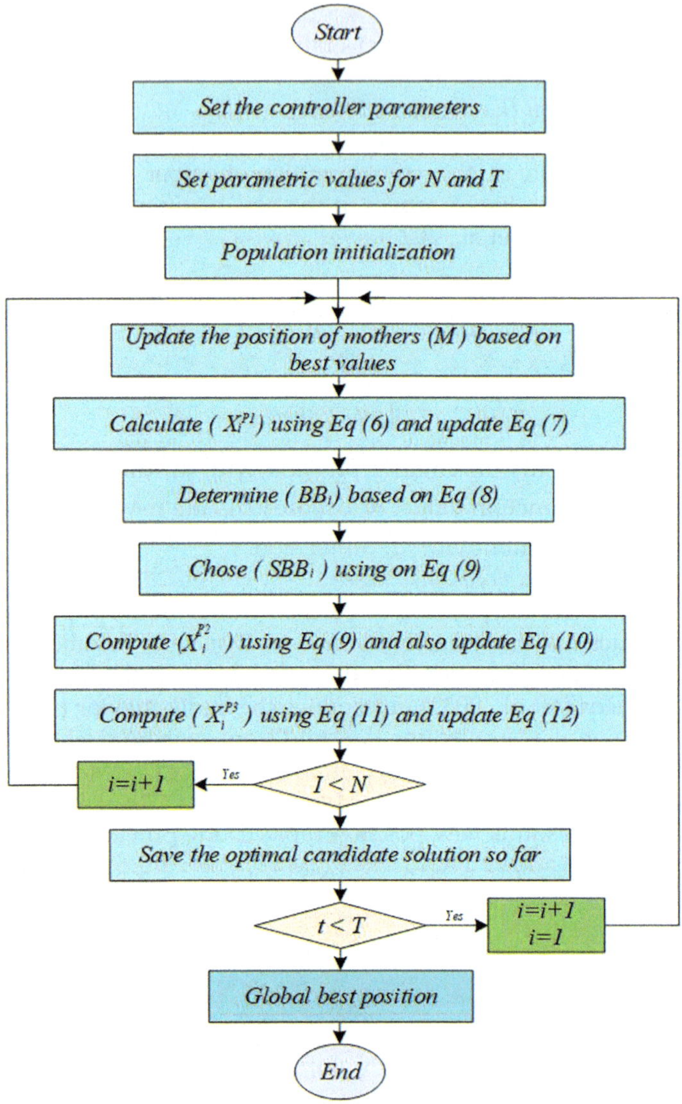

FIGURE 8.4 Flow chart of MOA.

replaced if the objective function value improves in the new position. The flowchart for MOA algorithm is shown in Fig. 8.4.

$$x_{i,j}^{P3} = x_{i,j} + (1 - 2.rand(0, 1).\frac{ub_j - lb_j}{t} \tag{8.11}$$

$$X_i = \begin{cases} X_i^{P3}, F_i^{P3} < F_i; \\ X_i, Otherwise \end{cases} \tag{8.12}$$

8.5.3 Controller performance evaluation

In this study, the performance of the FOI-TD controller was compared with other control approaches, including PI, PID, and FOPID, based on overshoot (Osh), minimum undershoot (Ush), and time settling (Ts), adjusting for changes in area-2, tie line, and area-1. In terms of settling time, overshoot, and undershoot, the transient performance analysis of the controllers (PI, PID, FOPID, and FOI-TD) reveals noticeable differences which has been shown in Fig. 8.5. Across all metrics, the FOI-TD controller consistently achieved the best results, displaying its ability to stabilize quickly with minimal oscillations. Among all parameters, it had the shortest settling times, measuring 2.22s for ΔF_1, 1.52s for ΔF_2, and 2.81s for ΔP_{tie}. This demonstrates its effectiveness in reducing oscillations and achieving steady-state conditions quickly. Similarly, FOPID controllers showed respectable settling times (2.34, 2.90, and 3.07s, respectively), while PID controllers had moderate performance, and PI controllers had the longest settling times, reflecting a more delayed response.

FOI-TD again excelled in overshoot, recording the lowest values across all parameters. It demonstrated precise control with minimal overshoot with values of 0.0,002,347 for ΔF_1, 0.0,002,201 for ΔF_2, and 0.0,003,508 for ΔP_{tie}, suggesting its ability to adhere to the target more closely. FOPID also performed well, maintaining low overshoot values, especially in ΔF_2 and ΔP_{tie}. A moderate overshoot was observed in the PID controller, while the PI controller showed the highest overshoot values, particularly in the ΔF_2 and ΔP_{tie} variables, suggesting greater response peaks and, therefore, less controlled tracking of the desired values. Undershoot analysis further highlighted FOI-TD's strong performance, as it achieved the smallest undershoot values, indicating a balanced and controlled response. During transient periods, it shows values of −0.0,006,158, −0.0,002,686, and −0.009,405 for ΔF_1, ΔF_2, and ΔP_{tie} respectively, which indicate efficient reductions of oscillatory behavior. It was found that the FOPID controller showed moderate undershoot values, while the PID controller showed higher undershoot, especially for ΔF_1 and ΔF_2, which may cause oscillations to increase. According to the PI controller's values, which are the highest among all controllers, there are significant deviations from target values, leading to less stability in the system's transient behavior.

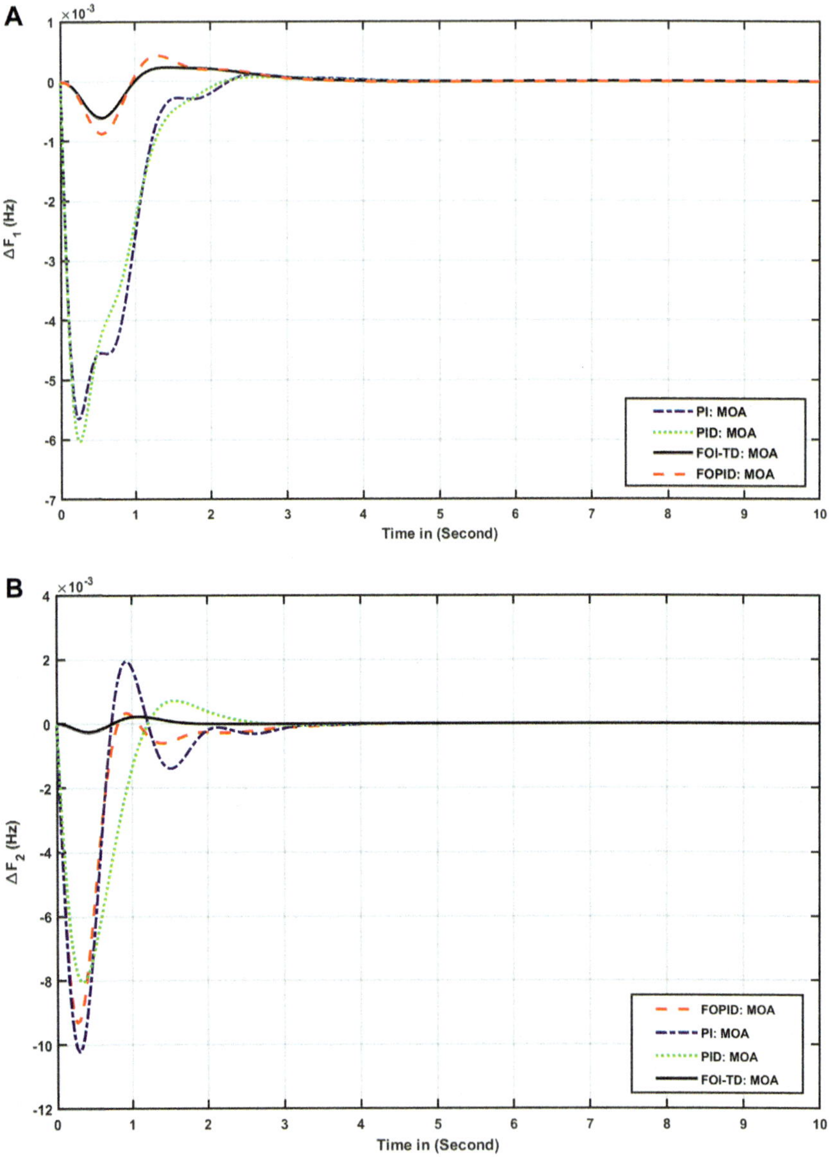
A
×10^-3
1
0
-1
-2
-3
-4
-5
-6
-7
ΔF_1 (Hz)
PI: MOA
PID: MOA
FOI-TD: MOA
FOPID: MOA
0 1 2 3 4 5 6 7 8 9 10
Time in (Second)
B
×10^-3
4
2
0
-2
-4
-6
-8
-10
-12
ΔF_2 (Hz)
FOPID: MOA
PI: MOA
PID: MOA
FOI-TD: MOA
0 1 2 3 4 5 6 7 8 9 10
Time in (Second)

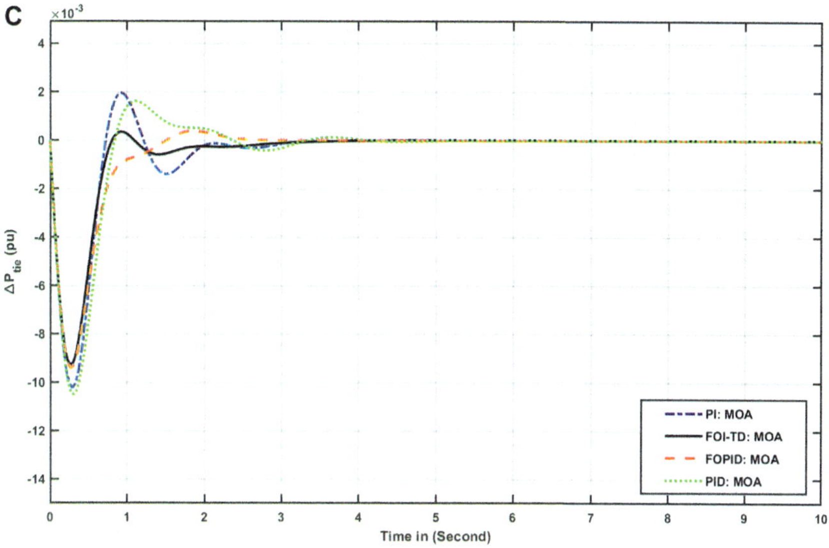

FIGURE 8.5 Analysis of the dynamic response of different controllers for (a) ΔF1; (b) ΔF2; (c) ΔPtie.

8.5.4 Algorithm performance evaluation

The results shown in Fig. 8.6 indicates that the mother optimization algorithm (MOA) with FOI-TD controllers outperforms both the gray wolf optimizer (GWO) and particle swarm optimization (PSO) in terms of transient response parameters, including settling time, overshoot, and undershoot. MOA obtained the shortest settling time (Ts) across all parameters (ΔF_1, ΔF_2, and ΔP_{tie}) with values of 2.69, 3.21, and 3.20s, respectively. This is a significant improvement over GWO, which had values of 2.86, 3.93, and 3.66s, and PSO, which had the longest settling times of 3.66, 4.09, and 3.71s for the same parameters. The reduction in settling time by MOA implies that the system will stabilize faster after a disruption. MOA demonstrated favorable overshoot findings, with ΔF_1, ΔF_2, and ΔP_{tie} values of 0.00,024, 0.00,023, and 0.00,007, respectively. This implies that MOA experienced less overshoot than GWO and PSO, which have higher overshoot values. PSO reported 0.00,007, 0.00,043, and 0.00,012 with the same settings. MOA continues to outperform in terms of undershoot, as indicated by the lowered undershoot values seen across all parameters ($\Delta F_1 = -0.00{,}061$, $\Delta F_2 = -0.00{,}026$, and $\Delta P_{tie} = -0.00{,}601$). Gray wolf optimization and particle swarm optimization have large undershoot values as compared to mother optimization algorithms. PSO values are −0.00,600, −0.00,062, and −0.00,803, while GWO values are −0.00,086, −0.00,035, and −0.00,564. From Fig. 8.6 it can be observed that

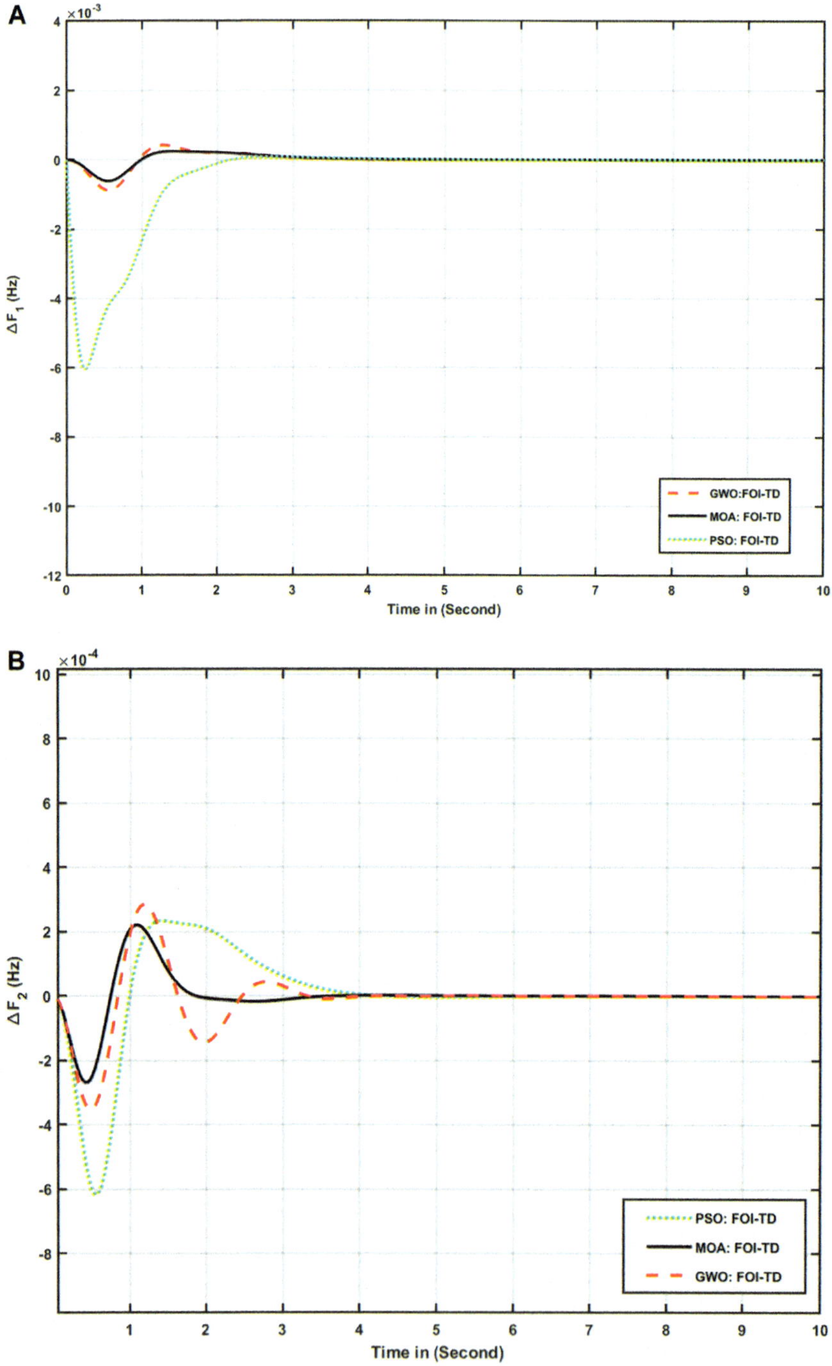
A
×10⁻³
4
2
0
-2
-4
-6
-8
-10
-12
ΔF₁ (Hz)
0 1 2 3 4 5 6 7 8 9 10
Time in (Second)
GWO:FOI-TD
MOA: FOI-TD
PSO: FOI-TD
B
×10⁻⁴
10
8
6
4
2
0
-2
-4
-6
-8
ΔF₂ (Hz)
1 2 3 4 5 6 7 8 9 10
Time in (Second)
PSO: FOI-TD
MOA: FOI-TD
GWO: FOI-TD

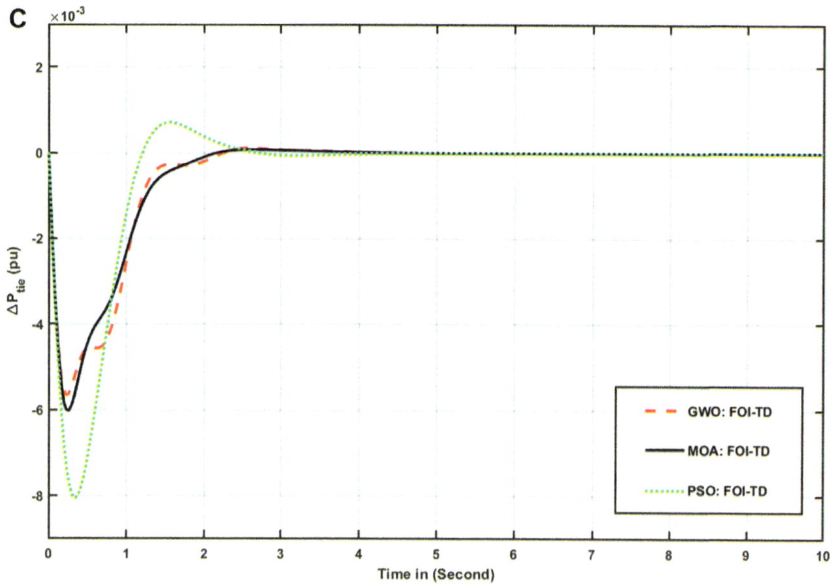

FIGURE 8.6 Analysis of the dynamic response of different algorithms for (a) ΔF1; (b) ΔF2; (c) ΔPtie.

mother optimization algorithms perform very well in respect of settling time, undershoot/overshoot as compared to gray wolf optimization and article swarm optimization for change in area 1, area-2 and change in tie-line power.

8.5.5 Future directions for multi-microgrid frequency control

The progression of multi-microgrid systems, introduces numerous research challenges and opportunities, regarding load frequency control under complex and dynamic circumstances. As energy demand raises and renewable energy incorporation increases, MMG system entail advance control mechanism to guarantee stability, resilience and real time adaptability. A promising track lies in the amalgamation of smart grid technologies within MMG frameworks. These technologies employ real time data from advance metering set-up, and distributed sensors to allow faster, more conversant response to frequency deviations. However, unlike conventional microgrids, MMG claim localized decision making, imposing development of decentralized control tactics that minimize communication delays while keeping global coordination across interconnected microgrids.

The used of machine learning and artificial intelligence offers a pathway to adaptive, predictive, and self-optimizing control strategies in MMG systems. By employing machine learning based approaches, one can incessantly analyze the performance of the system, permitting real-time optimization, which is predominantly appreciated in handling renewable energy sources. Based on antique and real-time statistics, artificial intelligence-based techniques can estimate variations in load and generation to enhance reliability and flexibility.

Cybersecurity is another critical domain for future MMG systems developments. A vigorous cyber-attack anticipation mechanism is obligatory to control the frequency of the MMG systems. As multi-microgrid systems progressively reliant on control systems and digital communication, robust cyber-attack detection and vindication mechanism must be implanted to ensure secure frequency stabilization and overall system integrity.

Additionally, improvement in energy storage systems especially in battery technology, will play an important role in mitigating supply fluctuations introduced by intermittent resources. The combination of energy storage with advanced control algorithms can significantly enhance the resilience of the LFC in multi-microgrids.

8.6 Conclusion

This chapter highlighted the crucial role of sophisticated load frequency control mechanisms in supporting multi-microgrid systems, which are essential in modern, renewable-dependent power grids. The MMG system integrates a variety of distributed generation resources in order to manage decentralized, variable energy sources. The comprehensive analysis presented in this chapter demonstrates the comparative efficacy of several control methodologies, focusing particularly on fractional order integral-tilt derivative controllers, which consistently outperform conventional controllers in terms of reducing frequency deviations, achieving rapid stabilization and improving resilience to fluctuations, among other factors. Mother optimization algorithms have dramatically improved LFC efficacy by using nature-inspired optimization algorithms. Furthermore, this chapter has covered a variety of coordination frameworks, such as hierarchical, decentralized, and hybrid approaches, that ensure system reliability by harmonizing the interactions between microgrids. MMG systems will benefit substantially from machine learning and artificial intelligence in the future, which will allow LFC applications to become more predictive and adaptable in real time. In addition, the MMGs' resilience and scalability will be further enhanced by the implementation of advanced energy storage solutions and the strengthening of their cybersecurity policies.

References

Abubakr, H., Guerrero, J.M., Vasquez, J.C., Mohamed, T.H., Mahmoud, K., Darwish, M.M.F., Dahab, Y.A., 2022. Adaptive LFC incorporating modified virtual Rotor to regulate frequency and tie-line power Flow in multi-area microgrids. IEEE Access 10, 33248–33268. https://doi.org/10.1109/ACCESS.2022.3161505.

Arya, Y., Guha, D., Sarangi, S., 2023. Special issue on: AI/data-based modeling, control, and operation of smart grid power systems. Int. J. Numer. Model. Electron. Network. Dev. Field. 36 (5). https://doi.org/10.1002/jnm.3156.

Boyanapalli, U.B., 2018. Implementation of Secure DNP3 Architecture of SCADA System for Smart Grids.

Can, O., Ozturk, A., Eroğlu, H., Kotb, H., 2022. A novel grey wolf optimizer based load frequency controller for renewable energy sources integrated thermal power systems. Elec. Power Compon. Syst. 49 (15), 1248–1259. https://doi.org/10.1080/15325008.2022.2050450.

Casavola, A., Franzè, G., Tedesco, F., 2022. Centralized Supervision and Coordination of Load/Frequency Control Problems in Networked Multi-Area Power Systems. Informa UK Limited, pp. 375–423. https://doi.org/10.1201/9781003293989-15.

Choudhary, R., Rai, J.N., Arya, Y., 2022. Cascade FOPI-FOPTID controller with energy storage devices for AGC performance advancement of electric power systems. Sustain. Energy Technol. Assessments 53, 102671. https://doi.org/10.1016/j.seta.2022.102671.

Daraz, A., 2023. Optimized cascaded controller for frequency stabilization of marine microgrid system. Appl. Energy 350, 121774. https://doi.org/10.1016/j.apenergy.2023.121774.

Daraz, A., Malik, S.A., Azar, A.T., Aslam, S., Alkhalifah, T., Alturise, F., 2022. Optimized fractional order integral-tilt derivative controller for frequency regulation of interconnected diverse renewable energy resources. IEEE Access 10, 43514–43527. https://doi.org/10.1109/ACCESS.2022.3167811.

Daraz, A., Malik, S.A., Basit, A., Aslam, S., Zhang, G., 2023. Modified FOPID controller for frequency regulation of a hybrid interconnected system of conventional and renewable energy sources. Fractal and Fractional 7 (1). https://doi.org/10.3390/fractalfract7010089.

Elkasem, A.H.A., Khamies, M., Hassan, M.H., Agwa, A.M., Kamel, S., 2022. Optimal design of TD-TI controller for LFC considering renewables Penetration by an improved Chaos Game optimizer. Fractal and Fractional 6 (4), 220. https://doi.org/10.3390/fractalfract6040220.

Fathollahi, A., Kargar, A., Yaser Derakhshandeh, S., 2022. Enhancement of power system transient stability and voltage regulation performance with decentralized synergetic TCSC controller. Int. J. Electr. Power Energy Syst. 135, 107533. https://doi.org/10.1016/j.ijepes.2021.107533.

Freitas, A.L., Fernandes, V., Coury, V., et al., 2022. Deep reinforcement learning-based secondary control for microgrids in islanded mode. Elec. Power Syst. Res. 212. https://doi.org/10.1016/j.epsr.2022.108315.

Guan, Y., Wei, B., Guerrero, J.M., Vasquez, J.C., Gui, Y., 2022. An overview of the operation architectures and energy management system for multiple microgrid clusters. iEnergy 1 (3), 306–314. https://doi.org/10.23919/IEN.2022.0035.

Gulzar, M.M., 2023. Designing of robust frequency stabilization using optimized MPC-(1+PIDN) controller for high order interconnected renewable energy based power systems. Protection and Control of Modern Power Systems 8 (1). https://doi.org/10.1186/s41601-023-00286-x.

Gulzar, M.M., Sibtain, D., Ahmad, A., Javed, I., Murawwat, S., Rasool, I., Hayat, A., Sharma, P., 2022. An efficient design of adaptive model predictive controller for load frequency control in hybrid power system. International Transactions on Electrical Energy Systems 2022, 1–14. https://doi.org/10.1155/2022/7894264.

Li, J., Zhou, T., 2024. Prior Knowledge incorporated large-scale Multiagent Deep reinforcement learning for load frequency control of Isolated microgrid considering multi-structure coordination. IEEE Trans. Ind. Inf. 20 (3), 3923–3934. https://doi.org/10.1109/TII.2023.3316253.

Matoušová, I., Trojovský, P., Dehghani, M., Trojovská, E., Kostra, J., 2023. Mother optimization algorithm: a new human-based metaheuristic approach for solving engineering optimization. Sci. Rep. 13 (1). https://doi.org/10.1038/s41598-023-37537-8.

Mehrizi-Sani, A., Iravani, R., 2010. Potential-function based control of a microgrid in islanded and grid-connected modes. IEEE Trans. Power Syst. 25 (4), 1883–1891. https://doi.org/10.1109/TPWRS.2010.2045773.

Mishra, D., Nayak, P.C., Prusty, R.C., 2020. PSO optimized PIDF controller for Load-frequency control of A.C Multi-Islanded-Micro grid system. In: Proceedings of the 2020 International Conference on Renewable Energy Integration into Smart Grids: A Multidisciplinary Approach to Technology Modelling and Simulation, ICREISG. Institute of Electrical and Electronics Engineers Inc., India, pp. 116–121. https://doi.org/10.1109/ICREISG49226.2020.9174552. http://ieeexplore.ieee.org/xpl/mostRecentIssue.jsp?punumber=9163073.

Mishra, S., Nayak, P.C., Prusty, R.C., Panda, S., 2023. Hybrid dragonfly algorithm and pattern search optimized adaptive fuzzy PID controller for frequency regulation of multi microgrid system using Henkel matrix based reduced-order model. Int. J. Numer. Model. Electron. Network. Dev. Field. 36 (1). https://doi.org/10.1002/jnm.3029.

Mohamed, A., Wazeer, E.M., El Masry, S.M., Abdel Ghany, A.M., Mosa, M.A., 2024. A novel scheme of load frequency control for a multi-microgrids power system utilizing electric vehicles and supercapacitors. J. Energy Storage 89, 111799. https://doi.org/10.1016/j.est.2024.111799.

Naderi, M., Khayat, Y., Shafiee, Q., Blaabjerg, F., Bevrani, H., 2023. Dynamic modeling, stability analysis and control of interconnected microgrids: a review. Appl. Energy 334, 120647. https://doi.org/10.1016/j.apenergy.2023.120647.

Nayak, P.C., Prusty, U.C., Prusty, R.C., Panda, S., 2025. Imperialist competitive algorithm optimized cascade controller for load frequency control of multi-microgrid system. Energy Sources, Part A Recovery, Util. Environ. Eff. 47 (1), 5538–5560. https://doi.org/10.1080/15567036.2021.1897710.

Pavan, K.N.V.S.R., Kiran Kurapati, K., Kumar Kundrapu, K., Srinamreddy, S., Praveena Pampana, R., 2023. Load frequency control of multi-microgrid system using world cup optimization. i-manager's Journal on Power Systems Engineering 11 (2), 1. https://doi.org/10.26634/jps.11.2.19962.

Prusty, U.C., Nayak, P.C., Prusty, R.C., Panda, S., 2022. An improved moth swarm algorithm based fractional order type-2 fuzzy PID controller for frequency regulation of microgrid system. Energy Sources, Part A Recovery, Util. Environ. Eff. https://doi.org/10.1080/15567036.2022.2038735.

Sabahi, K., Tavan, M., Amin, H., 2021. Adaptive type-2 fuzzy PID controller for LFC in AC microgrid. Soft Comput. 25 (11), 7423–7434. https://doi.org/10.1007/s00500-021-05704-6.

Saha, D., Bazmohammadi, N., Vasquez, J.C., Guerrero, J.M., 2023. Multiple microgrids: a review of architectures and operation and control strategies. Energies 16 (2). https://doi.org/10.3390/en16020600.

Sharma, G., Panwar, A., Arya, Y., Kumawat, M., 2020. Integrating layered recurrent ANN with robust control strategy for diverse operating conditions of AGC of the power system. Institution of Engineering and Technology, South Africa IET Generation, Transmission and Distribution 14 (18), 3886–3895. https://doi.org/10.1049/iet-gtd.2019.0935.

She, B., Li, F., Cui, H., Wang, J., Min, L., Oboreh-Snapps, O., Bo, R., 2023. Decentralized and coordinated V-f control for islanded microgrids considering DER inadequacy and demand control. IEEE Trans. Energy Convers. 38 (3), 1868–1880. https://doi.org/10.1109/TEC.2023.3258919.

Si, S., Sun, W., Wang, Y., 2024. A decentralized dispatch model for multiple micro energy grids system considering renewable energy uncertainties and energy interactions. J. Renew. Sustain. Energy 16 (1). https://doi.org/10.1063/5.0192716.

Singh, K., Arya, Y., 2023. Tidal turbine support in microgrid frequency regulation through novel cascade Fuzzy-FOPID droop in de-loaded region. ISA (Instrum. Soc. Am.) Trans. 133, 218–232. https://doi.org/10.1016/j.isatra.2022.07.010.

Siti, M.W., Mbungu, N.T., Tungadio, D.H., Banza, B.B., Ngoma, L., 2022. Application of load frequency control method to a multi-microgrid with energy storage system. J. Energy Storage 52, 104629. https://doi.org/10.1016/j.est.2022.104629.

Smith, C., Burrows, J., Scheier, E., Young, A., Smith, J., Young, T., Gheewala, S.H., 2015. Comparative Life Cycle Assessment of a Thai Island's diesel/PV/wind hybrid microgrid. Renew. Energy 80, 85–100. https://doi.org/10.1016/j.renene.2015.01.003.

Yakout, A.H., Kotb, H., Hasanien, H.M., Aboras, K.M., 2021. Optimal fuzzy PIDF load frequency controller for hybrid microgrid system using marine predator algorithm. IEEE Access 9, 54220–54232. https://doi.org/10.1109/ACCESS.2021.3070076.

Yoo, B.K., Yang, S.H., Yang, H.S., Kim, W.Y., Jeong, Y.S., Han, B.M., Jang, K.S., 2011. Communication architecture of the IEC 61850-based micro grid system. Journal of Electrical Engineering and Technology 6 (5), 605–612. https://doi.org/10.5370/JEET.2011.6.5.605SouthKorea.

Zhang, G., Daraz, A., Khan, I.A., Basit, A., Khan, M.I., Ullah, M., 2023. Driver Training based optimized fractional order PI-PDF controller for frequency stabilization of diverse hybrid power system. Fractal and Fractional 7 (4). https://doi.org/10.3390/fractalfract7040315.

Zohuri, B., 2017. Hybrid Renewable Energy Systems. Springer Science and Business Media LLC, pp. 1–38. https://doi.org/10.1007/978-3-319-70721-1_1.

CHAPTER

9

Hardware and real-time validation of microgrids

Preeti Gupta[1], *Vivek Pahwa*[2], *and Yajvender Pal Verma*[2]

[1] Department of Electrical Engineering, PEC Chandigarh, Chandigarh, India; [2] Department of Electrical & Electronics Engineering, UIET, Panjab University, Chandigarh, India

9.1 Introduction

Artificial Intelligence is revolutionizing the modern power systems by enabling intelligent decision-making and optimization across all levels of the utility grid. With the increasing complexity of power systems due to the integration of renewable energy sources, decentralized generation, electric vehicles, and smart grid technologies, it is providing the necessary tools to manage and operate these systems effectively (Srinivasan, 2022, Cardona et al., 2023). Despite of these advancements the result accuracy remains questionable because most of the studies have been carried out in an offline environment, and very less attempts have been made to evaluate the performance in real-time (RT).

Generally, MATLAB/SIMULINK is the most preferred choice of engineering researchers as it is an easy platform, which has matrix-based analysis. Conventionally, it involves immediate and approximate solution algorithms such as Newton's method (Garces, 2018), Runge-Kutta method (Pries and Hofmann, 2015), Lagrangian method Feng et al., (2020), and Newton–Raphson method (Liu et al., 2020). These algorithms provide an analytical solution, which is beneficial only when the complexities of the system are limited within a certain horizon (Bose, 1990; Bellini et al., 2002). Recently, it solves the power system equations using better solvers than earlier, like Dormand-Prince, Bogacki-Shampine, trapezoidal, etc, in variable time-step for analysis purposes

Frequency Control in Microgrids
https://doi.org/10.1016/B978-0-443-36476-1.00008-0

(Rahman and Rahim, 2016; Kusagur et al., 2010). The power system models compiled with a variable time-step is not suitable for RT operation study because they initiate triggering of the next cycle as soon as the previous iteration of the computation cycle gets over in a system, regardless of the real situation (Mirzahosseini and Iravani, 2019). Though offline is a platform that helps in developing and analyzing the system, it still lacks accuracy. Because receiving or sending the signals either before or after the required time does not make any difference in real power systems. Therefore, the accuracy of control action with a time frame is an essential requirement before implementing these controllers onto the actual power system.

Designing complex systems such as microgrids (MGs) in real time is another challenge for the platform. As this leads to the formation of a heavy admittance matrix of a system, which is a computational burden on the platform during each simulation time-step (Strasser et al., 2015). Therefore, controllers designed in an offline environment lead to inaccurate control actions when tested on the actual power system. To eliminate this computation burden, a few researchers have worked on algorithms that are designed for *field-programmable gate array* (FPGA) based systems, which helps in speeding up the calculations of a large system admittance matrix using parallel computation features (Mirzahosseini and Iravani, 2019; Kumashiro et al., 2020).

This FPGA feature is provided by many simulators in the market, namely, *OPAL-RT (ordinateur parallèle, applications logicielles temps-réel), RTDS (real-time digital system),* and *dSPACE (digital signal processing and control engineering).* These RT simulators provide features through a process for the analysis of the power system in real scenarios (Li et al., 2004; Breaz et al., 2014). The process ensures that the design of the model is in line with the real-world data. It also ensures that the characteristics, behavioral studies, and dynamic performances are subject to immediate adjustments and improvements as per the requirements (Wang et al., 2018; Shen and Lior, 2016). It possesses advanced solvers for calculating the large system matrix for instance, solvers: *artemis, artemis-ssn, e-megasim* solvers are provided by the OPAL-RT simulator through a software, RT-LAB (*real-time laboratory*). These solvers help give robust and accurate solutions within the prescribed time frame (Fu et al., 2018; Jonsdottir et al., 2015). With this, controllers developed are synchronized with the RT clock, and these can be easily implemented onto the actual power system (Bose, 1990; Bellini et al., 2002).

Therefore, the idea of developing a more realistic model is vital, as it can ensure the accuracy of simulation studies. With this, the performance validation of the models has become easier, and thus, it is an important step during design and development in industries. Due to the growing dependency, RT validation finds a large number of applications in various sectors like MGs, power systems, control development, electronics, etc..,

(Luna et al., 2018; Baghaee *et al.*, 2017). In MGs, it helps in energy management by validating the load and generation data (predicted and actual) (Abrishambaf et al., 2017). It helps in maintaining the voltage and frequency regulations within allowable limits and detects/isolates faults. It gives operational flexibility by supporting easy transition from on-grid mode to off-grid mode, thus ensuring MG's adaptability toward varying conditions (Li and Belanger, 2015; Dong et al., 2016). Apart from MGs, it is useful in fine-tuning electronic devices (Baghaee et al., 2024; Reis et al., 2019), accurately setting the dynamic response of the control systems (Wu et al., 2017a; Chou and Butler-Purry, (2012), design optimization, system testing (Li et al., 2004; Dong et al., 2016; Sirviö et al., 2020). Renewable integration with the grid (Strasser et al., 2015; Kumar and Bansal, 2019), performance study of energy storage devices, and quality assurance (Kumar and Bansal, 2019; Ensermu et al., 2019). Therefore, RT validation gives significant advantages in various sectors by enhancing accuracy, reliability, safety, and reducing the emergency cost of repairs. However, synchronization of different systems, latency, or implementation of advanced artificial intelligence is still a challenge with RT validation tools.

This chapter is divided into five sections. The first section gives an insight into the RT validation, its history, and its various simulating environments. The next section provides an overview of RT simulation, its modeling methods, and various platforms available in the market. A comparison between the latest available platforms is drawn, and the best platform i.e., OPAL-RT, is selected for further discussions. The third section introduces OPAL-RT, which includes its advantages, disadvantages, features, implementation procedure, and its utilization for MG validation. After this section, a brief case study on a photovoltaic–solid oxide fuel cell (PV-SOFC) grid-connected system is investigated to showcase the performance improvement of SOFC, its control development, artificial intelligence implementation, and use of the OPAL-RT platform for validation. At last, this chapter gives some technical challenges and possible solution approaches in hardware and RT validation.

9.1.1 History

The RT validation has evolved over the years, as shown in Fig. 9.1 (Li et al., 2015). During the 1940s, analog systems were used for validation purposes, which involved a huge setup, a large investment, and technical complexity. This resulted in a slow speed of simulation and reduced work efficiency. After a decade, the validation was executed using hybrid computers that used a combination of hardware and analog systems. Again, it involved a high cost of computation, a very slow speed for simulation, and the setup size was very large. These physical simulators, like EUROSTAG (Antone and Stubbe, 1992; Cole and Haut, 2013) were

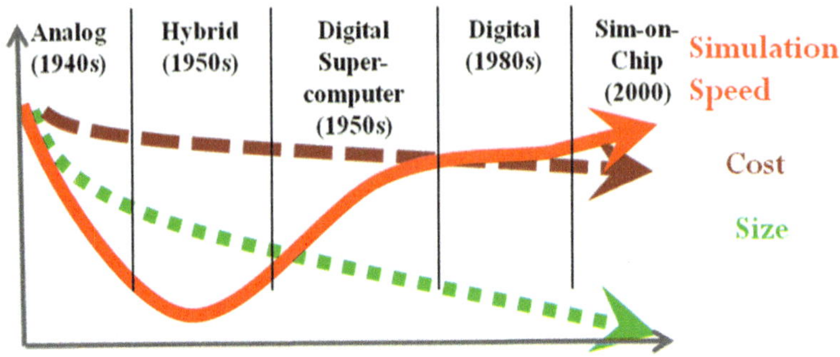

FIGURE 9.1 Comparison of various simulators.

complex technologies for analyzing the system in real conditions, due to which their usage reduced after the 1970s.

Thereafter, the use of complete digital simulators came into existence in the 1990s, such as ARENE, which uses simulation and coding software (Etxeberria-Otadui et al., 2002). For over 20 decades, a combination of these physical and digital simulations has played a role in analyzing the systems. Similarly, Siemens also developed an RT simulator, known as NETOMAC, for testing the protection devices. Again, it was a digital computer-based approach for testing modules under any kind of faulty conditions (Krebs and Ruhle, 2004; Lehn et al., 1995). However, such simulators give a smaller step time, which is not enough to study the instantaneous effects. After a decade or so, a new trend of digital signal processing (DSP) based RT simulators came into existence. They provided an ease of simulation study for the fundamental analysis of the models using a computer based platform called as "offline" and once the model is perfectly simulated then it is interfaced with the RT simulator, which is hardware based platform called as "online" for the study/analyzes of the real effects on the model (Li et al., 2004; Breaz et al., 2014).

9.1.2 Simulating environments

Once the modeling of any physical system has been designed for a particular application, the next challenge is to simulate the model of this overall system with a suitable simulation environment for evaluation of its performance under various scenarios to reduce cost and inadvertent design mistakes. In understanding, designing, and then the RT implementation of the developed model, an appropriate simulation environment plays a vital role. This section deals with the definitions of basic terms related to the simulation study in an environment, various simulating environments, and the selection of an appropriate environment as per the RT scenarios. Afterward, the detailed working procedure of this selected environment is also given.

A simulation is the implementation of the operation/scenario/behavior of the real-world process/physical system over time (Morris et al., 2019). It helps in studying the informational, organizational, and environmental changes and their effects on a system. It helps in gaining knowledge about the improvement of the system and finding important input parameters with changing simulation inputs (Sauro and Bergmann, 2010). Hence, the model construction in simulation with a conceptual framework describes the complete system. However, sometimes the time taken by the simulation in an environment to solve a system is either too less/more or rarely accurate as per the real clock. It is because a simulation study of any system consists of generating a solution of the mathematical equations of that system. These equations are solved successively at a given time step over a certain duration of time. If the number of equations is more than, slow will be the calculations; otherwise it will be fast enough, irrespective of the real scenario (Blanger and Paquin, 2010).Therefore, it is important to study the runtime accuracy of the system/controller design.

The basic terms related to any simulated environment are given below:

- *Real clock time:* The duration of time as per the physical clock is termed as real clock time, t_{rc}.
- *Build time*: The time taken by the solver to analyze the system and plan its strategy to solve the model correctly before the compilation is termed as build time, t_b.
- *Simulation time*: The time taken by the simulating tool in actually solving the model is termed as simulation time, t_s. It is the time consumed in computing all the equations of the system at a given time step.
- *Model execution time:* The total time taken by the environment in successfully executing the model is known as its model execution time, t_m (Fu et al., 2018; Jonsdottir et al., 2015). It is composed of build time, t_b and simulation time, t_s. Mathematically, the model execution time is given in Eq. (9.1),

$$t_m = t_b + t_s \tag{9.1}$$

In general, for the practical implementation of any developed system/control model, it is mandatory to achieve the condition as given below:

The simulation time must exactly match with real clock time, i.e., exactly match with real clock time in Eq. (9.2),

$$t_s = t_{rc} \tag{9.2}$$

Broadly, the simulating environments can be categorized in two ways as given below:

- Offline and
- Online.

9.1.2.1 *Offline environment*

Any computerized environment that simulates the power system model, irrespective of the duration of simulation, is an offline environment. A few examples of offline environments are NI, SCILAB, and MATLAB/SIMULINK. These are currently highly competitive at the industrial level for research and development purposes (Ruiz et al., 2020). These provide numerical computation of a system as per the application. However, MATLAB/SIMULINK is the most preferred choice of engineering researchers as it is an easy platform, which has matrix-based analysis.

Conventionally, it involves immediate and approximate solution algorithms such as Newton's method Garces, (2018), Runge-Kutta method (Pries and Hofmann, 2015), Lagrangian method (Feng et al., 2020), and Newton–Raphson method (Liu et al., 2020). This algorithm provides an analytical solution, which is beneficial only when the complexities of the system are limited within a certain horizon (Bose, 1990; Bellini et al., 2002). Recently, it solves the power system equations using better solvers than earlier, like Dormand-Prince, Bogacki-Shampine, trapezoidal, etc, in variable time-step for analysis purposes (Rahman and Rahim, 2016; Kusagur et al., 2010). For matching the practicality, it has several amendments for the mode of simulations, such as *normal*, *accelerator*, and *rapid accelerator* that may/may not help in developing accuracy, in the long run or during complex systems (Bose, 1990; Bellini et al., 2002).

9.1.2.2 *Online environment*

The most crucial aspect of designing or analyzing a power system as per real conditions is its RT modeling that will depict its physical model, for instance, the author in Breaz et al. (2014) presents the effectiveness of the modeling in the RT. The RT modeling is its mathematical interpretation based on the physical phenomenon for the critical analyzes. However, there is another way of discussing the mathematical modeling of a system i.e., system-based modeling, which is not so effective. As this leads to the formation of a heavy admittance matrix of a system that becomes a *computation burden*, *CB* on the platform during each simulation time-step, given by *T. Strasser* in (Strasser et al., 2015). It can be expressed mathematically using Eq. (9.3)

$$CB = \frac{t_m}{t_{rc}} \tag{9.3}$$

Few researchers, *Tatsuya Kamei* et al. in (Mirzahosseini and Iravani, 2019; Kumashiro et al. 2020) have worked on algorithms to reduce this burden that are designed for FPGA-based systems. These algorithms help in speeding up the calculations of a large system admittance matrix because FPGA-based RT simulators provide a parallel computation feature that enhances the overall performance of the complete power system (Li et al., 2004; Breaz et al., 2014; Mikkili et al., 2015).

9.2 Overview of RT simulation

To enhance the performance of the system, there are three ways of modeling a system, regardless of the version of the simulator used. These system models involve the numerical solving approach, and the model is operated at a fixed time-step for the precise firing of the signals.

The three ways of modeling are as given below:

- Software-in-loop (SIL)
- Hardware-in-loop (HIL)
- Rapid Control Processing (RCP)

9.2.1 Software-in-loop (SIL)

In this type of modeling, the developed model in an offline environment is simulated with the RT simulator as shown in Fig. 9.2 (Musleh et al., 2017).

In this figure, the offline model is with the host PC, which is connected via Ethernet cable to the RT simulator. The simulator has a single-core CPU, where all the model parameters are kept. These parameters are transferred to the FPGA board for parallel computation of the model parameters if possible, and then the output data of the model is retrieved from the core CPU as mat file. This modeling provides an ease of no inputs and outputs to/from the simulator, as a result, it helps in

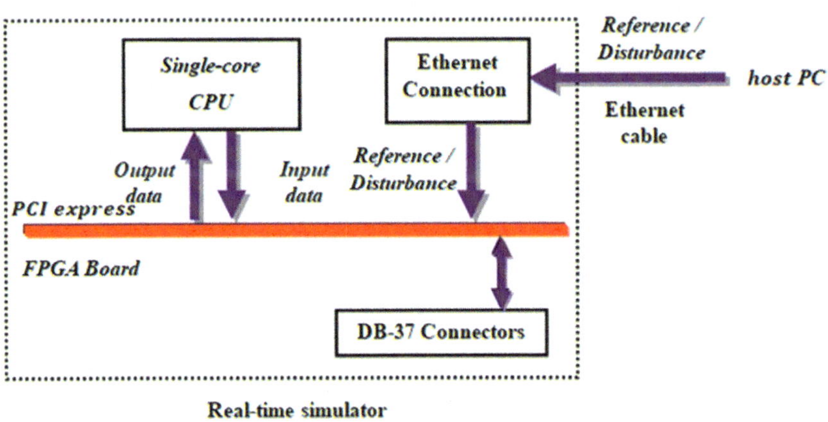

FIGURE 9.2 Software-in-loop.

maintaining the integrity of the signals. Thus, it is considered to be one of the best modeling approaches for validating the result (conceptual) accuracy and thus, it is considered as an elementary stage of RT simulation study (Musleh et al., 2017).

9.2.2 Hardware-in-loop (HIL)

In HIL type of modeling, the signals are transmitted from one system to another or back to the same system, which represents a connection between one physical system and one virtual system (Blanger and Paquin, 2010). It can be connected in two ways:

- virtual system and virtual controller present in different cores of the same simulator and connected via an external circuit as depicted in Fig. 9.3(a).
- virtual system in one simulator and virtual controller in another simulator connected via an external circuit as depicted in Fig. 9.3(b).

HIL gives the ease of testing the system under normal and adverse conditions that cannot be tested on a physical system. Since the physical layout of any power system is a costly affair, it is very difficult to test the designs on a physical system. On the other hand, HIL provides the reliable testing of the system designs at a very low cost (Gupta et al., 2020).

Broadly, HIL is further categorized in two ways:

- Controller hardware-in-loop (C-HIL) and
- Power hardware-in-loop (P-HIL)

In C-HIL, the controller is physically operated, and the power system is completely/partially executed virtually on an RT simulator. While in P-HIL, the power system is physically operated, and the controller is virtually executed on an RT simulator. These two categories are well-proven for the designing and testing of the system designs as IEEE and IEC standards have marked them superior in comparison to any offline-based simulations (Ropp, 2019; *IEEE Standard for the Testing of Microgrid Controllers*, 2018).

9.2.3 Rapid Control Processing (RCP)

In RCP type of modeling, the system controller is placed in the simulator and it is connected directly to the physical system Tawfique et al., (2016) as shown in Fig. 9.4. It helps in the faster implementation of a control prototype, which further helps in debugging the system errors faster. The control parameters can be tuned significantly in RT which helps in the effective design of the controller (Guillaud et al., 2015; Li and Belanger, 2015).

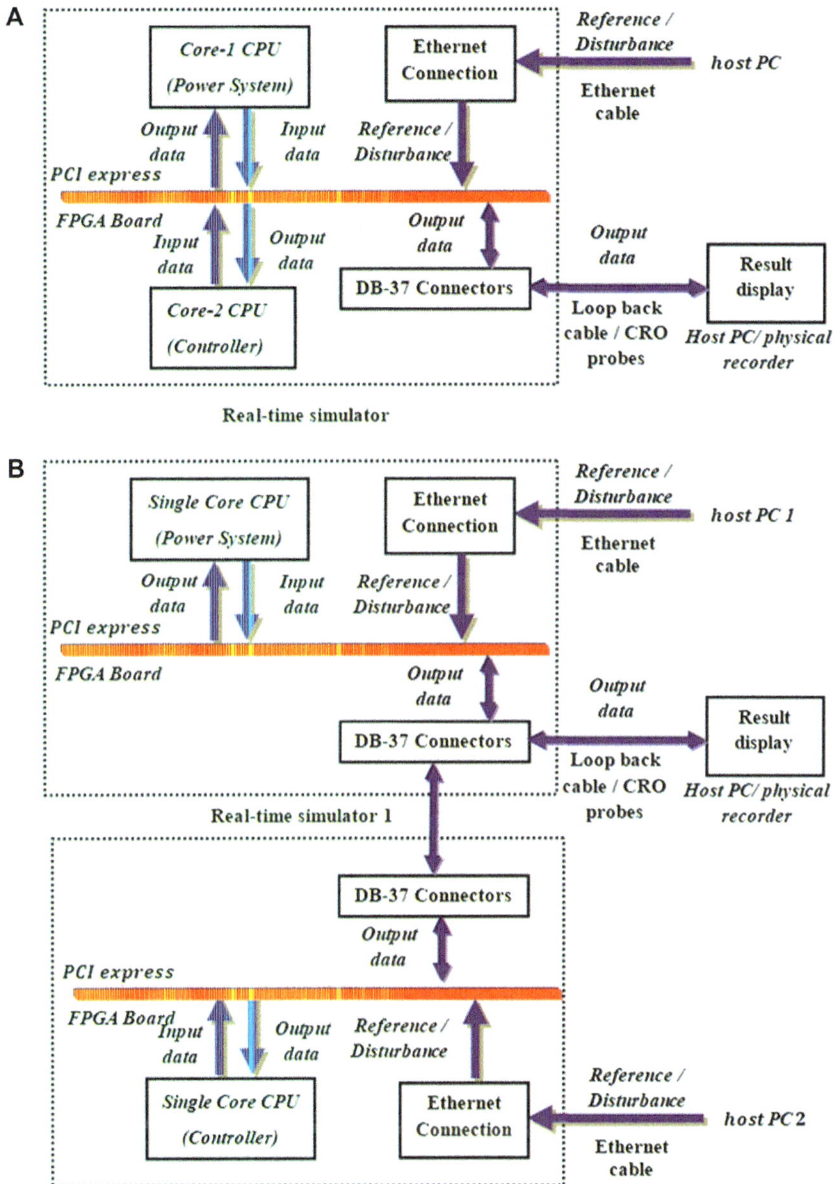

FIGURE 9.3 Hardware-in-loop.

Therefore, the online RT environment is globally accepted for the validation of the designs of the system/controllers as per IEEE and IEC standards, respectively. The verification of any model design is vital to demonstrate its accuracy in terms of time using SIL, HIL, or RCP type of modeling.

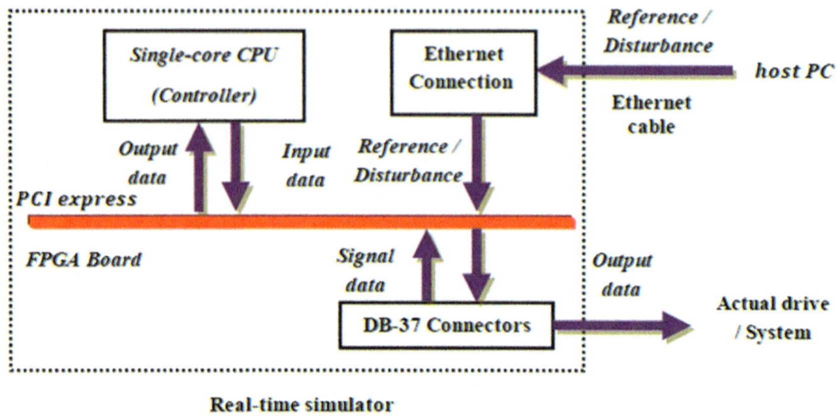

FIGURE 9.4 Rapid control processing.

9.2.4 DSP-based RT simulators

The DSP-based RT simulators are the foundation of today's studies. These consist of FPGA-based emulations that provide an excellent testing platform (Guillaud et al., 2015; Breaz et al., 2014; Mikkili et al., 2015). They transfer the RT data of the actual device under test, including system prototypes and actual devices, for the analyses as per RT conditions.

Apart from the ease of a simulation study, these are advantageous (Li et al., 2004; Guillaud et al. 2015) in many ways, as summarized below:

- They provide a time-critical simulating platform for the precise dynamic representation of the system (internal variables and control actions of the simulation) that ensures exact time length during execution, as actual power circuitry is involved.
- They are an automatic code-generating platform used in many engineering applications such as power systems, control algorithms, robotics, drives, aircraft, and grid protection for both designing and testing. This type of simulation provides a feature for analyzing the systems as per the RT scenarios.
- They provide the validation of the work as per the real conditions in the most secure manner, as it gives an easy and non-risky approach for testing, especially in case of faulty condition evaluations on a simulated model.
- They also help in using the compiled code of the model instead of the interpreted code, which is generally used by most of the simulation tools, thus enabling a rapid simulation study.

There are different RT simulators in the market, as given in Table 9.1 namely, ARENE, EMTP, NETOMAC, dSPACE, RTDS, OPAL-RT, and

TABLE 9.1 Comparison of different RT environments.

S. No.	Environment	Speed/Accuracy	Remarks
1.	**SimPowerSystem (SPS)**	Low No permission required	High computation time
2.	**Real-time**	High	• Simulations match real clock time • Reduced number of over-runs • Fast computations
	ARENE		
	EMTP		
	NETOMAC		
	dSPACE		
	RTDS		
	OPAL-RT		
	Typhoon HIL		• No PHIL integration

Typhoon HIL. Among these, the first three have high computational time in comparison to the others. dSPACE finds large applications in both development and testing, however, its P-HIL is a little cumbersome.

Only RTDS and OPAL-RT can handle the electrical systems in both C-HIL and P-HIL (Bose, 1990; Bellini et al., 2002). RTDS is generally preferred for electromagnetic transient studies and performance analysis of protection devices (Li et al., 2020). OPAL-RT provides an easy integration with the best-suited offline environment, i.e., MATLAB/ SIMULINK (Mikkili et al., 2015). It provides a step-size in the range of approximately 10 – 100μs (micro-seconds) for operation study, unlike Typhoon HIL (Reis et al., 2019). Apart from easy integration, it possesses certain features such as advanced solvers for calculating the large system matrix, for perfect analysis. These solvers provide an accurate solution within the prescribed time frame (Fu et al., 2018; Jonsdottir et al., 2015).

9.3 Overview of OPAL-RT

OPAL-RT is a modern DSP-based RT simulator. The hardware components used for validating any system are shown in Fig. 9.5 (Merabet et al., 2016). In this figure, a. OPAL-RT's Real-time simulator OP4510 b. DB-37 connector with Loop-back cable c. Host Personal Computer (PC) and Digital Storage Oscilloscope (DSO). It accurately simulates the response of an actual physical system (hardware) and has evolved over the years as a powerful tool in power system studies (Blanger and Paquin, 2010).

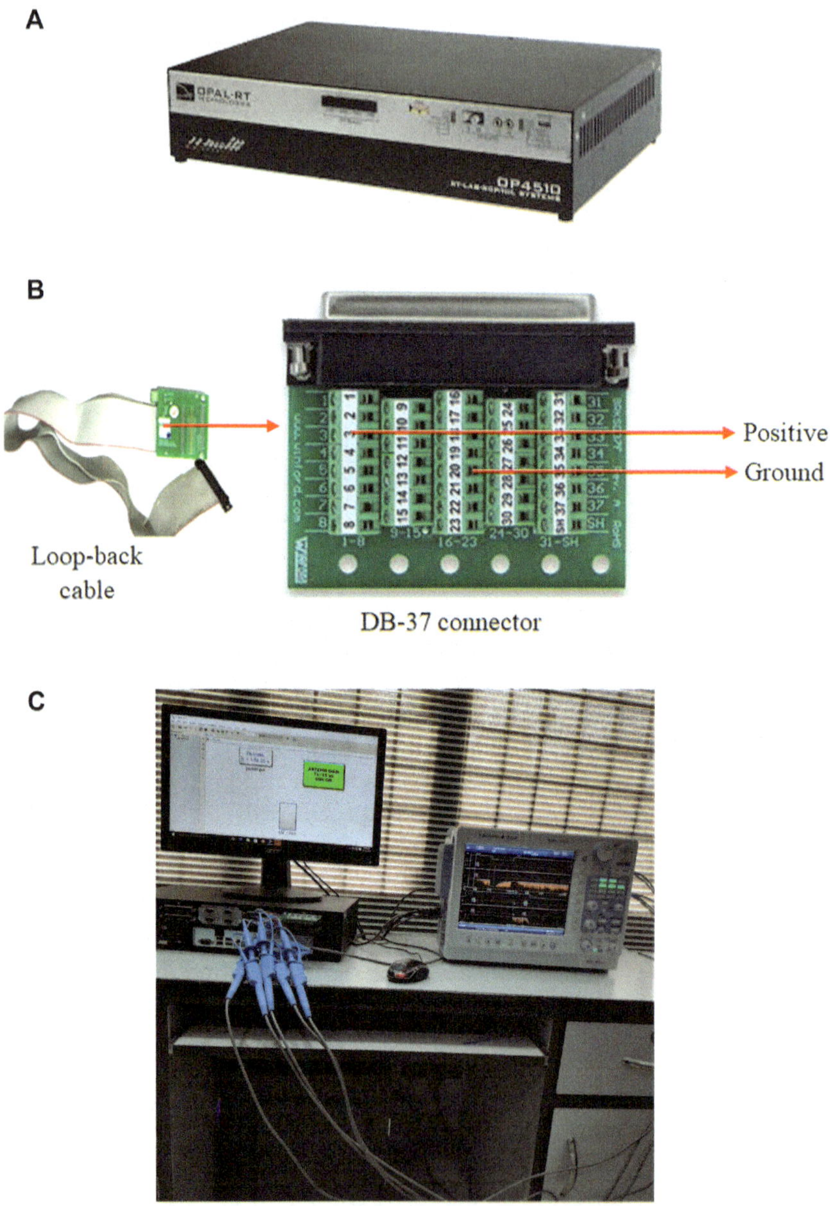

FIGURE 9.5 Hardware components used in HIL/RCP.

Table 9.2 gives a brief literature survey on different MG systems (isolated/grid-connected/multi-connected) that are using OPAL-RT for a certain application. It briefs about power generators, storage systems, types of controllers or optimization algorithms, and applications.

TABLE 9.2 Literature survey on different MG systems using OPAL-RT.

Author's Name/ Year	Type of MG system	Power generation sources	Storage systems	Controller type/ optimization algorithm	Application	Ref. No.
Wu et al. (2017a)	Isolated	Fuel cell	Battery	Adaptive	RT control development	Wu et al. (2017a)
Amine Yamane et al. (2015)	Grid-connected	Wind and photovoltaic	–	–	Distribution system	Amine Yamane et al. (2015)
Ginbar Ensermu et al. (2019)	Grid-connected	Wind and photovoltaic	Battery	G-VSC	Smart DC microgrid	Ginbar Ensermu et al. (2019)
Adel Merabet et al., 2016	Grid-connected	Wind	–	ANN	Multivariable controller in RT	Merabet et al. (2016)
Khadanga et al. (2018)	Isloated	Micro-Turbine (MT), DEG, WTG, PV, FC	Battery energy storage system (BESS), FESS	Multi-stage PID controller	Islanded microgrid	Khadanga et al. (2018)

It has a number of features that help in enhancing the simulation studies as given below:

1. *Monitoring Time:* It allows the user to monitor the time accuracy required for the precise RT simulations (Parizad et al., 2019). There is a number of timing information of the model such as "computation time", "real step size", "idle/wait time", and "number of overruns". In the library of the OPAL-RT's platform, a monitoring toolbox has an *OpMonitor* block as shown in Fig. 9.6 (Riquelme-Dominguez et al., 2022).

It can be seen that the block allows the user to view all the timing information of the model during runtime as given below:

- *Computation time:* The time taken by the model in its complete execution is called computation time in this block. It is to be noted here that it is the same timing information that is referred to as model execution time, t_m.
- *Real step-size:* The size of the step-time required for the collection of data during discrete running of the model is termed as real step-size, t_{step}.
- *Idle/wait time:* The time during which the simulator is idly, waiting for the response action to occur in the model is termed as idle/wait time, t_w. If the system takes less time in computation than the specified time-step, then the firing pulse waits for the exact time to trigger. Else, if the system takes more time, then the solver of the simulator utilizes the previous idle/wait time to compute the iterations of the mathematical equations of the system in order to trigger the control action on time. In this way, the online execution utilizes efficient ways of computing the large system admittance matrix with precision.
- *Nb overruns:* It is called the number of overruns, n_{over} and it is a verification parameter. While the system is online, the model is actually checked for the RT synchronization through this parameter

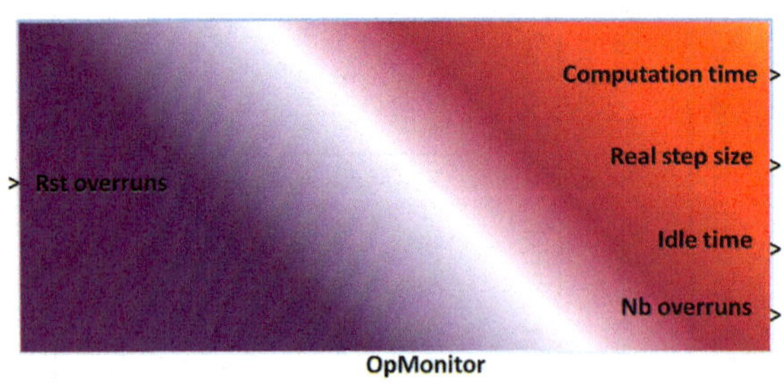

FIGURE 9.6 OpMonitor block.

(Strasser et al., 2015). If any overrun, n_{over} in the system model exists, it means that the system is not synchronized with the actual clock. Hence, the analysis of the system is not efficient in computing the results with respect to time, and the control strategy designed cannot be implemented in a real power system. Such a type of system modeling is known as erroneous in RT. Thus, for the practical implementation, the condition is given in Eq. (9.4),

$$n_{over} = 0 \tag{9.4}$$

On the other hand, if no overrun occurs, it means the system is synchronized (Blanger and Paquin, 2010). This helps in defining whether the model is running on time or not.

2. *Probe Control*: During HIL for a multi-variable system, it is vital to manage the data acquisition and transmission for perfect tuning in the synchronized execution mode. It is because if too much data is transmitted, then the display will update in visible bursts, and if too little, then there will be a risk of overloading the computation node as well as loss of data. Thus, a probe control helps in determining how much data should be gathered before it is sent.

For this control, the acquisition and transmission parameters are set per "acquisition group" in the "*OpComm*" block of the console subsystem. Other parameters are the "decimation factor", "number of samples per signal", and "duration":

- The decimation factor, dm determines how often on the computation node we collect a data point (for each requested signal) to put into the memory buffer.
- The number of samples per signal n_s represents the size of the acquisition window. It defines how many data points are gathered in the buffer before being sent to the command station.
- The duration is another way of selecting the number of samples, per signal, to gather data before sending to the command station. It specifies the window size in milliseconds as per Eq. (9.5),

$$\text{Duration} = dm \times n_s \times Ts \tag{9.5}$$

where Ts is the model sample time, for instance, the duration for the two decimation factor and two number of samples per signal will be 4 ms when the sampling time of the model is 1 ms.

3. *Data loss control:* In this type of control feature, the loss of data can be easily dealt with, irrespective of the cause of data loss, in many ways:
   - If the target/host processor is busy, then prefer to use a larger step-size for the model.

- If the communication network bandwidth is higher than prefer to use a larger decimation factor.

Despite of all the attempts to correct the modeling, if there is a data loss condition then RT-LAB has a synchronization algorithm that inserts dummy data points in order to maintain the RT clock for the execution of the model. It also interpolates the data between the last and the new data points for the smooth curve.

4. *Data logging:* RT-LAB provides direct data acquisition to a file instead of going through a network to the console subsystem. It has an "*Opwrite*""block Riquelme-Dominguez et al., (2022) as shown in Fig. 9.7 that helps in writing the data acquisition to.mat file in the workspace of the console processor.
5. *Dynamic signal acquisition:* A gain RT-LAB provides the selection of signals for the dynamic visualization at run time. This gives a user-friendly control/manipulation of the variables at run time. It has an "*OpCtrl*" block (Riquelme-Dominguez et al., 2022).
6. *Services:* OPAL-RT has well-organized products and services for the ease of operation. It gives software toolboxes, special simulators, and application-based hardware as given in Table 9.3 below.
   - The *software toolboxes* are beneficial as per the requirements. ARTEMiS is a CPU-based software used in electrical systems,

FIGURE 9.7 OpWrite block.

TABLE 9.3 OPAL-RT services.

Software toolboxes	Special Simulators
• ARTEMiS • eHS • eFPGASIM • ePHASORISM • RT-XSG • TestView • ScopeView	• OP1200 • OP1300 • OP1420

eHS, eFPGASIM, ePHASORSIM, and RT-XSG are FPGA-based software generally used in power electronics, power systems, electro-mechanical phasors, and input/output & communication (I/O & Comm) respectively, TestView is again CPU-based software used in test automation, and ScopeView is used in graph and data processing (Strasser et al., 2015).
- There are many simulators for the design and development, such as OP4510, etc. However, there are special simulators for effective and faster computation, like OP1200 can be used for multi-level converters, OP1300 for power electronics, and OP1420 for the MG PHIL system.

Therefore, the benefits of OPAL-RT's simulator indicate that it is one of the best platforms for executing/validating the designs of the developed models.

9.3.1 RT implementation procedure using OPAL-RT

The RT implementation is a process of executing the designed models of any system as per RT conditions. It is also a verification or testing stage of the design process industry, where the validation of the work is demonstrated (Blanger and Paquin, 2010; Dufour and Bélanger, 2004). OPAL-RT's online environment provides software, named as RT-LAB, for drafting the RT compatible model from the developed models in the offline environment. The component architecture of this drafting involves three stages of modeling as shown in Fig. 9.8. It consists of the following:

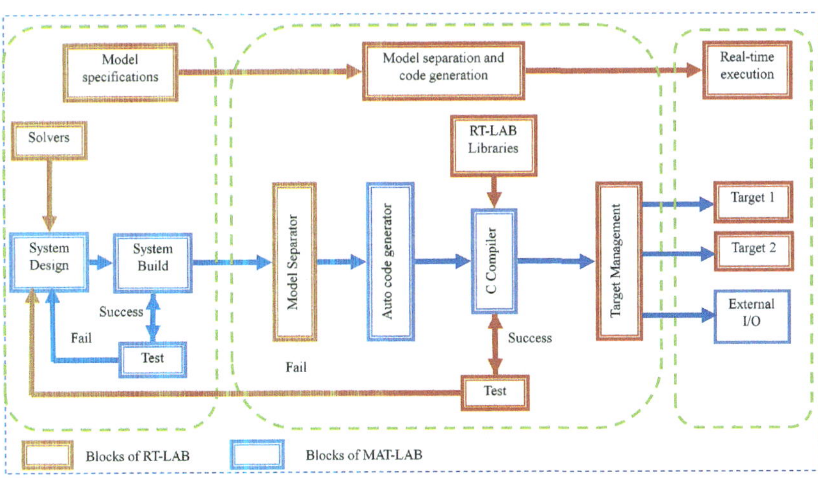

FIGURE 9.8 Components of the system architecture modeling.

- *Model specifications*: At this stage, the user defines the system design in MATLAB/SIMULINK or any other offline environment. After the successful development, it is ensured that there is no error in the offline execution of the model. Then, re-grouping is done in which the model consists of two groups named as "*master*" and "*console*". The naming syntax of these groups is SM_name and SC_name for the master and the console subsystems, respectively, where name can be user user-defined model name. At this stage, the data retrieval process, offline model testing, and solver selection from RT-LAB are done.
- *Model separation and code generation:* At this stage, the model is made compatible for the RT execution in the RT-LAB (Merabet et al., 2016) by adding RT blocks such as "*OpComm*" and by setting RT parameters such as step size. The model is separated into different cores of the simulator for parallel execution, and its code is auto-generated. Further, it is compiled using a C compiler for faster execution.
- *RT execution:* The model RT parallel execution takes place.

In Fig. 9.8 (Strasser et al., 2015), the most important is the model testing at two stages, one during offline model designing in MATLAB/ SIMULINK stage, where the model is tested after the system is built. And other testing is performed during RT execution, where the model is actually checked for the RT synchronization through the number of overruns, n_{over} (Strasser et al., 2015). Otherwise, the model is edited again using different computing solvers available in the RT-LAB, so that the computation time of the master subsystem matches the clock. In case the model size is complex and huge than the model is separated into parts to verify these stages; otherwise, a different concept of "*master and slave*" subsystems is also available (Merabet et al., 2016).

9.3.2 Utilization of RT platform for MG validation

After developing an understanding of OPAL-RT, the validation of various aspects of MG is easier. OPAL-RT provides a wide range of powerful tools for simulating complex systems, integrating controls, and evaluating performances under varying scenarios as discussed in previous sections. This makes the system compatible for industry deployment. Here's a structured approach for utilizing the RT simulator set-up (Patra et al., 2025; Maged et al., 2024) for MG validation :

- ***Define Objectives and Requirements***
  - **Determine Scope:** Identify the specific task of the MG to be validated, such as its control strategies, reliability, stability, or response to disturbances.

  - **Set Objectives:** Define the objective, such as evaluation of performance metrics, etc.
- *Develop the MG Model*
  - **Create Models:** Develop the detailed model of MG components that may include any generation source (solar, wind, diesel), storage device (batteries or super-capacitors), and load profiles in MATLAB/SIMULINK or an offline environment.
  - **Integrate Components:** Use RT-LAB software to integrate this offline model into an RT model using any technique SIL, HIL, or RCP.
- *Design and Implement Control Strategies*
  - **Control Algorithms:** Design controllers for the specific power source, manage the operation of the MG using energy management systems (EMS), set the demand response, and develop coordination between AC and DC systems in an MG.
  - **Simulation Integration:** Integrate these control strategies into the OPAL-RT's simulation environment for the evaluation of the performance in real-time scenarios.
- *Set up RT simulation Testing*
  - **Hardware Integration:** Connect real hardware components such as host PC, controllers, oscilloscopes, drives, systems, sensors, or actuators to the simulator. Select as per the model development is done, i.e., SIL, HIL, or RCP.
  - **Test RT Interaction:** Ensure that the hardware setup and simulation models are well-synchronized in time and are accurately reflecting the responses in actual time.
- *Execute simulations and Analyze*
  - **Initial Execution:** Perform preliminary simulations of the MG under normal operating conditions.
  - **Test Scenarios:** Test various scenarios like normal operational changes or fault conditions to evaluate the MG's robustness and control strategies.
  - **Performance Evaluation:** Evaluate the performance based on metrics such as system stability, response time, efficiency, and reliability.
  - **Identify Issues:** Detect any issues or areas for improvement in the MG design, control algorithms, real-time connections, or execution stages.
- *Process, modify, and reiterate*
  - **Modify Models and Controls:** Perform necessary changes to the system, control or connections based on the analysis.
  - **Re-iterate:** Perform model execution and analyses again and again until the desired results are obtained.

9.4 Case study

The modern power system is currently adopting advanced technologies to satisfy the increasing power demand. Accordingly, every branch of the power system, including generation, designing, protection, controlling, management, etc.., is going through a rapid change to meet it (Lasseter and Paigi, 2004; Tedesco et al., 2015). In the present scenario, the power generation from renewable sources has gained momentum owing to the fast depletion of fossil fuel, increasing pollution levels, and the pressing need for a sustainable energy solution to meet ever-increasing energy demand. Renewable sources such as wind, PV, etc.., alone are not reliable as the availability and the amount of production of energy highly depend on the climatic conditions. Thus, these are highly variable and show intermittent characteristics, which is not acceptable to consumers (Shen and Lior, 2016). This intermittent nature can be solved by connecting these via the grid to the supply load. Thus, it is important to penetrate the various renewable resources of generations to the grid that can form an MG (Wang et al., 2015; Denholm et al., 2011). However, renewable sources such as SOFC are not dependent on climate, and it is growing in demand due to their high efficiency and myriad of applications (Srinivasan, 2006; Alhassan and Garba, 2006) thus it is a reliable option for power generation in an MG (Zhu and Tomsovic, 2002a). Further, it can help in compensating for the power imbalance caused by PV (Wu et al., 2017b). Therefore, among all the available sources, PV and SOFC are an attractive option to form an MG.

The SOFC technology has accordingly been exploited in designing the layout/structure Buonomano et al., (2015), discussion on material stress Fischer and Seume (2012) physical modeling (Zhu and Tomsovic, 2002b; Marra et al., (2016) and various controls of SOFC (Wächter et al., 2006; Wang and Nehrir, 2007). In controlling, the variables like voltage, fuel input, partial pressures of oxygen-hydrogen, air excess ratio, utilization factor, temperature, etc.., of SOFC are exploited (Sun et al., 2018; Yu et al., 2017). Significant work has been reported where the variables are being controlled for the improvement of the performance of SOFC, but coordinated control between a set of DC-sources (PV and SOFC) and an inverter has not been researched much. The SOFC's response is very slow when a load demand change is applied (Gupta et al., 2020; Li et al., 2015; Han et al., 2018). While the inverter's time constant is low, making it a fast responding device (Sun et al., 2018; Li et al., 2005). Therefore, the essential coordination of these slow and fast responding devices generates a control action during any load change, making a complex control problem.

Earlier reports have suggested a feed-forward current control loop for this coordination analysis (Li et al., 2005, 2007). The observations reveal that it could not produce the expected dynamic performance. Further, some works provide the performance indices of SOFC in the grid and the micro-grid system using a PI-based controller. This controller works in either of the two different coordinated control strategies, which are based on fast load tracking and constant fuel utilization factor (Sun et al., 2018). It was not sufficient to control the two variables (1) fuel input and (2) utilization factor, simultaneously. Additionally, the time gap considered between the two consecutive demand power changes is generally small. This gap does not help in the study of the complete dynamics of such a slow-acting device, as it may not be a realistic situation. Thus, it necessitates the investigation of advanced control techniques that can control both fuel input and its utilization factor. Such controllers are indeed being designed, but these are either based on stochastic (Lagrange method, parent optimization, etc.,) or on meta-heuristic methods (genetic algorithm, evolutionary algorithm, particle swarm optimization, etc.,), which are not able to maintain a constant utilization factor and do not discuss the fuel input of SOFC (Wu et al., 2017b; Ramadhani et al., 2017; Chatterjee et al., 2014). Hence, a new technology to control both these variables efficiently seems to be a hybrid combination of *particle swarm optimization* (PSO) Sandre-Hernandez et al., (2015) and *adaptive neural network fuzzy inference system* (ANFIS) (Krasopoulos et al., 2018; Wu et al., 2018) termed as PSO-ANFIS. A controller designed using this combination is efficient in providing solutions for several applications in different power systems (Kumar et al., 2011; Semero et al., 2018).

Inspite of these advancements in designing controllers using a combination of various meta-heuristic methods, the result accuracy is not satisfactory because most of the controlling studies are carried out in an offline environment. To maintain this accuracy, it is important to design the system with more precision and reality. RT simulation is one such way that can help study large and complex systems.

This section presents an investigation on MG consisting of PV and SOFC in a grid-connected power system (Sun et al., 2018) as shown in Fig. 9.9 (Bai et al., 2016). A grid connection is used to provide a balance in a system (as it gives support during transients). It is necessary to keep the system balanced because an MG can get imbalanced either due to intermittent nature of renewable resource or any major load variations in a system (Sun et al., 2018). A PV array represents the intermittent nature while SOFC compensates for the power deficits and ultimately reduces the impact on grid. The current case study only focuses on the detailed study of SOFC, its advanced modeling and control development for the overall enhanced performance of the system. As, literature already reports lot of work in managing PV (Kumar and Bansal, 2019; Semero et al., 2018; Mahmud et al., 2017; Martin et al., 2018), inverter (Bellini et al., 2002; Mahmud et al., 2017; Bahrami et al., 2020; Fei et al., 2014) and grid-side (Li and Wang, 2021; Mitra et al., 2015).

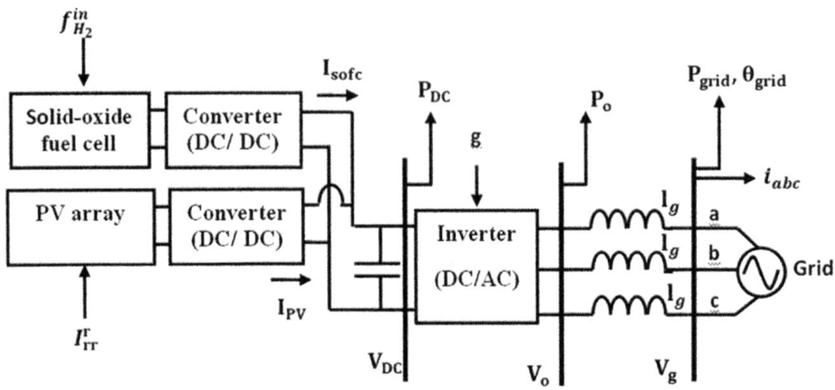

FIGURE 9.9 Block diagram of PV-SOFC grid-connected system.

9.4.1 SOFC

The basic functioning of the SOFC is that it undergoes a chemical reaction between hydrogen and oxygen to form water in the gaseous state and releases some energy (Costamagna et al., 1998; Gebregergis et al., 2009). Its dynamic model is given in Fig. 9.10 (Gupta, Pahwa and Verma, 2021a).

The rate at which it consumes fuel, f_i defines its fuel utilization factor, U_f. The change in fuel utilization factor, ΔU_f in SOFC, which is the ratio of reacted flow rate of H_2 gas, $\Delta f^r_{H_2}$ to the input flow rate of H_2 gas, $\Delta f^{in}_{H_2}$ depends on the feedback of SOFC current and input flow rate of H_2 gas. It should be within operating limits for the secure operation of the SOFC (Li et al., 2007). Thus, SOFC has defined a feasible operating area (FOA) for (a) operating temperature, T from 1173 to 1273K, (b) active power produced, P_{sofc} within 0.1–1 per unit and (c) utilization factor, U_f in the range 0.7 –0.9.

With FOA, the behavior of the output voltage, V_{sofc} with respect to the change in processed input fuel flow rate of hydrogen, $\Delta f^{in}_{H_2}$ is generally examined for analyzing the dynamic response of SOFC (Li et al., 2005). It is accelerated with the control action of a fuel controller and processor that provides a feedback control loop of fuel input. These reduce the time taken by the SOFC in processing the fuel (Sun et al., 2018). However, to further accelerate the dynamics of the SOFC system, it must coordinate with the inverter.

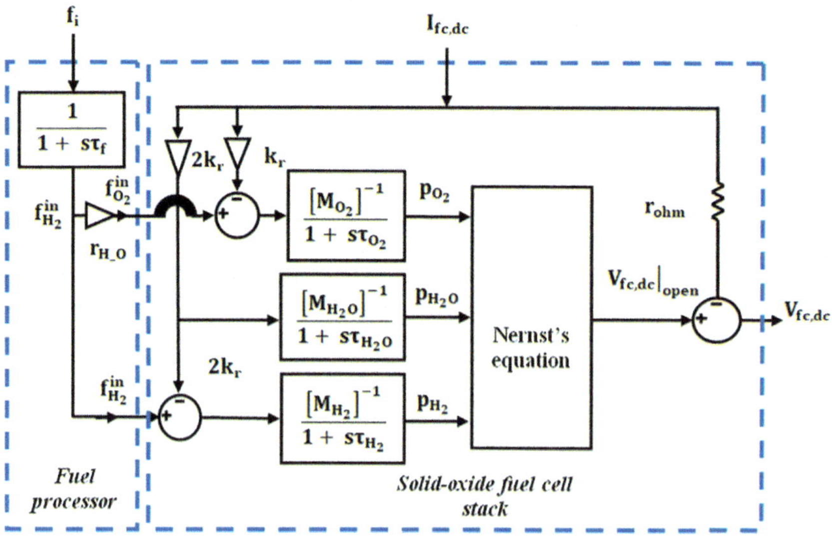

FIGURE 9.10 Dynamic model of the SOFC.

9.4.2 Coordinated control between SOFC and inverter

For appropriate coordination between the SOFC and the inverter, a closed-loop control circuitry as shown in Fig. 9.11 is developed. It consists of an intelligent fuel-valve controller, a refined inverter controller, and a fuel processor with correct positioning of the fuel into the stack.

- *Intelligent fuel-valve controller and fuel processor:* During the load change, the change in fuel utilization factor, ΔU_f given in (4.8)

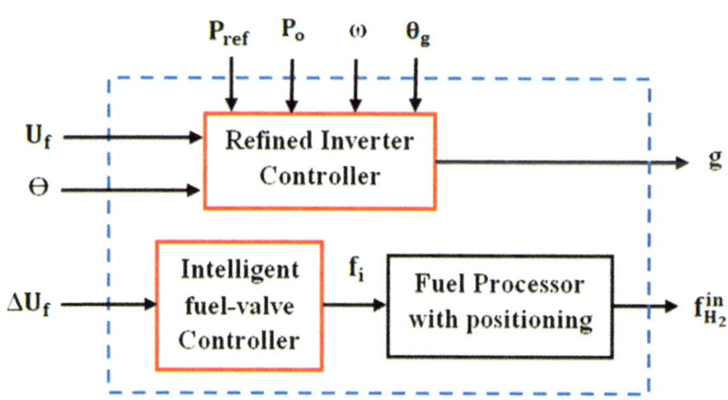

FIGURE 9.11 Block diagram for coordinated control operation.

determines the rate of change of the fuel consumption in cells, which should be within the operating limits as per FOA.

$$\Delta U_f = \frac{2k_r}{\Delta f_{H_2}^{in}} \tag{4.8}$$

The control action of an intelligent fuel-valve controller and fuel processor provides a feedback control loop of the fuel flow of hydrogen, $f_{H_2}^{in}$ which reduces the time taken by the fuel processor in Fig. 9.10 in processing the fuel, f_i. Thus, it helps in accelerating the dynamics of the SOFC system (Sun et al., 2018). This change in the utilization factor, ΔU_f is subject to $0.7 \leq U_f(t) \leq 0.9$.

- Refined inverter controller: When applying the input active reference power, P_{ref} to the complete system, the inverter operates as per the gate pulses, g obtained from RIC as shown in Fig. 9.12. In this figure, the switching block utilizes a control algorithm that helps in processing the SOFC power, P_o. When the fuel utilization factor is within the safe limits then the power adjustment will feed the corresponding SOFC power, $P_o^* = \begin{cases} 0.7P_o \forall U_f < 0.7 \\ 0.9P_o \forall U_f > 0.9 \\ P_o \forall 0.7 \leq U_f \leq 0.9 \end{cases}$ for further processing, otherwise, it will wait for the SOFC's response. Next, the output SOFC power, P_o^* is being compared with the reference power, P_{ref} to regulate the error, P_ε^* using an integral control. Therefore, with this closed-loop control of the SOFC completes its coordination with an inverter in a grid-connected system.

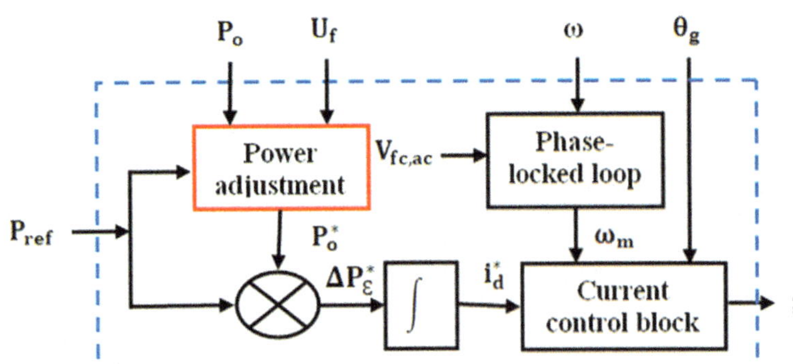

FIGURE 9.12 Block diagram of refined inverter control.

9.4.3 Design of PSO-ANFIS based intelligent fuel controller

To regulate the fuel input to SOFC, a PSO-ANFIS-based IFC is shown in Fig. 9.13. This IFC utilizes the optimization ability of particle swarm that identifies the global best, the power of neural network that can adapt, and the discipline of fuzzy rules. It is an efficient way of handling the dynamically changing variables due to its ability to learn and adapt as per the situation. It consists of the following:

- ISE block for calling the model,
- PSO algorithm for k_p and k_i determination,
- PSO trained PI controller for generating the set of input-output data and
- ANFIS block for final optimization.

In ISE block, u is the change in utilization factor, ΔU_f which is optimized by PSO algorithm. The optimal solution for the $k_p = 0.9978$ and $k_i = 5.0021$ of PSO-based PI helps the system to get tuned efficiently (Ensermu et al., 2019). With this solution, the execution of the model in the MATLAB/SIMULINK environment leads to the generation of the desired output data for the controller i.e., a set of input-output data which is $\left(\Delta U_f - f_i\right)$ data. This input-output $(\Delta U_f - f_i)$ data is loaded in the ANFIS toolbox (MATLAB/SIMULINK environment), and the controller is re-iterated until the parameters fall within limits.

Here, the two inputs are time, t (seconds), and ΔU_f where $i = 1$ for $u_1 = t$ and $i = 2$ for $u_2 = \Delta U_f$ respectively. The network is a Takagi-Sugeno model with two inputs and one output function consisting of a 2:6:9:9:1 architecture as shown in Fig. 9.14 (a). In layer 2, the membership functions, j selected for each input by hit and trial method are 3, and their type is "*pimf*". Therefore, the function *u* is f{x, [pqrs]}, where *p*, *q*, *r*, *s* are the "*pimf*" parameters that are generated by the neural network itself. Thereafter, the unified rules of FIS are applied that form layer 3 in the network architecture

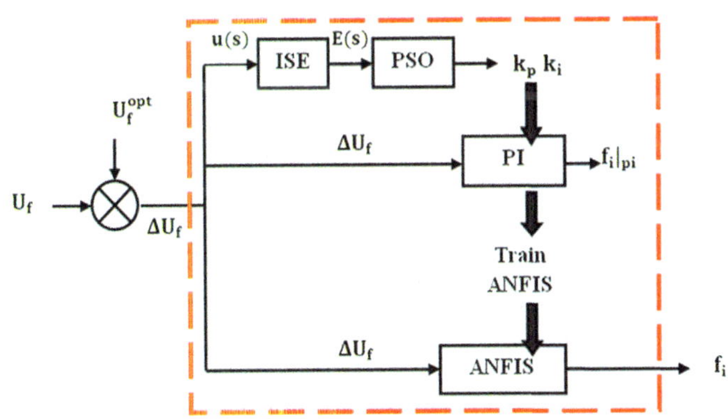

FIGURE 9.13 Block diagram of PSO-ANFIS based IFC.

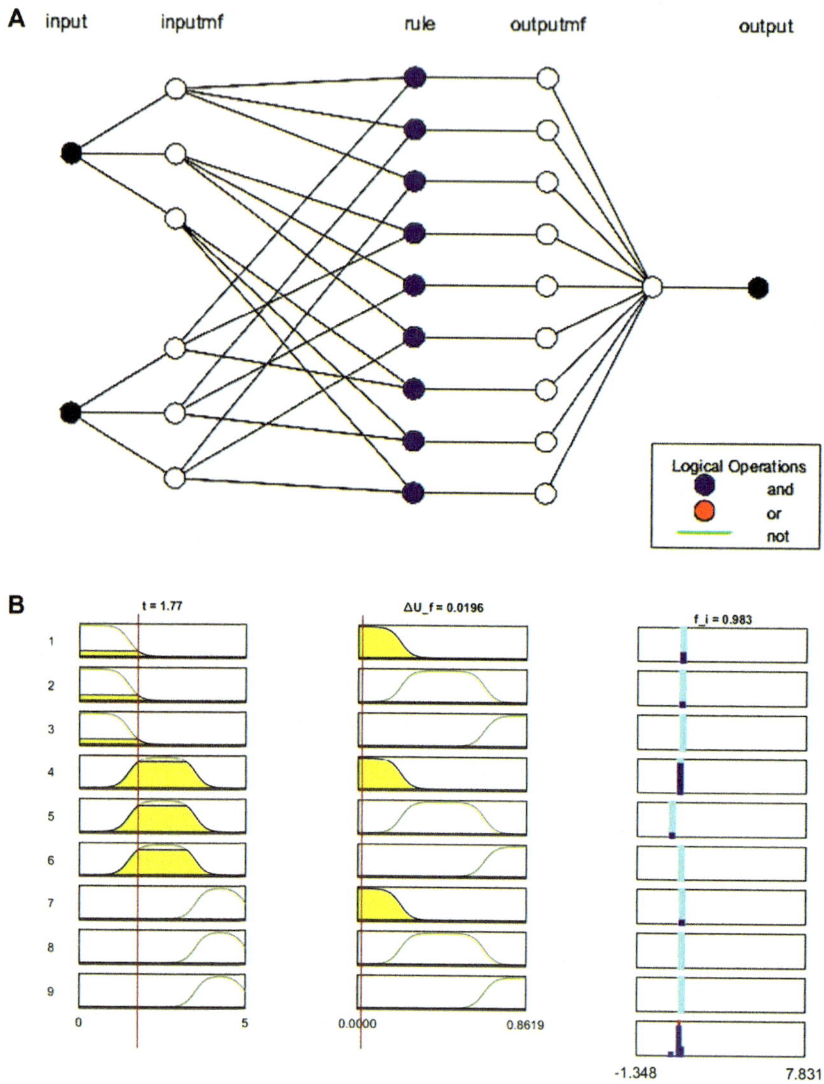

FIGURE 9.14 Network and layer 3 of ANFIS for the two inputs, and time, t (seconds) and one output, fuel input, (a) network architecture(b) rule viewer.

(Fig. 9.14(a)). These rules are represented with the help of rule viewer as shown in Fig. 9.14(b). In this figure, the output, y_j and the de-fuzzified output, y of FIS using "wataver and aggMethod" are given below:

$$y_j = \begin{matrix} [19.8684, 19.9348, 19.9861, 19.8720, \\ 19.9362, 19.9896, 19.8744, 19.9391, 19.9876] \end{matrix} \text{ and } y = 0.985 \text{mol/sec}.$$

With this, the crisp result, i.e., the controlled output, $f_i = 0.985$ mol/s is obtained after incorporating the appropriate limiter.

Finally, the training of approximately 30 epochs is carried out to test the proposed controller until the error is reduced upto 0.0035 as shown in Fig. 9.15(a). This small error indicates that the developed controller is behaving in the same pattern as the initial loaded data.

After the number of iterations and error reduction, the dependency of fuel input on the change in utilization factor and time is checked with the help of a control surface viewer, as shown in Fig. 9.15(b). In this figure, a

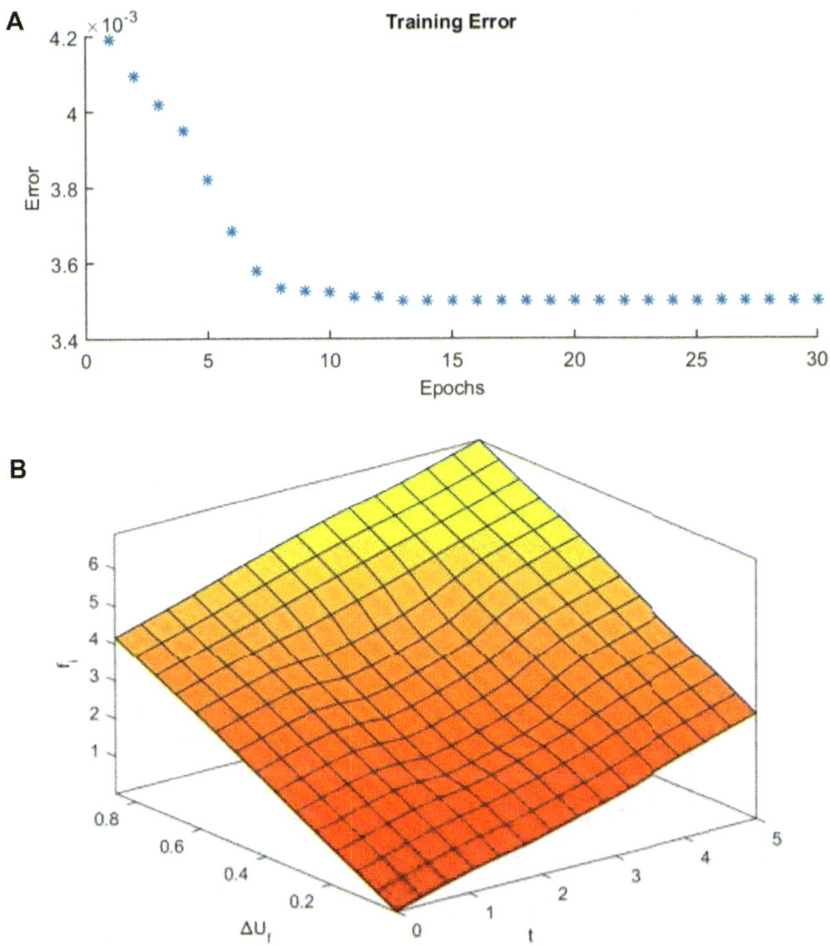

FIGURE 9.15 Layer 5 of ANFIS for the fuel input (a) error-checking curve (b) control surface.

three-dimensional plot is generated where the x-axis is time, t in seconds, the y-axis is the change in utilization factor, and z-axis is the fuel input (net fuzzy output). The fuel input is linear with respect to the change in utilization factor whereas it is nearly constant with respect to time. Hence, the proposed controller is itself handling the requirement of the fuel with respect to its utilization, irrespective of the time.

9.4.4 RT simulation of grid-connected microgrid

The RT environment is well-suited for the analysis of complex system operations and process studies. It provides a platform to verify the system as per RT domain (Ramezani et al., 2017; Baghaee et al., 2024; Merabet et al., 2016). The simulator used for investigation of the model under study is the OP4510 module of the OPAL-RT simulator, which is composed of Intel Xeon 4-core processors and Xilinx Kintex 7 FPGA. The standard configuration includes 32 digital inputs-outputs and 16 analog inputs and outputs (Blanger and Paquin, 2010; Dufour and Bélanger, 2004). For the RT model development, first, the system shown in Fig. 9.9 with its associated control circuits (given in further sub-sections) is developed in MATLAB/Simulink. Secondly, after the successful development of the model, it is re-grouped in the RT-LAB environment. At this stage, the model consists of two groups named as "master" and "console" as mentioned. The groups are named as per the syntax i.e., "SM_name" and "SC_name" used for master and console subsystems, respectively. Thereafter, the model is made compatible for RT execution in RT-LAB (Merabet et al., 2016). The model is first executed offline, and if it is error-free then it is perfectly designed in RT; otherwise, it is re-evaluated.

Now, the connections are made for the experimental setup as shown in Fig. 9.16. As shown, it consists of a host computer where RT-LAB is

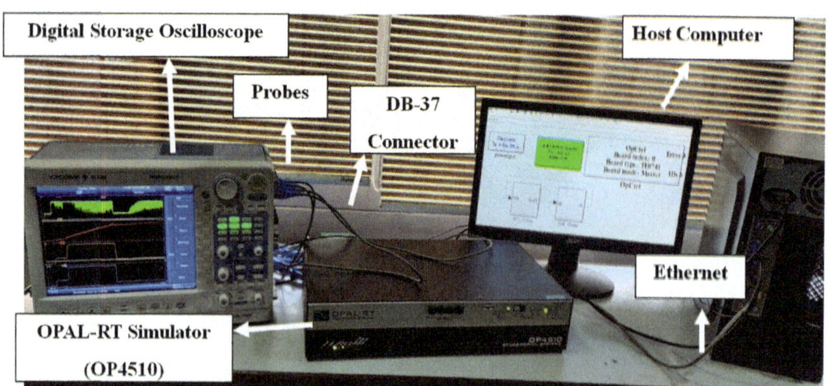

FIGURE 9.16 Experimental HIL set-up.

installed. This host computer is connected to the simulator OP4510 via Ethernet. The simulator acts as a target computer, which behaves as the master subsystem during execution. After the connections, the model is "built" "loaded", and "executed" onto the simulator. This execution is known as an online simulation. At this stage, while the system and the controller are online, the model is actually checked for synchronization through the parameter called "number of overruns". If it checks for zero, then the model is perfectly synchronized with RT clock. Otherwise, the model is edited again using different computing solvers available in RT-LAB, so that the computation time of the master subsystem matches the clock.

Further, the communication can be completed by monitoring the signals on a digital storage oscilloscope (DSO) via a DB37 connector. The DSO can either store the results as.mat file (viewed and plotted in MATLAB) or capture the runtime scenario as.jpeg file. This completes hardware-in-loop because DSO demonstrates as a monitoring unit to observe parameters, the simulator as an actual site (where the power plant and fuel controller are installed), and the host computer as a console (where change in load demand is made).

9.4.5 Results and discussion

The grid-connected microgrid taken into consideration has been analyzed for fuel input to the SOFC and the impact of its utilization on the system during any load demand power change. The developed controller has a control loop of the fuel flow of input (hydrogen) to the SOFC, and it utilizes control mechanisms based on PI and PSO-ANFIS for the effective comparison of the proposed controller. Later, the case study analyzes the SOFC's stability using a Bode plot and presents total harmonic distortions using Fast Fourier Transform analysis of the system with/without an intelligent controller in MATLAB/SIMULINK environment only.

The simulation study proceeds with the following: PV array comprises 50 parallel strings, with each string consisting of 7 SunPower SPR-415E modules connected in series. The system has a total rated power output of 135 kW. A switching function based inverter is used to convert the DC output to AC (reason to use this inverter model are referred to (Gupta et al., 2021b). To ensure optimal energy harvesting, a Maximum Power Point Tracking (MPPT) controller is implemented using the Perturb and Observe (P&O) algorithm system. The grid of 440 V operates at 60 Hz supply. The capacity of SOFC is taken as 100 kW at an operating temperature of 1273K with 450 number of series cells in a stack with base voltage of 530 V (1.18×450), and its other parameters are given in Table 9.4.

The system load has been changed at 5, 15, and 55 seconds from 0.0 p.u. to 0.3 p. u., 0.3 p. u. to 1 p. u. and, 1 p. u. to 0.1 p. u. respectively as shown

TABLE 9.4 Data of SOFC (Sun et al., 2018).

SOFC parameters	Symbol	Value	Unit
Faraday's constant	F	96.487×10^6	C/kmol
Molar constants of H_2	M_H	8.343×10^{-4}	kmol/s-atm
Molar constants of H_2O	M_{H_O}	2.81×10^{-4}	kmol/s-atm
Molar constants of O_2	M_O	2.52×10^{-4}	kmol/s-atm
Resistive loss	r_{ohm}	3.2813×10^{-4}	ohms
Reaction constant	k_r	1.166×10^{-9}	kmol/s-A
Stoichiometric-ratio	r_{H_O}	1.145	–
Stack area	A	654	cm^2
Standard cell potential	V_0^o	1.18	volts
Time constant of fuel	τ_f	5	s
Time constants of H_2	τ_{H_2}	26.1	s
Time constants of H_2O	τ_{H_2O}	78.3	s
Time constants of O_2	τ_{O_2}	2.91	s
Universal gas constant	R	8.314	J/kmol-K

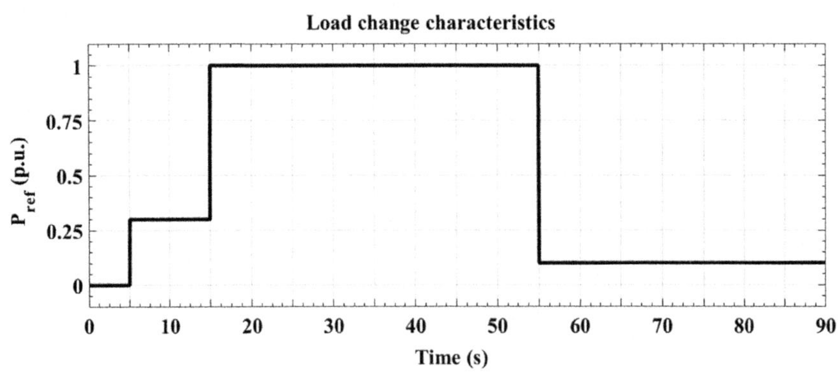

FIGURE 9.17 Load change characteristics.

in Fig. 9.17. In this figure, it is interesting to note that the time gap of 5, 10, 40, and 35 seconds between the two consecutive load power changes has been intentionally provided for observing the behavior of the SOFC. Since the SOFC takes 5–10 seconds to reach the steady-state under disturbed conditions (Sun et al., 2018), a long time gap has been considered for the analysis. The complete system is investigated for 90 seconds to observe the dynamic performance of the SOFC system.

9.4.5.1 Dynamic responses in SIL

During this mode of execution, the system executes without any risk to showcase the dynamic responses of the SOFC. It mainly consists of the fuel utilization factor, U_f, the amount of fuel input, f_i (mol/sec) and the output power, P_o (p.u.). These comparative dynamic responses are shown in Fig. 9.18 using conventional PI and proposed POS-ANFIS based fuel

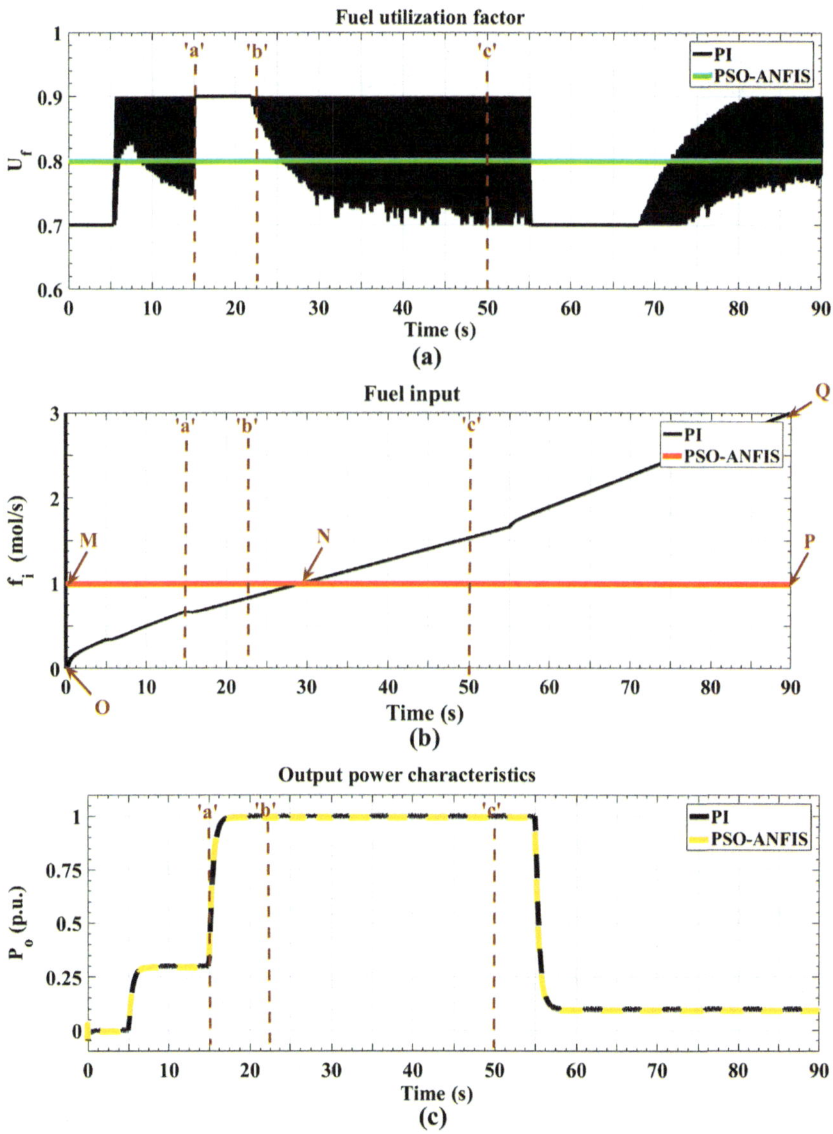

FIGURE 9.18 Output responses of the system with PI and PSO-ANFIS based fuel controllers in SIL (a) fuel utilization factor, (b) fuel input, (mol/sec) (c) output power, (p.u.).

valve controllers. The proposed control mechanism enhances the dynamic response of the SOFC by controlling the fuel utilization factor, U_f and fuel input, f_i as shown in Fig. 9.18(a) and (b). In Fig. 9.18(a), despite the full load change at 55 s, the response of U_f is 0.8 using the proposed controller, whereas the PI controller gives a sluggish response. The utilization factor is edging the cells of the stack to work on limits. Since the load on any real power system changes frequently, therefore accordingly U_f will change with a conventional PI-based fuel controller. Further, depending upon the magnitude of the load change, the fuel input, f_i to the SOFC, as well as its chemical reactions, will take place at a faster or slower rate accordingly, making it and its associated control circuitry less efficient. This ultimately leads to the reduced life of the SOFC system. In contrast, U_f is optimized with PSO-ANFIS-based IFC, making the SOFC and its associated control circuitry more efficient.

In Fig. 9.18(b) it is noteworthy that the response of the fuel input, f_i with PI-based controllers is increasing continuously, and in case of PSO-ANFIS-based IFC, it is constant. Although the developed controller takes more fuel at the start but in the end it is low. It can be understood like this, as the fuel input, f_i the area inside the points "MNO" is less in comparison to the area inside the points "QNP". This means that the average fuel consumed by the SOFC with the proposed controller is less, representing the superior performance of it because the system experiences jerks at the time of load change with the PI-based controller, which highly affects the life of the SOFC.

This continuously increasing fuel input is not that effective when observed for minor gaps, i.e., 5 or 10 seconds, but it is immensely affecting the life of the SOFC when considered for large time gap between the two consecutive load changes. The developed PSO-ANFIS-based IFC helps in maintaining the fuel input, f_i at 0.995 mol/s, showcasing its robust design (Sun et al., 2017). Moreover, the load tracking capability of the SOFC system for the same reference power changes in Fig. 9.17 is unaffected because RIC, along with IFC are guiding the fuel cell to respond as per the demand power requirements as depicted in Fig. 9.18(c). In this figure, the developed output power, P_o is the same with both the controllers. In other words, the power tracking with both controllers is the same. Hence, the overall performance of the system is enhanced with the help of the proposed controller.

9.4.5.2 Dynamic responses in HIL

The same study is carried out further in HIL as depicted in Fig. 9.19. In this figure, (a) fuel utilization factor, U_f (b) fuel input, f_i (mol/sec), (c) output, P_o (p.u.) and (d) reference, P_{ref} (p.u.) powers to the SOFC are shown with the requisite amount of gain parameters on the y-axis. The

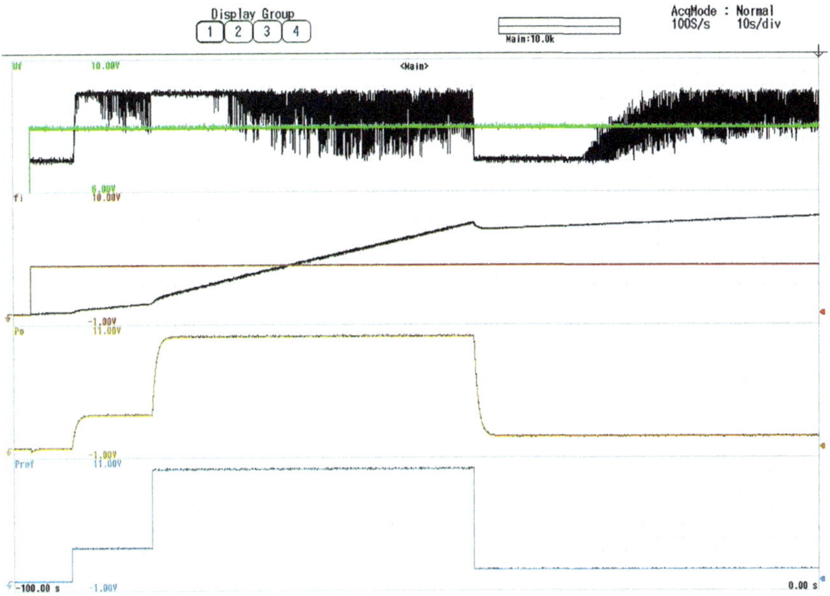

FIGURE 9.19 Output response of the system in HIL with PI and PSO-ANFIS based fuel controllers on DOS (from top to bottom).

gain is necessary because the real-time simulator can only transfer voltage signals in the range of ±16 V.

9.4.5.3 Stability analysis

As discussed, the SOFC is a slow-acting device and its dynamic behavior cannot be analyzed in the time domain properly. Therefore, its analysis is also done in s-domain using the MATLAB built-in program. It is linearized and its Bode in three types of control operations is plotted with PSO-ANFIS, PI, and without any controller as shown in Fig. 9.20. In this figure, the operating points "a" and "d" show the instant at which the step change is applied and any arbitrary point during the steady-state period, respectively. The operating points "b" and "c" are two arbitrary points representing the transient period. $|x|$, $|y|$ and $|z|$ represents the phase angle distance from the reference phase of −90 degrees to the response of PSO-ANFIS-based IFC, PI-based fuel controller, and without any controller, respectively. The comparison of the controllers under consideration is given in Table 9.5 based on the above Bode plot. Three steps are applied on SOFC system (refer to Fig. 9.17). Out of these, the step-change occurring at 15 seconds is considered because it has maximum amplitude (3-1 p.u.) and the time gap (40 s) respectively for analyzing the SOFC's performance in the s-domain. It can be observed in

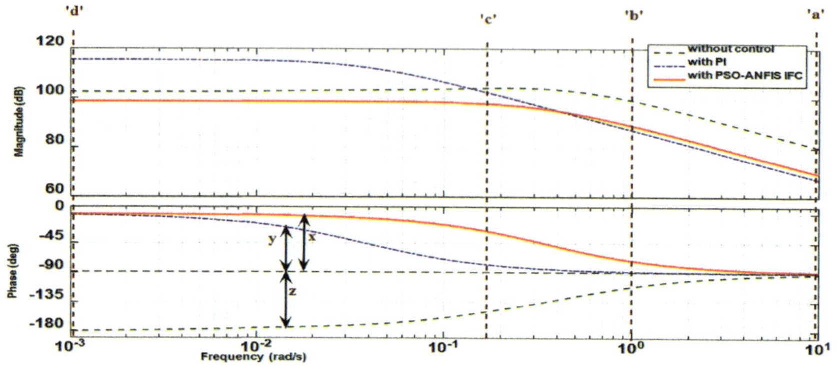

FIGURE 9.20 Bode plot of SOFC with both the controllers and without controller operation.

Fig. 9.20 and Table 9.6 that at operating points "a" and "d", the effects of both the controllers are the same on the SOFC, i.e.., the phase angle differences, $|z-x|$ and $|y-x|$ are in the range of $0.2-0.5$ (approximately equal to zero).

Whereas, during the transient period i.e.., at operating points "b" and "c", the control efforts generated by both the controllers are different. The phase angle difference, $|z-x|$ due to the proposed controller is 1.5 degrees, which is nearly equal to zero. This effect is shown in the time-domain plot of the utilization factor, which clearly indicates that for any change the utilization factor settles down to its optimum value of "0.8", which results into a smooth control. In contrast to this, the phase angle difference, $|y-x|$ due to the conventional PI based controller is in the range of 35 degrees – 47 degrees, which is quite high. This results into settling of the utilization factor in the range of 0.7–0.9, shows inferior performance of the conventional controller.

9.4.5.4 THD

Out of many power quality parameters, total harmonic distortion (THD) is one of the main parameters. The THD of the current, I_a is 3.73%. This is within limits as per the latest grid code as shown in Fig. 9.21. As the system inverter utilizes the hysteresis current loop and thereby defines the switching signal for an inverter, the overall performance of the grid-connected SOFC system maintains the power quality.

9.4.5.5 Computation time analysis with overruns

To analyze the developed model using OPAL-RT, the *OpMonitor* block, as shown in Fig. 9.6 is used. The analysis is performed to display the controller's robustness and accuracy in an online environment, in contrast

TABLE 9.5 Comparison of controllers.

Operating point	Time (sec)	Condition of SOFC	Without fuel controller	PI-based fuel controller			PSO-ANFIS based IFC		
			$\lvert z\rvert$	$\lvert y\rvert$	$\lvert z-y\rvert$	*Response*	$\lvert x\rvert$	$\lvert z-x\rvert$	*Response*
"a"	15	Step-change	39.5 degrees	0.16 degrees	39.34 degrees	-NA-	39 degrees	0.5 degrees	*Smooth*
"b"	21	Transient	45.2 degrees	8.7 degrees	36.5 degrees	Sluggish	43.7 degrees	1.5 degrees	
"c"	50		59.5 degrees	12.85 degrees	46.55 degrees		60.96 degrees	1.54 degrees	
"d"	6298.1	Steady	89 degrees	88.47 degrees	0.53 degrees	Smooth	88.88 degrees	0.01 degrees	

TABLE 9.6 Comparison of online and offline environment on the basis of time (sec).

Environment	Simulation time (Sec)	Number of overruns	Real-time clock (Sec)	Clock synchronization	Computation burden	Practical implementation (Difficult/Easy)
	t_s	n_{over}	t_{rc}			
Offline	280	65,413	90	$t_s \neq t_{rc}$	$n_{over} \neq 0$	Difficult
Online	90	00	90	$t_s = t_{rc}$	$n_{over} = 0$	*Easy*

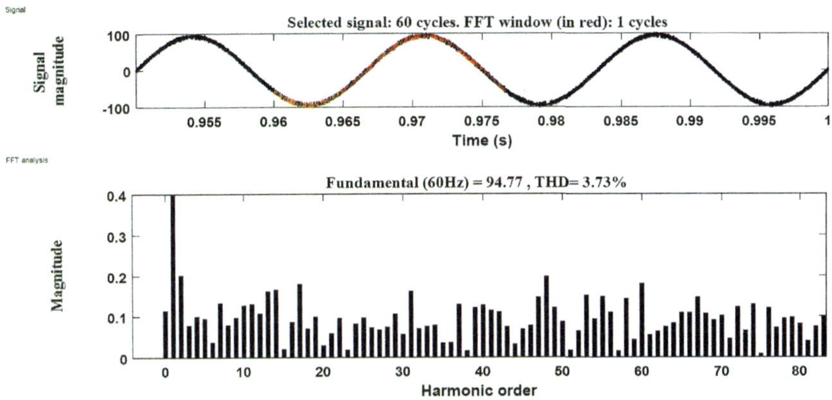

FIGURE 9.21 FFT analysis of grid-connected SOFC system.

to offline, in terms of the model execution time, t_m (sec), and the number of overruns, n_{over} as given in Table 9.6. As observed from the data given in table, for the same simulation time under consideration, the offline model takes more time in computing the results while the online model is computing within the exact time as represented with the real-time clock, t_{rc}. It is because the online environment utilizes an *artemis* solver of the OPAL-RT simulator which helps in providing the parallel computation across the same core of the processor. This enables accuracy of the proposed control actions. It is also observed that the model designed in the MATLAB (offline mode) has high number of overruns, n_{over} while the same model that has been re-designed in the RT-LAB (online mode) using an *artemis* solver contains zero overruns. This represents large computation burden on the offline environment due to which accuracy of the control actions is questionable. Thus, the computation time analysis with overruns of the system with its associated complex control circuitry reveals accurate control actions in real-time, and the proposed controllers are ready for the actual power system operations.

9.5 Challenges and solutions in hardware and real-time validation

Real-time validation is now an integral part of many industrial (like automotive, aerospace, robotics, and autonomous vehicles) simulation study processes. It helps in designing and testing the performance without any actual setup deployment, thus saving extra cost and dangerous experiments. However, there are several challenges faced during validation using real-time simulation (2015). This section discusses the challenges and then provides some possible solutions.

9.5.1 Technical challenges in practical validation

There can be several technical challenges as per the scenario, system requirements, application, resource used, and process of operation (Abrishambaf et al., 2017; Lugaresi and Matta, 2018; Menghal and Laxmi, 2012). Here are a few common challenges faced during any kind of RT operation:

- ***Latency:*** Any system must respond simultaneously to every input and generate the corresponding output without any delay or buffer. This delay creates lag in computation speed and causes inaccuracy in the simulation model. Thus, it makes an ineffective approach for validating the dynamic behavior of any system.
- ***Model Fidelity and Accuracy:*** Generally, practical systems are complex to simulate as they lead to a high computation burden. On the other hand, simple models do not provide an actual behavior study. Therefore, achieving high fidelity and accuracy in simulation models is very difficult.
- ***Synchronization:*** The Time taken by each generator, system, controller, sensor, etc.., is different, which creates an error during the physical integration of the models and hardware setups.
- ***Implementation of Artificial intelligence or Machine learning:*** Designing system controllers based on AI or ML that replicate the complexity of real-world scenarios in an RT simulation can be difficult. It may not capture all variables, resulting in incomplete or misleading validation.
- ***Re-iterate or Debugging errors:*** In complex systems, the validation process becomes difficult for non-linear systems when studying dynamically, as runtime issues are difficult to debug.

9.5.2 Addressing challenges

The above-mentioned challenges are stubborn in the case of complex systems, and it becomes very difficult to solve practically. Here are some of the possible solution approaches (Sirviö et al., 2020; Riquelme-Dominguez et al., 2022; Ibarra et al., 2017; Dufour and Belanger, 2014) that might be effective:

- ***Optimization:*** Use of high-performance methods to refine the system variables and find best best-suited outputs in less time can be an effective way. For example, a control design based on modern techniques such as fuzzy logics, particle swarm, genetic, etc.., (Bermudez et al., 2019; Maged et al., 2023).
- ***Enhanced methods of modeling:*** There are several methods to improve system modeling, such as model reduction methods, multi-

scale modeling, etc. This helps in maintaining computational efficiency without affecting accuracy.

- ***Co-simulation:*** This approach gives the advantage of parallel simulation using two or more platforms where a real-time simulating model executes alongside the hardware components to exchange data (Wang et al., 2021). The platforms that allow co-simulation with external hardware are Simulink, MATLAB, Simscape, etc. In addition, industrial standard communication protocols such as Ethernet, FlexRay, etc, are used during co-simulation, which ensures smooth interaction (on-time exchange of information rather than any delay) (Riquelme-Dominguez et al., 2022).
- ***Cloud Computing:*** This is another advanced way of managing information. This gives flexibility in utilizing hardware, storage, and services that can help in scaling the simulation study dynamically.

9.6 Summary and future scope

The RT validation has evolved over the years, and it has made a remarkable impact in the design and development industries. It is giving a process to test systems in various environments like SIL, HIL, and RCP that provide ease of dynamic studies, including lower cost and damage operation. These environments are as per IEEE and IEC standards. This chapter discusses the various simulators that provide such environments for validation purposes. After the comparison drawn between these simulators, it can be stated that OPAL-RT provides ease of integration with existing software and hardware, i.e., flexibility, high accuracy, and faster computation for the complex systems. Thus, this chapter discusses this platform in detail, giving insights about its history, features, implementation procedure, and a case study on it.

The case study is conducted on a PV-SOFC grid-connected system for which an ANFIS-based fuel controller is designed. This controller regulates the flow of hydrogen gas to the SOFC stack in an optimized manner. The complete system is first developed on MATLAB/SIMULINK, and then it is made compatible with the OPAL-RT platform. The model is executed on SIL and HIL environments using the OPAL-RT simulator, and it has been showcased that the technical result's accuracy is unaffected. Additionally, a computational analysis is carried out to discuss the time accuracy of the developed controller, which indicates that it is ready for deployment after HIL execution.

References

Abrishambaf, O., Faria, P., Gomes, L., Spínola, J., Vale, Z., Corchado, J.M., 2017. Implementation of a real-time microgrid simulation platform based on centralized and distributed management. Energies 10 (6). https://doi.org/10.3390/en10060806.

Alhassan, M., Garba, M.U., 2006. Design of an alkaline fuel cell. LEJPT 99–106.

Antone, J.P., Stubbe, M., 1992. EUROSTAG, software for the simulation of power system dynamics, its application to the study of a voltage collapse scenario. In: IEE Colloquium on Interactive Graphic Power System Analysis Programs, pp. 1–4.

Baghaee, H.R., Mirsalim, M., Gharehpetian, G.B., Talebi, H.A., Niknam-Kumle, A., 2024. Notice of violation of IEEE publication principles: a hybrid ANFIS/ABC-based online selective harmonic elimination switching pattern for cascaded multi-level inverters of microgrids. IEEE Trans. Ind. Electron. 1. https://doi.org/10.1109/tie.2017.2694403.

Bahrami, A., Norambuena, M., Narimani, M., Rodriguez, J., 2020. Model predictive current control of a seven-level inverter with reduced computational burden. IEEE Trans. Power Electron. 35 (6), 5729–5740. https://doi.org/10.1109/TPEL.2019.2952533.

Bai, W., Abedi, M.R., Lee, K.Y., 2016. Distributed generation system control strategies with PV and fuel cell in microgrid operation. Control Eng. Pract. 53, 184–193. https://doi.org/10.1016/j.conengprac.2016.02.002.

Bellini, A., Bifaretti, S., Costantini, S., 2002. A new approach to hysteresis modulation techniques for NPC inverters. In: IEEE International Symposium on Industrial Electronics, vol. 3. Institute of Electrical and Electronics Engineers Inc., Italy, pp. 844–849. https://doi.org/10.1109/isie.2002.1025843.

Bermudez, M., Gomozov, O., Kestelyn, X., Barrero, F., Nguyen, N.K., Semail, E., 2019. Model predictive optimal control considering current and voltage limitations: real-time validation using OPAL-RT technologies and five-phase permanent magnet synchronous machines. Math. Comput. Simulat. 158, 148–161. https://doi.org/10.1016/j.matcom.2018.07.005.

Blanger, J., Venne, Paquin, J.N., 2010. The What, Where and Why of Real-Time Simulation IEEE PES General Meeting, pp. 37–49.

Bose, B.K., 1990. An adaptive hysteresis-band current control technique of a voltage-fed PWM inverter for machine drive system. IEEE Trans. Ind. Electron. 37 (5), 402–408. https://doi.org/10.1109/41.103436.

Breaz, E., Gao, F., Paire, D., Tirnovan, R., 2014. Fuel cell modeling with dSPACE and OPAL-RT real time platforms. In: IEEE Transportation Electrification Conference and Expo: Components, Systems, and Power Electronics - from Technology to Business and Public Policy, ITEC 2014. Institute of Electrical and Electronics Engineers Inc., Romania https://doi.org/10.1109/ITEC.2014.6861813.

Buonomano, A., Calise, F., d'Accadia, M.D., Palombo, A., Vicidomini, M., 2015. Hybrid solid oxide fuel cells-gas turbine systems for combined heat and power: a review. Appl. Energy 156, 32–85. https://doi.org/10.1016/j.apenergy.2015.06.027.

Cardona, M.A., Rodriguez, R.J., Ishmael, K., 2023. Artificial Intelligence and the Future of Teaching and Learning, pp. 1–71.

Chatterjee, K., Shankar, R., Kumar, A., 2014. Fuzzy logic based controller for a grid-connected solid oxide fuel cell power plant. J. Fuel Cell Sci. Technol. 11 (5). https://doi.org/10.1115/1.4027709.

Chou, H.M., Butler-Purry, K.L., 2012. Real-time wide area power control studies using PXI and RTDS. In: North American Power Symposium, NAPS 2012. https://doi.org/10.1109/NAPS.2012.6336369.

Cole, S., Haut, B., 2013. Robust modeling against model-solver interactions for high-fidelity simulation of VSC HVDC systems in EUROSTAG. IEEE Trans. Power Syst. 28 (3), 2632–2638. https://doi.org/10.1109/TPWRS.2012.2230194.

Costamagna, P., Costa, P., Antonucci, V., 1998. Micro-modelling of solid oxide fuel cell electrodes. Electrochim. Acta 43 (3–4), 375–394. https://doi.org/10.1016/S0013-4686(97)00063-7.

Denholm, P., Ela, E., Kirby, B., Milligan, M., 2011. The Role of Energy Storage with Renewable Electricity Generation Energy Storage: Issues and Applications. Nova Science Publishers, Inc., pp. 1–58. https://www.novapublishers.com/catalog/product_info.php?products_id=21441

Dong, Y., Pan, L., Qiu, D., Tian, J., Weihua, W., Li, W., 2016. IEEE Computer Society China Hardware-in-the-loop simulation and test of a control and protection system for MMC-based UPFC. In: IEEE Power and Energy Society General Meeting. https://doi.org/10.1109/PESGM.2016.7741817.

Dufour, C., Bélanger, J., 2004. A PC-based real-time parallel simulator of electric systems and drives. In: International Conference on Parallel Computing in Electrical Engineering: Workshop on System Design Automation, SDA, PARELEC, pp. 105–113.

Dufour, C., Belanger, J., 2014. On the use of real-time simulation technology in smart grid research and development. IEEE Trans. Ind. Appl. 50 (6), 3963–3970. https://doi.org/10.1109/TIA.2014.2315507.

Ensermu, G., Bhattacharya, A., Panigrahy, N., 2019. Real-time simulation of smart DC microgrid with decentralized control system under source disturbances. Arabian J. Sci. Eng. 44 (8), 7173–7185. https://doi.org/10.1007/s13369-019-03842-6.

Etxeberria-Otadui, I., Manzo, V., Bacha, S., Baltès, F., 2002. Generalized average modelling of FACTS for real time simulation in ARENE. IECON Proceedings (Industrial Electronics Conference) 2, 864–869. https://doi.org/10.1109/IECON.2002.1185385.

Fei, Z., Xiaolin, Z., Junjun, Z., Jingsheng, H., 2014. Hardware-in-the-loop simulation, modeling and close-loop testing for three-level photovoltaic grid-connected inverter based on RT-LAB International Conference on. Power Syst. Technol. 20–22.

Feng, F., Zhang, J., Jin, J., Na, W., Yan, S., Zhang, Q.J., 2020. Efficient FEM-based em optimization technique using combined Lagrangian method with Newton's method. IEEE Trans. Microw. Theor. Tech. 68 (6), 2194–2205. https://doi.org/10.1109/TMTT.2020.2979982.

Fischer, K., Seume, J.R., 2012. Thermo-mechanical stress in tubular solid oxide fuel cells: Part II - operating strategy for reduced probability of fracture failure. IET Renew. Power Gener. 6 (3), 194–205. https://doi.org/10.1049/iet-rpg.2011.0109.

Fu, X., Seye, S.M., Mahseredjian, J., Cai, M., Dufour, C., 2018. A comparison of numerical integration methods and discontinuity treatment for EMT simulations. In: 20th Power Systems Computation Conference, PSCC 2018. Institute of Electrical and Electronics Engineers Inc., Canada https://doi.org/10.23919/PSCC.2018.8442452.

Garces, A., 2018. On the convergenc e of Newton's method in power flow studies for dc microgrids. IEEE Trans. Power Syst. 33 (5), 5770–5777. https://doi.org/10.1109/TPWRS.2018.2820430.

Gebregergis, A., Pillay, P., Bhattacharyya, D., Rengaswemy, R., 2009. Solid oxide fuel cell modeling. IEEE Trans. Ind. Electron. 56 (1), 139–148. https://doi.org/10.1109/TIE.2008.2009516.

Guillaud, X., Omar Faruque, M., Teninge, A., Hariri, A.H., Vanfretti, L., Paolone, M., Dinavahi, V., Mitra, P., Lauss, G., Dufour, C., Forsyth, P., Srivastava, A.K., Strunz, K., Strasser, T., Ali, D., 2015. Applications of real-time simulation technologies in power and energy systems. IEEE Power Energy Technol. Syst. J. 2 (3), 103–115. https://doi.org/10.1109/jpets.2015.2445296.

Gupta, P., Pahwa, V., Verma, Y.P., 2020. Load tracking enhancement of a grid connected SOFC system using an advanced controller in real time. In: 1st IEEE International Conference on Measurement, Instrumentation, Control and Automation, ICMICA 2020. Institute of Electrical and Electronics Engineers Inc., India https://doi.org/10.1109/ICMICA48462.2020.9242674.

Gupta, P., Pahwa, V., Verma, Y.P., 2021a. Efficient control of a nonlinear solid oxide fuel cell system using an adaptive model predictive controller. Int. Trans. Electr. Energy Syst. 31 (11). https://doi.org/10.1002/2050-7038.13095.

Gupta, P., Pahwa, V., Verma, Y.P., 2021b. Switching function based inverter modeling for a grid-connected SOFC system in real time. In: IOP Conference Series: Materials Science and Engineering, vol 1033. IOP Publishing Ltd, India. https://doi.org/10.1088/1757-899X/1033/1/012023.

Hamid, R.B., Mirsalim, M., Gharehpetian, G.B., Talebi, H.A., 2017. A decentralized power management and sliding mode control strategy for hybrid AC/DC microgrids including renewable energy resources. IEEE Trans. Ind. Inf. 1. https://doi.org/10.1109/TII.2017.2677943.

Han, S., Sun, L., Shen, J., Pan, L., Lee, K.Y., 2018. Optimal load-tracking operation of grid-connected solid oxide fuel cells through set point scheduling and combined L1-MPC control. Energies 11 (4). https://doi.org/10.3390/en11040801.

Ibarra, L., Rosales, A., Ponce, P., Molina, A., Ayyanar, R., 2017. Overview of real-time simulation as a supporting effort to smart-grid attainment. Energies 10 (6). https://doi.org/10.3390/en10060817.

IEEE Standard for the Testing of Microgrid Controllers, 2018,

Jonsdottir, G.M., Almas, M.S., Baudette, M., Palsson, M.P., Vanfretti, L., 2015. IEEE Computer Society Sweden RT-SIL performance analysis of synchrophasor-and-active load-based power system damping controllers. In: IEEE Power and Energy Society General Meeting. https://doi.org/10.1109/PESGM.2015.7286372.

Kassa Semero, Y., Zhang, J., Zheng, D., 2018. PV power forecasting using an integrated GA-PSO-ANFIS approach and Gaussian process regression based feature selection strategy. CSEE J. Power Energy Syst. 4 (2), 210–218. https://doi.org/10.17775/CSEEJPES.2016.01920.

Khadanga, R.K., Padhy, S., Panda, S., Kumar, A., 2018. Design and analysis of multi-stage PID controller for frequency control in an islanded micro-grid using a novel hybrid whale optimization-pattern search algorithm. Int. J. Numer. Model. Electron. Network. Dev. Field. 31 (5). https://doi.org/10.1002/jnm.2349.

Krasopoulos, C.T., Beniakar, M.E., Kladas, A.G., 2018. Multicriteria PM motor design based on ANFIS evaluation of EV driving cycle efficiency. IEEE Trans. Transp. Electrif. 4 (2), 525–535. https://doi.org/10.1109/TTE.2018.2810707.

Krebs, R., Ruhle, O., 2004. NETOMAC real-time simulator - a new generation of standard test modules for enhanced relay testing. IEE Conf. Publ. 2, 669–674. https://doi.org/10.1049/cp:20040212.

Kumar, R., Bansal, H.O., 2019. Real-time implementation of adaptive PV-integrated SAPF to enhance power quality. Int. Trans. Electr. Energy Syst. 29 (5). https://doi.org/10.1002/2050-7038.12004.

Kumar, G.S., Kumar, B.K., Mishra, M.K., 2011. Mitigation of voltage sags with phase jumps by UPQC with PSO-Based ANFIS. IEEE Trans. Power Deliv. 26 (4), 2761–2773. https://doi.org/10.1109/TPWRD.2011.2165301.

Kumashiro, S., Kamei, T., Hiroki, A., Kobayashi, K., 2020. An efficient and accurate time step control method for power device transient simulation utilizing dominant time constant approximation. IEEE Trans. Comput. Aided Des. Integrated Circ. Syst. 39 (2), 451–463. https://doi.org/10.1109/TCAD.2018.2889673.

Kusagur, A., Kodad, S.F., Sankar Ram, B.V., 2010. Modeling, design and simulation of an adaptive neuro-fuzzy inference system (ANFIS) for speed control of induction motor. Int. J. Comput. Appl. 6 (12), 29–44. https://doi.org/10.5120/1123-1472.

Lasseter, R.H., Paigi, P., 2004. Microgrid: a conceptual solution. In: PESC Record - IEEE Annual Power Electronics Specialists Conference, vol 6, pp. 4285–4290. https://doi.org/10.1109/PESC.2004.1354758.

Lehn, P., Rittiger, J., Kulicke, B., 1995. Comparison of the ATP version of the EMTP and the NETOMAC program for simulation of HVDC systems. IEEE Trans. Power Deliv. 10 (4), 2048–2053. https://doi.org/10.1109/61.473344.

Li, W., Belanger, J., 2015. An FPGA-based real-time HIL test bench for full-bridge modular multilevel STATCOM controller. In: 17th European Conference on Power Electronics and Applications, EPE-ECCE Europe 2015. Institute of Electrical and Electronics Engineers Inc., Canada https://doi.org/10.1109/EPE.2015.7309244.

Li, X., Wang, S., 2021. Energy management and operational control methods for grid battery energy storage systems. CSEE J. Power Energy Syst. 7 (5), 1026–1040. https://doi.org/10.17775/CSEEJPES.2019.00160.

Li, Y., Vilathgamuwa, D.M., Loh, P.C., 2004. Design, analysis, and real-time testing of a controller for multibus microgrid system. IEEE Trans. Power Electron. 19 (5), 1195–1204. https://doi.org/10.1109/TPEL.2004.833456.

Li, Y.H., Choi, S.S., Rajakaruna, S., 2005. An analysis of the control and operation of a solid oxide fuel-cell power plant in an isolated system. IEEE Trans. Energy Convers. 20 (2), 381–387. https://doi.org/10.1109/TEC.2005.847998.

Li, Y.H., Rajakaruna, S., Choi, S.S., 2007. Control of a solid oxide fuel cell power plant in a grid-connected system. IEEE Trans. Energy Convers. 22 (2), 405–413. https://doi.org/10.1109/TEC.2005.853756.

Li, Y., Wu, Q., Zhu, H., 2015. Hierarchical load tracking control of a grid-connected solid oxide fuel cell for maximum electrical efficiency operation. Energies 8 (3), 1896–1916. https://doi.org/10.3390/en8031896.

Li, F., Wang, Y., Wu, F., Huang, Y., Liu, Y., Zhang, X., Ma, M., 2020. Review of real-time simulation of power electronics. J. Mod. Power Syst. Clean Energy 8 (4), 796–808. https://doi.org/10.35833/MPCE.2018.000560.

Liu, Z., Zhang, X., Su, M., Sun, Y., Han, H., Wang, P., 2020. Convergence analysis of Newton-raphson method in feasible power-flow for dc network. IEEE Trans. Power Syst. 35 (5), 4100–4103. https://doi.org/10.1109/TPWRS.2020.2986706.

Lugaresi, G., Matta, A., 2018. Real-time simulation in manufacturing systems: challenges and research directions 2018-. In: Proceedings - Winter Simulation Conference. Institute of Electrical and Electronics Engineers Inc., Italy, pp. 3319–3330. https://doi.org/10.1109/WSC.2018.8632542.

Luna, E.G., Becerra, J.E.C., Sáenz, E.M., 2018. Current status and future trends in protection, control and communications testing in electrical grids using real-time simulation. J. Eng. Sci. Technol. Rev. 11 (4), 204–214. https://doi.org/10.25103/jestr.114.25.

Maged, N.A., Hasanien, H.M., Ebrahim, E.A., Tostado-Véliz, M., Turky, R.A., Jurado, F., 2023. Optimal Real-time implementation of fuzzy logic control strategy for performance enhancement of autonomous microgrids. Int. J. Electr. Power Energy Syst. 151. https://doi.org/10.1016/j.ijepes.2023.109140.

Maged, N.A., Hasanien, H.M., Ebrahim, E.A., Tostado-Véliz, M., Jurado, F., 2024. Optimal super twisting sliding mode control strategy for performance improvement of islanded microgrids: validation and real-time study. Int. J. Electr. Power Energy Syst. 157. https://doi.org/10.1016/j.ijepes.2024.109849.

Mahmud, N., Zahedi, A., Mahmud, A., 2017. A cooperative operation of novel PV inverter control scheme and storage energy management system based on ANFIS for voltage regulation of grid-tied PV system. IEEE Trans. Ind. Inf. 13 (5), 2657–2668. https://doi.org/10.1109/TII.2017.2651111.

Marra, D., Pianese, C., Polverino, P., Sorrentino, M., 2016. Models for Solid Oxide Fuel Cell Systems, first ed. . Springer.

Martin, A.D., Vazquez, J.R., Cano, J.M., 2018. MPPT in PV systems under partial shading conditions using artificial vision. Elec. Power Syst. Res. 162, 89–98. https://doi.org/10.1016/j.epsr.2018.05.005.

Menghal, P.M., Laxmi, A.J., 2012. India Real time simulation: recent progress & challenges. Int. Conf. Power Signals Controls Comput. EPSCICON 2012. https://doi.org/10.1109/EPSCICON.2012.

Merabet, A., Tanvir, A.A., Beddek, K., 2016. Speed control of sensorless induction generator by artificial neural network in wind energy conversion system. IET Renew. Power Gener. 10 (10), 1597–1606. https://doi.org/10.1049/iet-rpg.2016.0285.

Mikkili, S., Panda, A.K., Prattipati, J., 2015. Review of real-time simulator and the steps involved for implementation of a model from MATLAB/SIMULINK to real-time. J. Inst. Eng. India Ser. B 96 (2), 179–196. https://doi.org/10.1007/s40031-014-0128-6.

Mirzahosseini, R., Iravani, R., 2019. Small time-step FPGA-based real-time simulation of power systems including multiple converters. IEEE Trans. Power Deliv. 34 (6), 2089–2099. https://doi.org/10.1109/TPWRD.2019.2933610.

Mitra, P., Hanning, J., Köhlström, J., Larsson, T., Danielsson, J., 2015. IEEE Computer Society Sweden DC grid system behavior: a real-time case study. In: IEEE Power and Energy Society General Meeting. https://doi.org/10.1109/PESGM.2015.7286123.

Morris, T.P., White, I.R., Crowther, M.J., 2019. Using simulation studies to evaluate statistical methods. Stat. Med. 38 (11), 2074–2102. https://doi.org/10.1002/sim.8086.

Musleh, A.S., Muyeen, S.M., Al-Durra, A., Kamwa, I., 2017. Testing and validation of wide-area control of STATCOM using real-time digital simulator with hybrid HIL-SIL configuration. IET Gener. Transm. Distrib. 11 (12), 3039–3049. https://doi.org/10.1049/iet-gtd.2016.1557.

Parizad, A., Mohamadian, S., Iranian, M.E., Guerrero, J.M., 2019. Power system real-time emulation: a practical virtual instrumentation to complete electric power system modeling. IEEE Trans. Ind. Inf. 15 (2), 889–900. https://doi.org/10.1109/TII.2018.2837079.

Patra, P.C., Panigrahi, P.K., Satpathy, A., Nayak, N., 2025. Assessing stability in renewable microgrid using a novel-optimized controller for PV-battery based micro grid with OPAL-RT based real-time validation. Results Eng. 26. https://doi.org/10.1016/j.rineng.2025.104290.

Pries, J., Hofmann, H., 2015. Steady-state algorithms for nonlinear time-periodic magnetic diffusion problems using diagonally implicit runge-kutta methods. IEEE Trans. Magn. 51 (4). https://doi.org/10.1109/TMAG.2014.2344005.

Rahman, M.M.A., Rahim, A.H.M.A., 2016. Performance evaluation of ANN and ANFIS based wind speed sensor-less MPPT controller. In: 5th International Conference on Informatics, Electronics and Vision, ICIEV 2016. Institute of Electrical and Electronics Engineers Inc., Bangladesh, pp. 542–546. https://doi.org/10.1109/ICIEV.2016.7760062.

Ramadhani, F., Hussain, M.A., Mokhlis, H., Hajimolana, S., 2017. Optimization strategies for solid oxide fuel cell (SOFC) application: a literature survey. Renew. Sustain. Energy Rev. 76, 460–484. https://doi.org/10.1016/j.rser.2017.03.052.

Ramezani, M., Li, S., Sun, Y., 2017. Combining droop and direct current vector control for control of parallel inverters in microgrid. IET Renew. Power Gener. 11 (1), 107–114. https://doi.org/10.1049/iet-rpg.2016.0107.

Reis, A.W., Guarda, F.G.K., Gastaldini, C.C., 2019. Simulation of a phasor measurement unit in real time using Typhoon virtual HIL. In: IEEE PES Conference on Innovative Smart Grid Technologies, ISGT Latin America 2019. Institute of Electrical and Electronics Engineers Inc., Brazil https://doi.org/10.1109/ISGT-LA.2019.8895261.

Riquelme-Dominguez, J.M., Gonzalez-Longatt, F., Melo, A.F.S., Rueda, J.L., Palensky, P., 2022. Cyber-physical testbed Co-simulation real-time: system frequency response. In: IEEE International Conference on Environment and Electrical Engineering and 2022 IEEE Industrial and Commercial Power Systems Europe, EEEIC/I and CPS Europe 2022. Institute of Electrical and Electronics Engineers Inc., Spain https://doi.org/10.1109/EEEIC/ICPSEurope54979.2022.9854592.

Ropp, M., 2019. Guide to the IEEE 1547-2018 Standard and its Impacts on Cooperatives National Rural Electric Cooperative Association (NRECA).

Ruiz, A.P., Rivas, M.A., Harbour, M.G., 2020. Non-blocking synchronization between real-time and non-real-time applications. IEEE Access 8, 147618–147634. https://doi.org/10.1109/ACCESS.2020.3015385.

Sandre-Hernandez, O., Morales-Caporal, R., Rangel-Magdaleno, J., Peregrina-Barreto, H., Hernandez-Perez, J.N., 2015. Parameter identification of PMSMs using experimental measurements and a PSO algorithm. IEEE Trans. Instrum. Meas. 64 (8), 2146–2154. https://doi.org/10.1109/TIM.2015.2390958.

Sauro, H.M., Bergmann, F.T., 2010. Software Tools for Systems Biology Systems Biomedicine. Elsevier Inc., United States, pp. 289–314. https://doi.org/10.1016/B978-0-12-372550-9.00012-2.

Shen, P., Lior, N., 2016. Vulnerability to climate change impacts of present renewable energy systems designed for achieving net-zero energy buildings. Energy 114, 1288–1305. https://doi.org/10.1016/j.energy.2016.07.078.

Sirviö, K.H., Mekkanen, M., Kauhaniemi, K., Laaksonen, H., Salo, A., Castro, F., Babazadeh, D., 2020. Accelerated real-time simulations for testing a reactive power flow controller in long-term case studies. J. Electr. Comput. Eng. 1–17. https://doi.org/10.1155/2020/8265373.

Srinivasan, S., 2006. Fuel Cells: From Fundamentals to Applications Fuel Cells: From Fundamentals to Applications. Springer US, pp. 1–691. https://doi.org/10.1007/0-387-35402-6.

Srinivasan, V., 2022. AI & learning: a preferred future. Comput. Educ. Artif. Intell. 3. https://doi.org/10.1016/j.caeai.2022.100062.

Strasser, T., Lauss, G., Jalili-Marandi, V., Forsyth, P., Dufour, C., Dinavahi, V., Monti, A., Kotsampopoulos, P., Martinez, J.A., Strunz, K., Saeedifard, M., Wang, X., Shearer, D., Paolone, M., Brandl, R., Matar, M., Davoudi, A., Iravani, R., 2015. Real-time simulation technologies for power systems design, testing, and analysis. IEEE Power Energy Technol. Syst. J. 2, 63–73. https://doi.org/10.1109/JPETS.2015.2427370.

Sun, L., Hua, Q., Shen, J., Xue, Y., Li, D., Lee, K., 2017. A combined voltage control strategy for fuel cell. Sustainability 9 (9), 1517. https://doi.org/10.3390/su9091517.

Sun, L., Wu, G., Xue, Y., Shen, J., Li, D., Lee, K.Y., 2018. Coordinated control strategies for fuel cell power plant in a microgrid. IEEE Trans. Energy Convers. 33 (1), 1–9. https://doi.org/10.1109/TEC.2017.2729881.

Tawfique, K.A., Merabet, A., Ibrahim, H., Beguenane, R., 2016. Standalone wind energy conversion system using OPAL-RT real-time HIL/RCP laboratory. In: International Conference on Industrial Informatics and Computer Systems, CIICS 2016. Institute of Electrical and Electronics Engineers Inc., Canada https://doi.org/10.1109/ICCSII.2016.7462404.

Tedesco, F., Mariam, L., Basu, M., Casavola, A., Conlon, M.F., 2015. Economic model predictive control-based strategies for cost-effective supervision of community microgrids considering battery lifetime. IEEE J. Emerg. Sel. Top. Power Electron. 3 (4), 1067–1077. https://doi.org/10.1109/JESTPE.2015.2446894.

Wächter, C., Lunderstädt, R., Joos, F., 2006. Dynamic model of a pressurized SOFC/gas turbine hybrid power plant for the development of control concepts. J. Fuel Cell Sci. Technol. 3 (3), 271–279. https://doi.org/10.1115/1.2205360.

Wang, C., Nehrir, M.H., 2007. A physically based dynamic model for solid oxide fuel cells. IEEE Trans. Energy Convers. 22 (4), 887–897. https://doi.org/10.1109/TEC.2007.895468.

Wang, T., O'Neill, D., Kamath, H., 2015. Dynamic control and optimization of distributed energy resources in a microgrid. IEEE Trans. Smart Grid 6 (6), 2884–2894. https://doi.org/10.1109/tsg.2015.2430286.

Wang, B., Wang, Q., Wei, Y.M., Li, Z.P., 2018. Role of renewable energy in China's energy security and climate change mitigation: an index decomposition analysis. Renew. Sustain. Energy Rev. 90, 187–194. https://doi.org/10.1016/j.rser.2018.03.012.

Wang, Z., Qi, D., Mei, J., Li, Z., Wan, K., Zhang, J., 2021. Real-time controller hardware-in-the-loop co-simulation testbed for cooperative control strategy for cyber-physical power system. Glob. Energy Interconnect. 4 (2), 214–224. https://doi.org/10.1016/j.gloei.2021.05.004.

Wu, C., Chen, J., Xu, C., Liu, Z., 2017a. Real-time adaptive control of a fuel cell/battery hybrid power system with guaranteed stability. IEEE Trans. Control Syst. Technol. 25 (4), 1394–1405. https://doi.org/10.1109/TCST.2016.2611558.

Wu, G., Lee, K.Y., Sun, L., Yali, X., 2017b. Coordinated fuzzy logic control strategy for hybrid PV array with fuel-cell and ultra-capacitor in a microgrid. IFAC-PapersOnLine 50 (1), 5554–5559. https://doi.org/10.1016/j.ifacol.2017.08.1098.

Wu, J., Wu, H., Song, Y., Zhang, T., Zhang, J., Cheng, Y., 2018. Adaptive Neuro-fuzzy inference system based estimation of EAMA elevation joint error compensation. Fusion Eng. Des. 126, 170–173. https://doi.org/10.1016/j.fusengdes.2017.11.025.

Yamane, A., 2015. Real-time Simulation of Distributed Energy Systems and Microgrids International Conference on Sustainable Mobility Applications, Renewables and Technology, pp. 1–6.

Yu, S., Fernando, T., Chau, T.K., Iu, H.H.C., 2017. Voltage control strategies for solid oxide fuel cell energy system connected to complex power grids using dynamic state estimation and STATCOM. IEEE Trans. Power Syst. 32 (4), 3136–3145. https://doi.org/10.1109/TPWRS.2016.2615075.

Zhu, Y., Tomsovic, K., 2002. Development of models for analyzing the load-following performance of microturbines and fuel cells. Elec. Power Syst. Res. 62 (1), 1–11. https://doi.org/10.1016/s0378-7796(02)00033-0.

CHAPTER

10

Microgrid from perspective of issues of cybersecurity, ethics and regulation

Rahul Saini and Yajvender Pal Verma

Department of Electrical & Electronics Engineering, UIET, Panjab University, Chandigarh, India

10.1 Introduction

The concept of microgrid originated for the utilization of distributed generation sources to meet local demand of a region. Formation of a microgrid requires integration of small-scale generation plants with nearby loads and this connected system acts like a miniature grid. With the popularization and commercialization of renewable energy sources to act as small-scale power generation plants, concept of microgrid can be implemented practically on the ground. Microgrids would change the structure and roles in the grid significantly over time. Regulations, which were created for the current system, would need to be updated to incorporate changes induced due to microgrid (Yalta et al., 2018).

Power system is going through a rapid change in its basic structure. New technologies are coming up and are getting integrated with the system. It can start from the small rooftop solar plant or EV in household to big windfarms and PMUs based grid monitoring system. Whenever something new get introduced to system it has multiple effects on the system. Cybersecurity is new factor which needs to be considered when adding something to power system. Ethical and regulatory factors are more of important but non-technical factors that need keen attention for the sustainability of the system.

Power grids are moving toward smart grids, in which most of the redundant decisions will be taken automatically by algorithms. Hence,

Frequency Control in Microgrids
https://doi.org/10.1016/B978-0-443-36476-1.00011-0

the system will become a lot more dependent on these control algorithms. These need to be designed in such a way that both ethical and regulatory factors need to be considered in their design stage only. Cybersecurity, ethical issues, and regulatory considerations will be further discussed individually in subsequent sections.

10.2 Cybersecurity threats and vulnerabilities

According to National Institute of Standards and Technology (NIST), the term cybersecurity is termed as, "Prevention of damage to, protection of, and restoration of computers, electronic communications systems, electronic communications services, wire communication, and electronic communication, including information contained therein, to ensure its availability, integrity, authentication, confidentiality, and nonrepudiation (cybersecurity - Glossary | CSRC, 2024)."

Cyber vulnerabilities are the weaknesses in the microgrid system that can be exploited to cause cyber-attack, like zero-day, bad cyber-security policy, no access control, etc., Through these vulnerable points in the microgrid, unauthorized access to the control system can be gained. Once a person with malicious intent gets into the system can disrupt the healthy working state.

Cybersecurity works for the protection of system against cyberattacks that can hinder, availability of the system when it is required to serve, can compromise the integrity of data (or action) that system produce, confidentiality of data (system information), to ensure the authentication and with nonrepudiation (Fig. 10.1). Integrating of electric vehicles and renewable energy sources to microgrid introduces new potential attack surfaces. These are mostly dispersed among multiple customers, and everyone has a different type of EV with different components and operating on different protocol. This diversity also leads to cyber vulnerabilities

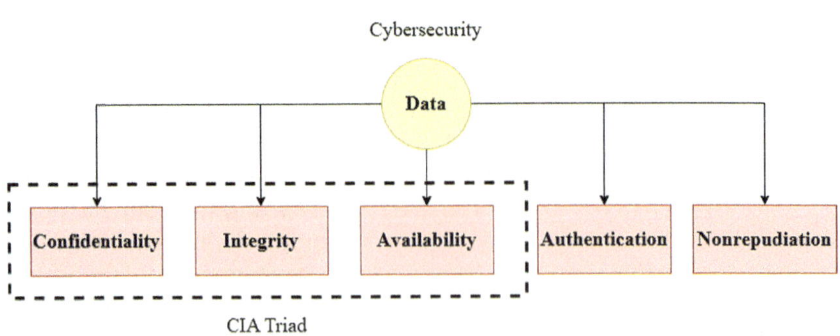

FIGURE 10.1 **Basic principles of cybersecurity.** The core principles are show in form of a classification diagram.

which can be used to execute cyber-attack on to the system. The energy management system of microgrid requires data collected for its working. Data needs to be authenticated to ensure the data integrity before the data is communicated as a sensor signal (Cui et al., 2021).

There are a variety of cybersecurity threats that can impact the health of microgrid systems. Some of the possible types of cyber-attacks on LFC are sensor and actuator attacks, denial of service attacks, false data injection attacks, covert attack, data integrity attacks, deception attacks, ramp attack, pulse attacks, random attacks, scaling attack, etc.., (Mohan et al., 2020) (Table 10.1).

For its operation microgrid need to communicate various control and status parameters with as well as to the external grid. Multiple communication links open potential for cyber-attack on the communication channel. For the simplification of microgrid communication open communication infrastructure is required (Rasolomampionona et al., 2024). So, new methods can be developed based on openly available information, and by doing so, the cybersecurity of microgrid can be enhanced.

Centralized control architecture requires data to be sent to the control present at the centralized location, from where it will make decisions for the entire system. These create multiple communication links, which in turn open multiple potential cyber vulnerabilities (Rasolomampionona et al., 2024). Attacks on control signals by hackers and introducing false control commands into the system can cause a catastrophic impact on the frequency stability of the microgrid. Hacker can alter the sensor measurement of supply and demand, which will create a mismatch in the system. In the worst case, the system might lose complete stability and may get isolated (Nejabatkhah et al., 2021). Microgrids are prone to time-delay attacks on their communication network.

A cybersecurity incident can completely or partially disturb the healthy microgrid operation. Cybersecurity incidents mainly target the communication channel, which in turn will cause transmission delay, data packet losing, and complete outage (Wang et al., 2019). Cybersecurity incidents can impact the economy, create issues related to technical and practical factors. Financial loss risk is one of the direct economic losses that a microgrid can face during operations. Cyber-attacks will affect the stability of microgrid operation (Nejabatkhah et al., 2021). Once microgrid operation is disrupted due to any factor will cause direct economic impact and loss of availability of power to the customers. Fig. 10.2 shows the direct and indirect impact of a cybersecurity incident on microgrid operations, with flow diagram.

TABLE 10.1 Some attacks on microgrid.

Type of attack		Basic about attack		Potential target component of microgrid	Potential impact on microgrid
Attack on availability	Denial of service attack	User is denied service, even when capacity is available.		Microgrid communication network	Users will not be able to get power.
Attack on integrity	False data injection attack	False data is injected into healthy data flowing across the system		Communication channel and sensors	Due to false data into the system stability will be impacted and wrong control decisions will be made.
		FDI attack templates			
		Ramp attack	Gradually changing ramp function for attack	Sensor output measurements	
		Pulse attack	Modifying data with random pulses		
		Random attack	Changing data based on random function		
		Scaling attack	Changing scaling parameters		
	Replay attack		In this attack, system's healthy state data is recorded and replayed during unhealthy condition to fool the system monitoring.	Monitoring system	Real state of microgrid cannot be estimated and due to that controlling will become difficult.
	Covert attack		Attack in which system parameters are altered to hide the presence of other attack.	Sensors and communication channel together	Attack can remain in system without detection and could potentially collect data.

10.3 Cybersecurity measures for microgrid LFC

Cybersecurity of microgrid LFC can be improved by limiting the vulnerabilities in the systems and mitigating them as these were detected. The majority of issue of cyber security is concern with data protection and access control. First, the data needs to be protected from the point of collection, then at the storage location, then while processing, and later safe disposal or archiving of the data for future use cases. Secondly, access control in other factor to only allow access to authorize person at each level. Cybersecurity of microgrid can be enhanced by doing continuous monitoring of the system for early detection of any threat, firewall can be deployed with custom rules for restricting unauthorized access to microgrid communication network, proper authentication protocols can be deployed to check access, data encryption technology can be used to secure data flow, human needs to be trained for cybersecurity, and system's backup need to be taken to start again system after cyber incident (shown in Fig. 10.3).

False data injection attacks on microgrids are first detected and then mitigated under a unified framework. This framework consists of a distance-dependent MMD (Maximum mean discrepancy)- based index for collected measurement samples to locate the attack, then using an unknown input observer to estimate the parameters and backstepping integrated sliding mode control to mitigate the impact of the attack by injecting reverse control (De and Sodhi, 2024). Cyber attacks on microgrid's communication lines can cause instability by disturbing the LFC control parameters, which will cause frequency deviations. Hence, a type 2 fuzzy logic control-based LFC scheme is designed to make LFC resilient to control deviations that are caused by cyber-attacks (Yıldız et al., 2024). There are certain types of dynamic attacks that can reside in a system called rootkit attacks. These can bypass static defenses by remaining silent for longer period and start functioning at a later point in time. A technique was proposed based on a Markov game to counter such type of dynamic attacks. A deep reinforcement learning-based defense strategy was proposed to detect and mitigate dynamic rootkit attacks (Rath et al., 2024). An enhanced $H\infty$ controller was proposed for the mitigation of the dynamic impact of simultaneous cyber-attacks on secondary control for microgrid (Kerdphol et al., 2024).

Microgrids need to operate smoothly in standalone mode as well, in cases where grid support is not available all the time. At that time there are more chances of cyber attacks and microgrid need to withstand those. For this situation, a method is proposed based on correcting and smoothing the reference voltage profiles in d and q axes, based on a defined control benchmark (Mahvash et al., 2024). A two-step strategy was proposed for

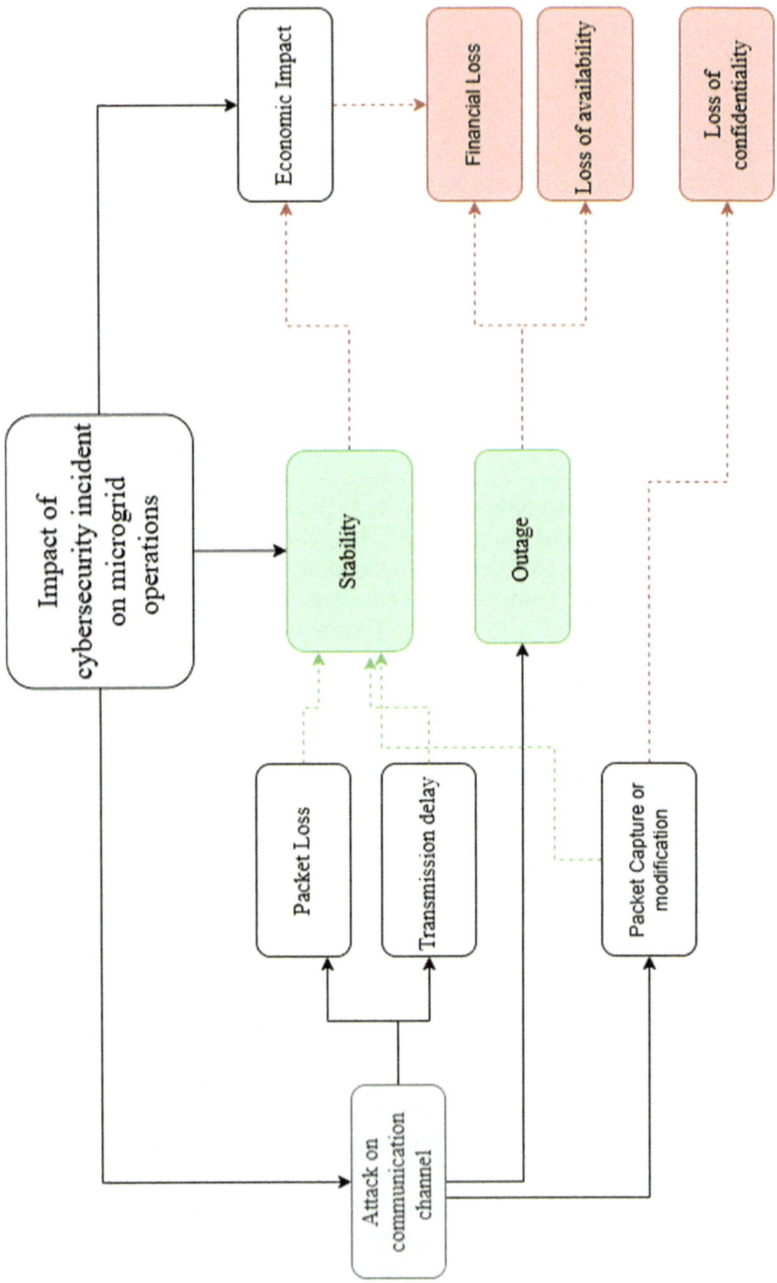

FIGURE 10.2 **Some of impact of cybersecurity incident on microgrid operation.** Figure explains the how cybersecurity issues can impact microgrid with logical diagram.

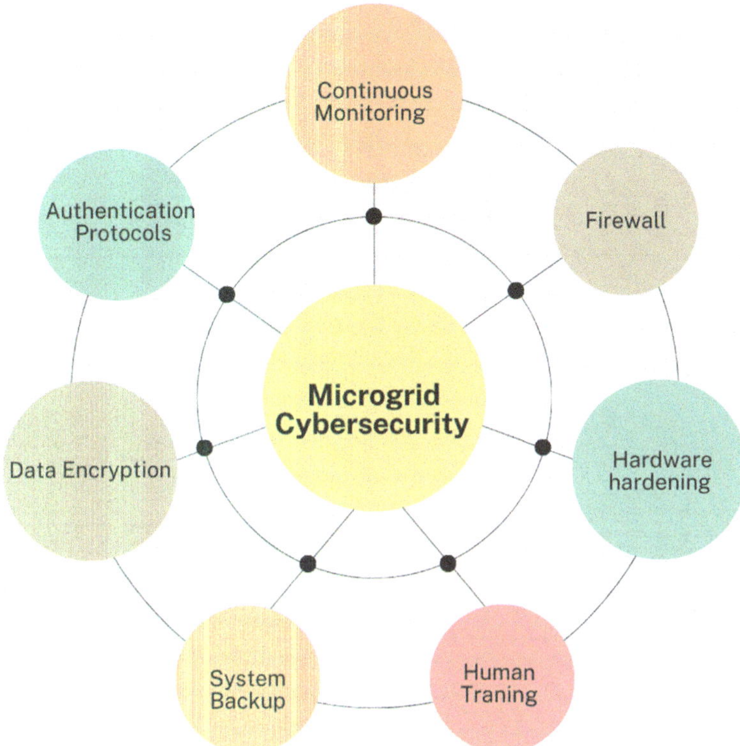

FIGURE 10.3 **Microgrid cybersecurity.** This figure explain about major methods that are used to enhance the microgrid's cybersecurity.

managing cyber anomalies in islanded microgrid. A machine learning-based SVM (Support Vector Machine) algorithm was used to detect the attack first and then the reinforcement learning technique was used to mitigate the impact of the cyber-attack (Tabassum et al., 2024). Cyber-attacks on DC microgrids are detected and mitigated based on robust state estimators using Kalman Filters (El-Ebiary et al., 2024). A brief comparison of the above-discussed cybersecurity measures is provided in Table 10.2.

10.3.1 Authentication mechanisms

For the authentication of source information of distribution synchronous phasors, a hybrid approach is developed based on self-adaptive mathematical morphology (SAMM) and time frequency (TF) techniques. In this first part SAMM is employed to consider the local environmental characteristics, and later TF to extract information signatures (Cui et al., 2021). A distributed identity authentication system based on

TABLE 10.2 Some of the recent work for mitigating cyber threats in microgrids.

Mitigation technique	References	System under consideration	Type of attack
Maximum mean discrepancy (MMD)-based index	De and Sodhi (2024)	IEEE-13 bus Modeled in RSCAD with RTDS	False data injection attack
DIT2FLC-FOTID controller with Circle-Inspired optimization algorithm	Yıldız et al. (2024)	Hydrogen and renewable sources based isolated microgrid in MATLAB Simulink	False data injection (FDI) and denial of service (DoS) attacks, on LFC controllers
By correcting the reference voltage profiles in the d and q axes sending to the PWM modulation in the RSC of each DFIG	Mahvash et al. (2024)	AC microgrid (ACMG) in MATLAB/SIMULINK	False data injection attack (FDIA), hijack attack (HjA) and denial of service (DoS)
24 node – 2-layer neural network	Rath et al. (2024)	4-DG virtual microgrid testbed in the MATLAB environment	Rootkit attacks
SVM based data-driven detection mechanism with RL agent control	Tabassum et al. (2024)	Matlab/SIMULINK	False data injection (FDI) attacks
Two mitigation layers based on Kalman filter with adaptive Proportional Integral (API) and Fractional API (FAPI) controllers	El-Ebiary et al. (2024)	DC microgrid MATLAB/Simulink software and Hardware in the Loop Typhoon HIL-402 with dSPACE Mircolab box RTI-1202	Voltage and current cyber-attacks (FDIAs)
Enhanced robust H∞ technique	Kerdphol et al. (2024)	Load frequency control model with microgrid model in MATLAB/Simulink	Combined false data injection (FDI), denial of service (DoS), and controller hijacking attacks

FISCO BCOS consortium blockchain technology to support multiple center is designed for a microgrid system (Guanqun et al., 2021). Considering the development of quantum computing, a lightweight authentication protocol is designed for fog-based microgrid (Lu and Li, 2021).

10.3.2 Encryption techniques

A hybrid encryption approach based on Advanced Encryption Standard (AES) and Rivest Cipher 4 (AES), is used to encrypt data before sending it to a wireless sensor network, which is used for monitoring and control in microgrid system (Murugan and Vijayarajan, 2023). The attribute-based encryption (ABE) technique is used to provide privacy protection and integrity for data of users of microgrid power trading system (Huang et al., 2024). A homomorphic encryption-based communication scheme is used for privacy protection of data used in a distributed optimization algorithm in microgrid (Yuan et al., 2024).

10.3.3 Data integrity measures

To protect data integrity, multi-dimensional and multi-functional data aggregation is used, considering both theoretical and practical aspects of privacy enhancement (Lu, 2016). To protect data of energy network against eavesdropping, data spoofing, and tampering attacks, a secure and privacy-preserving energy trading system (SPETS) is proposed based on blockchain technology (Samy et al., 2021) and a distributed privacy-preserving anonymous data collection method is proposed, which is based on tangle distributed ledger network (Alavi et al., 2023).

10.3.4 Intrusion detection and prevention systems (IDPS)

IDPS is used to continuously monitor command data and detect any false data in the energy management system of the microgrid. When malicious commands are detected by IDPS, it will raise an alarm to the operator for further action (Ali and Malik, 2022). A cloud-based Intrusion Detection and Prevention System (CB-IDPS) is proposed in which sensor data is routed to the cloud based on software-defined networking, in cloud, it is fully inspected for any malicious data presence (Brugman et al., 2019).

10.4 Ethical considerations in microgrid LFC

Ethics is one of the last things that is considered while developing a microgrid. Meeting the load demand, healthy and stable working are some of the main concerns of the microgrid operator. In addition, keep making profit out of this whole operation (Tally et al., 2019).

In microgrid systems, the controller that takes operational decisions lies near the customers and is also familiar with them. It needs to follow ethical practices while implementing LFC strategies. As to maintain frequency load needs to be shed under certain conditions; it need to be very ethical while deciding which loads to be shed. It needs to follow standard practices. Rule-based decision making needs to be adopted.

10.4.1 Fairness, transparency, and accountability in LFC operations

As LFC operations are based on data collected form the user and they have first right over the data, it needs to be a standard practice to get approval of usage of data from customer needs to be done. There needs to be a proper transparent mechanism through which customer can get idea how their data will be used, by whom, up to what time it will be stored and what are further procedure for destruction of that data. There needs to be some mechanism given so that people who want to opt-out from this data sharing can move out of this along with his pervious data.

Fairness of LFC operation can be observed from three distributive fairness principles. These are equality, means every participant member will get equal share of goods and burdens, min-max fairness, means that maximum allocation of resource will take place to every member and meritocracy, which means allocation of resources and burdens will be based on their contribution and merit (Soares et al., 2024). System needs to be reliable in doing this function then only fairness can be assured. Methods need to be developed that can work reliably to provide fairness in LFC operations. Stake holders can trust the fairness in decision making reliably over the entire period of operations (Wang et al., 2019). Accountability in LFC operation means microgrid operator need to protect data as well as data minimization, accuracy, integrity and confidentiality (Chiarini and Compagnucci, 2022). It will be accountable for these issues and can be checked at any time by the regulators.

10.4.2 Ethical challenges in data collection, analysis, and utilization

In smart microgrid to take decisions data from customers is required in large amount regularly, to take control actions based on this data (Tally et al., 2019). This collected data need to be carefully used without compromising the privacy of the customers. Additionally, data need to be stored and shared with stakeholders considering ethical and regulatory standards.

Energy usage data of customer is collected on regular interval, which can be further analyzed to generate customer behavior profiles, customer habits, to deduce their personal and social as well as economic information, their equipment usage patterns, their staying at home pattern, etc.., (Review of Smart Meter Data Analytics, 2024). There are further lot more information that can be analyzed and utilized to impact the life on customers.

Microgrid system being not that much financially big, managing proper data center and control center will cause addition financial burden on microgrid. To follow standard data collection and utilization methods, it requires experts to be deployed which understand and make sure these were properly followed at every level and throughout the microgrid system. Data collected from users can be later used to charge extra from them, if peaks are studied from consumption data to fulfill the profit desires of microgrid operators. These types of issues need to be addressed by regulators keeping interested of both stakeholders in mind.

10.5 Regulatory framework for microgrid LFC

Microgrid have multiple sub-systems working together in synchronism. For stable working microgrid needs to comply with standard of each of its subsystem.

10.5.1 Overview of regulatory standards and guidelines: NERC CIP standards, IEEE 2030.7, IEC 61,810

The North American Electric Reliability Corporation (NERC) Critical Infrastructure Protection (CIP) standards are a set of mandatory security regulations that protect the North American power grid from cyber threats (Reliability Standards, 2024). These standards deal with issues relating to cybersecurity, security management controls, personnel and training, system security management, configuration change management & vulnerability assessments, information protection, and other directives and guidelines (Project 2014-02, 2024).

The IEEE 2030.7 4 standard offers the most comprehensive technical process for describing the functions of a microgrid controller. Although aimed at the controller, these functional modes serve as a convenient way to actually specify the full microgrid (Danley, 2019). This standard defines six states of operation for a microgrid: two steady state modes and four transition modes to reduce the complexity around this issue of microgrid state. This standard also defines various equipment and control systems for microgrid. It is also disclosed in this standard that microgrid control system needs to fully ensure data protection against any cybersecurity threat (Danley, 2019).

IEC 61,800 primarily focuses on the adjustable speed electrical power drive systems, including their components, control methods, and applications. A microgrid has multiple components that have adjustable drives into it, with which components can vary the load for the best optimized result. These drives help in balancing load in microgrid environments, this standard deals with all aspects related to adjustable speed drives (2024).

10.5.2 Compliance requirements for microgrid LFC systems: Reporting, auditing, certification

Auditing ensured system performance optimization, effective resource allocation, and the identification of areas for improvement. The role of auditing is to assess energy efficiency, identify energy losses, optimize energy production, reduce operational cost, and to minimize environmental impacts (Gopinath et al., 2024). For microgrid doing an audit is a challenging task as different components follow different reporting methods, and there are very few parameter metrics for the relevant comparisons. For example: generation from a generator set cannot be compared with generation from a solar or wind plant with in the microgrid.

10.5.3 International and national regulatory perspectives: Comparison and harmonization efforts

International regulatory efforts are focused on operation, reliability, security, standardization, protection, and interoperability primarily. For the national regulatory impact of large adoption of microgrids on the stability of the national grid is a big concern. Load dispatch in case of large-scale microgrid connection to the main grid to meet their load. Local environmental considerations and national laws are major concerns for national regulators. At the national level regulators need to consider long-term planning and forecasting as well, which is not the perspective of international regulatory frameworks. Standards need to be developed, providing flexibility to accommodate the national regulatory framework for microgrid operations.

10.6 Case studies and best practices

10.6.1 Real-world examples of cybersecurity incidents in microgrid LFC

There are very few direct reported cases of cybersecurity incidents in microgrid. But there are a significant number of incidents in which pointing toward cyberattack on the microgrid system. For example, in case of a petrochemical plant of Saudi Arabia, a rogue code was discovered during later audit. In an incident like this, it is difficult to track when exactly the attack occurred and how much loss is done, it requires detail study. Similar incidents occur with Brazilian state-owned utility when it comes under ransomware attack in 2021 (Inside the Breach, 2024), which lead to the leaking of sensitive data online. In this type of attacks data of microgrid operating under these conditions will also get breached. Center for Strategic and International Studies (CSIS) published a comprehensive report highlighting major cyber incidents since 2006 (Significant Cyber Incidents, 2024), some of the recent and exclusive on the energy sector have been discussed in Table 10.3.

10.6.2 Ethical dilemmas and regulatory challenges faced by microgrid operators

Microgrid operators have multiple challenges like keeping in mind Community Wellbeing & Quality of Life versus Economic Benefits (Privatization), Community electric service & Sustainability versus Electric Power Company Vision and Community Socio-Economic Development versus Economic Benefits (Privatization) (Carrion et al., 2018). Operator needs to keep in mind the needs of the members participating in microgrid as well as the profits and interests of the entity operating the microgrid. The ambiguous legal definition of microgrids and associated obligations and liabilities present difficulties. Interconnection, island mode regulations, and tariff setting present both technical and financial difficulties. In case of some natural emergencies, responsibilities during the event and after the event and raising finances for the rebuilding process (Yalta et al., 2018).

TABLE 10.3 Some of recent incident of cyber-attacks on energy system.

System under attack	Year	References	Type of attack	Comment
Nuclear energy industry in China	2023	Robinson (2024)	Social engineering with phishing attacks by Bitter APT	Attack is detected by third party actual impact of attack is not known.
Energy research organization	2023	North Korean hackers (2024)	APT attack by North Korean Lazarus hacking group	100 GB of data was stolen
Ukraine energy SCADA system (MicroSCADA products)	2022	Sandworm Disrupts (2024)	Sandworm threat actor (APT attack) used CADDYWIPER	Caused two-time power disruption
Italian renewable energy source company	2022	BlackCat (2024)	BlackCat ransomware	700 GB of files from the Italian energy agency's servers.
Italian energy company	2022		BlackCat ransomware	Impacts on its operations
Luxembourg electricity network operator	2022		BlackCat ransomware	Customer portal become unavailable

10.6.3 Best practices for enhancing cybersecurity, ethics, and regulatory compliance in microgrid LFC

Study of the impact of cyber threats is important, and later, based on this study, taking action to counter the potential threat is one of the best cybersecurity practices. This type of approach comes under proactive cybersecurity measures, which will enhance the cybersecurity of the system against threats beforehand. An analytical method was proposed to calculate the impact of cyber incidents on the operational reliability of a microgrid (Wang et al., 2019). Defensive strategies, which are based on the detection of cyber-attacks and then taking mitigation action, have two types of detectors based on their characteristics. Static detectors are those that detect attacks on the steady-state stability of the microgrid, whereas dynamic detectors are those that detect attacks based on dynamic information of the microgrid (Nejabatkhah et al., 2021). Proper financial incentives need to be provided to customers following the microgrid operation policies and penalties to the violator of those.

10.7 Future trends and emerging issues

Future Trends will be closely aligned with the development of smart grid technologies. Microgrid will become more responsive, and faster decisions can be made with less human interaction in decision-making (Yusuf et al., 2025). Advancement of AI and data processing will be used in microgrids to increase the efficiency and profitability of operators. Load and supply can be balanced with the help of faster computation and using patterns from previously collected data. More use of IoT and 5G communication networks will provide more high-quality data to the controller for better decisions. Microgrids can interact with energy markets and earn profits using better prediction algorithms for forecasting. Machine learning and data analytics are being used for predictive maintenance, pattern recognition, and optimization in response to changing conditions (Gopinath et al., 2024).

Surplus power produced by renewable energy sources during the peak of supply and lower demand level is converted to hydrogen with help of an electrolyzer and later utilized as backup (Yıldız et al., 2024). This will help in making more sustainable and self-sufficient microgrids in the future. Type 3 generators in configuration, a doubly fed induction generators (DFIGs) is being coupled with marine current turbines (MCTs) for converting tidal energy to electricity (Mahvash et al., 2024). This technology will support microgrids on islands in a sustainable manner as tidal energy is a renewable energy source.

10.7.1 Evolving cybersecurity threat landscape in microgrid LFC

The concept of microgrid is moving from centralized control to more decentralized control structure. More players are coming into system and plays active role both in terms of generation and well as power selling. Due to the opening of the system, more players are now aware of inner workings of these system. Vulnerabilities of system can be easily exploited before any active measures are taken.

Verifying the source of data will be a big challenge and a big cyber threat. Proper authentication mechanisms are hard to implement at the microgrid level and most of component is present in the field, which make it more difficult to prevent devices from physical tampering. Over dependency on data and communication channels will be a big area of concern for cybersecurity-related threats to microgrid.

10.7.2 Ethical and regulatory considerations in next-generation LFC technologies: AI, blockchain

These new technologies both AI and blockchain, are hard to regulate. Blockchain has been changing the domain of energy trading which provides automated execution of smart contracts in peer-to-peer energy trading (Blockchain, 2024).

Establishing trust over the decision taken by an AI-based system will be a challenge for next generation of LFC technologies. AI technologies depend on datasets on which they are being trained. Data for training is collected from the microgrid system only. Hence, this factor of how that training data is being collected, stored and it needs to be properly regulated, who can have access to that data. Stored data will always be prone to cyberattacks which also needs consideration for regulation. AI algorithms are prone to bias and, based on the quality of data can cause discrimination (Parikh et al., 2019), this is concerning and requires regulation.

Blockchain is a new technology; there is very little availability of experience on the practical implementation of it into microgrids. But an idea can be taken from other domain regarding ethical and regulatory challenges that might come up, with its implementation into real-world microgrids. Some of the issues are system limitations of scalability, speed, data processing and storing costs, reporting and auditing mechanisms (Barceló et al., 2023).

10.7.3 Potential impacts of emerging technologies on microgrid security and compliance

Impact of emerging technologies will be in the direction of increasing speed of decision making, further minimizing the size of components, availability of large amounts of data, automation of redundant tasks, and increase in participation of customers as prosumers. As per security concerns more data is collection, create more potential for cybersecurity risk. Use of AI for decision making based on collected data will be affected of potential bias in the data, which might be visible in long-term implementation.

It will be a big challenge to balance out the finances for the adoption of new technologies on smaller microgrids (Altaf et al., 2022). It can be confidently assumed that it will be impossible for microgrids to be up to date with every cybersecurity vulnerability that will show up. It will take a long time for mitigation techniques to reach the small microgrids. Backward compatibility will be a big issue in incorporating newer technology into the older microgrids. Along with interoperability with different standards followed by different vendors for various components (Choudhury, 2020).

10.8 Conclusion

10.8.1 Summary of key cybersecurity, ethical, and regulatory considerations

The concept of microgrid is maturing with time, and new challenges are also coming up. Some of the major challenges that need urgent attention of every stakeholder participating in a microgrid environment, apart from technical and operational, are related to cybersecurity, regulatory, and ethical. Data is at the prime focus in the working of microgrid; proper frameworks are required for this. Optimization of every operational parameter for generating maximum profit generation need not to be the sole role of the microgrid operator. Customer satisfaction, maximum profit sharing with prosumers, and environmental factors are to be considered as well. Cybersecurity-related vulnerabilities need to be addressed using available solutions, before it turns into cyber threat. Regulations need to be designed based on the latest worldwide adopted standards like IEEE 2030.7, IEC 61,810, etc.

10.8.2 Recommendations for addressing challenges and enhancing security and compliance in microgrid LFC

Integrating regulation objectives with new best practices and technological developments to increase the generation of renewable energy and reduce environmental impact (T et al., 2024). Environmental and social factors need to be considered while framing regulatory policies for the microgrids. For this, consultation should be taken from every stakeholder while designing the policy. Training and awareness of stakeholders need to be done for updated cybersecurity practices.

References

Alavi, S.A., Javadipour, M., Rahimian, A., Mehran, K., 2023. A novel distributed privacy-preserving control and data collection method for IoT-centric microgrids. IET Gener. Transm. Distrib. 17 (10), 2249–2259. https://doi.org/10.1049/gtd2.12803.

Ali, S., Malik, T.N., 2022. Intrusion detection and prevention against cyber attacks for an energy management system. Mehran Univ. Res. J. Eng. Technol. 41 (1), 202–219. https://doi.org/10.22581/muet1982.2201.20.

Altaf, M.W., Arif, M.T., Islam, S.N., Haque, M.E., 2022. Microgrid protection challenges and mitigation approaches-a comprehensive review. IEEE Acc. 10, 38895–38922. https://doi.org/10.1109/ACCESS.2022.3165011.

Barceló, E., Dimić-Mišić, K., Imani, M., Spasojević Brkić, V., Hummel, M., Gane, P., 2023. Regulatory paradigm and challenge for blockchain integration of decentralized systems: example—renewable energy grids. Sustainability 15 (3), 2571. https://doi.org/10.3390/su15032571.

BlackCat Ransomware Claims Attack on Italian Energy Agency. (2024).

Blockchain Technology in the Energy Sector: A Systematic Review of Challenges and Opportunities - ScienceDirect. (2024).

Brugman, J., Khan, M., Kasera, S., Parvania, M., 2019/11/01. Cloud based intrusion detection and prevention system for industrial control systems using software defined networking. In: Proceedings - 2019 Resilience Week, RWS 2019. Institute of Electrical and Electronics Engineers Inc, United States, pp. 98–104. https://doi.org/10.1109/RWS47064.2019.8971825.

Carrion, G.A., Cintron, R.A., Rodriguez, M.A., Sanabria, W.E., Reyes, R., Orneill-Carrillo, E., 2018. Community microgrids to increase local resiliency. In: International Symposium on Technology and Society, Proceedings. Institute of Electrical and Electronics Engineers Inc, Puerto Rico, pp. 1–7. https://doi.org/10.1109/ISTAS.2018.8638269.

Chiarini, A., Compagnucci, L., 2022. Blockchain, data protection and P2P energy trading: a review on legal and economic challenges. Sustainability 14 (23). https://doi.org/10.3390/su142316305.

Choudhury, S., 2020. A comprehensive review on issues, investigations, control and protection trends, technical challenges and future directions for microgrid technology. Int. Trans. Electr. Energy Syst. 30 (9). https://doi.org/10.1002/2050-7038.12446.

Cui, Y., Bai, F., Yan, R., Saha, T., Ko, R.K.L., Liu, Y., 2021. Source authentication of distribution synchrophasors for cybersecurity of microgrids. IEEE Trans. Smart Grid 12 (5), 4577–4580. https://doi.org/10.1109/TSG.2021.3089041.

Cybersecurity - Glossary. CSRC. (2024).

Danley, D.R., 2019. Defining a Microgrid Using IEEE 2030.

De, S., Sodhi, R., 2024. A unified cyber attack detection and mitigation framework for an islanded AC microgrid. IEEE Trans. Syst. Man Cybern.: Systems 54 (9), 5270–5282. https://doi.org/10.1109/TSMC.2024.3403749.

El-Ebiary, A.H., Attia, M.A., Awad, F.H., Marei, M.I., Mokhtar, M., 2024. Kalman filters based distributed cyber-attack mitigation layers for DC microgrids. IEEE Trans. Circ. Syst. I: Reg. Pap. 71 (3), 1358–1370. https://doi.org/10.1109/TCSI.2023.3348928.

Gopinath, T., Kesavan, J., Srividhya, P.K., Ramkumar, P., 2024. Towards greener horizons: a comprehensive review of energy auditing in hybrid solar and wind systems. Interactions 245 (1). https://doi.org/10.1007/s10751-024-02049-y.

Guanqun, Y., Yin, L.I.U., Hao, X.U., Hongwei, Jianhui, Z., Entang, L.I., 2021. Credible distributed identity authentication system of microgrid based on blockchain. Chin. J. Network Informat. Sec. 7 (6), p88. https://doi.org/10.11959/j.issn.2096-109x.2021054.

Huang, K., Hu, X., Zhou, R., Xiang, D., 2024. Optimization and privacy protection of microgrid power trading system based on attribute encryption technology. EAI Endor. Trans. Energy Web 10, 1–9. https://doi.org/10.4108/ew.5431.

Inside the Breach. Unraveling Iconic Data Breaches in the Energy Industry. (2024).

Kerdphol, T., Ngamroo, I., Surinkaew, T., 2024. Enhanced robust frequency stabilization of a microgrid against simultaneous cyber-attacks. Elec. Power Syst. Res. 228. https://doi.org/10.1016/j.epsr.2023.110006.

Lu, R., 2016. In: Privacy-Preserving Multifunctional Data Aggregation. Springer Science and Business Media LLC, pp. 85–110. https://doi.org/10.1007/978-3-319-32899-7_5.

Lu, S., Li, X., 2021. Quantum-resistant lightweight authentication and key agreement protocol for fog-based microgrids. IEEE Acc. 9, 27588–27600. https://doi.org/10.1109/ACCESS.2021.3058180.

Mahvash, H., Taher, S.A., Guerrero, J.M., 2024. Detecting and mitigating cyber-attacks in AC microgrid composed of marine current turbine DFIGs to improve energy management system. E-Prime - Adv. Electr. Eng. Electr. Energy 7. https://doi.org/10.1016/j.prime.2024.100464.

Mohan, A.M., Meskin, N., Mehrjerdi, H., 2020. A comprehensive review of the cyber-attacks and cyber-security on load frequency control of power systems. Energies 13 (15). https://doi.org/10.3390/en13153860.

Murugan, G., Vijayarajan, S., 2023. IoT based secured data monitoring system for renewable energy fed micro grid system. Sustain. Energy Technol. Assessments 57. https://doi.org/10.1016/j.seta.2023.103244.

Nejabatkhah, F., Li, Y.W., Liang, H., Ahrabi, R.R., 2021. Cyber-security of smart microgrids: a survey. Energies 14 (1). https://doi.org/10.3390/en14010027.

North Korean Hackers Stole Research Data in Two-Month-Long Breach. (2024).

Parikh, R.B., Teeple, S., Navathe, A.S., 2019. Addressing bias in artificial intelligence in health care. JAMA, J. Am. Med. Assoc. 322 (24), 2377–2378. https://doi.org/10.1001/jama.2019.18058.

Project 2014-02. Critical Infrastructure Protection Standards Version 5 Revisions. (2024).

Rasolomampionona, D.D., Połecki, M., Zagrajek, K., Wróblewski, W., Januszewski, M., 2024. A comprehensive review of load frequency control technologies. Energies 17 (12). https://doi.org/10.3390/en17122915.

Rath, S., Das, T., Sengupta, S., 2024. Improvise, adapt, overcome: dynamic resiliency against unknown attack vectors in microgrid cybersecurity games. IEEE Trans. Smart Grid 15 (4), 4245–4258. https://doi.org/10.1109/TSG.2024.3372643.

Reliability Standards. (2024),

Review of Smart Meter Data Analytics. Applications, methodologies, and challenges | IEEE Journals & magazine. IEEE Xplore. (2024).

Robinson, R., 2024. Phishing Campaign Targets Chinese Nuclear Energy Industry.

Samy, A., Yu, H., Zhang, H., Zhang, G., 2021. Spets: Secure and privacy-preserving energy trading system in microgrid. Sensors 21 (23). https://doi.org/10.3390/s21238121.

Sandworm Disrupts Power in Ukraine Using a Novel Attack against Operational Technology Google Cloud Blog. (2024).

Significant Cyber Incidents. Strategic Technologies Program | CSIS. (2024).

Soares, J., Lezama, F., Faia, R., Limmer, S., Dietrich, M., Rodemann, T., Ramos, S., Vale, Z., 2024. Review on fairness in local energy systems. Appl. Energy 374. https://doi.org/10.1016/j.apenergy.2024.123933.

Tabassum, T., Lim, S., Khalghani, M.R., 2024. Artificial intelligence-based detection and mitigation of cyber disruptions in microgrid control. Elec. Power Syst. Res. 226. https://doi.org/10.1016/j.epsr.2023.109925.

Tally, H., Rodrigues, R., Wright, D., 2019. Smart grids and ethics. Orbit J. 2 (2), 1–28. https://doi.org/10.29297/orbit.v2i2.108.

Wang, C., Zhang, T., Luo, F., Li, F., Liu, Y., 2019. Impacts of cyber system on microgrid operational reliability. IEEE Trans. Smart Grid 10 (1), 105–115. https://doi.org/10.1109/TSG.2017.2732484.

Yalta, J., Makinen, S., Kotilainen, K., Jarventausta, P., Mendes, G., 2018/12/10. Comparison of regulatory challenges faced by different microgrid ownership models. In: Proceedings - 2018 IEEE PES Innovative Smart Grid Technologies Conference Europe, ISGT-Europe 2018. Institute of Electrical and Electronics Engineers Inc, Finland. https://doi.org/10.1109/ISGTEurope.2018.8571890.

Yıldız, S., Yildirim, B., Özdemir, M.T., 2024. Enhancing load frequency control and cybersecurity in renewable energy microgrids: a fuel cell-based solution with non-integer control under cyber-attack. Int. J. Hydrogen Energy 75, 438–449. https://doi.org/10.1016/j.ijhydene.2024.02.145.

Yuan, Z.P., Li, P., Li, Z.L., Xia, J., 2024. A fully distributed privacy-preserving energy management system for networked microgrid cluster based on homomorphic encryption. IEEE Trans. Smart Grid 15 (2), 1735–1748. https://doi.org/10.1109/TSG.2023.3309405.

Yusuf, S.S., Kunya, A.B., Abubakar, A.S., Salisu, S., 2025. Review of load frequency control in modern power systems: a state-of-the-art review and future trends. Electr. Eng. 107 (5), 5823–5848. https://doi.org/10.1007/s00202-024-02828-4.

CHAPTER

11

Future trends and research directions in microgrid LFC

Ark Dev[1], *Vineet Kumar*[2], *Vikash Kumar Verma*[3], *and Sumant Anand*[4]

[1] Department of Electronics and Communication Engineering, SR University, Warangal, Telangana, India; [2] Department of Electrical Engineering, Netaji Subhas University of Technology (NSUT), Delhi, New Delhi, India; [3] Department of Electrical Engineering, Parul Institute of Engineering and Technology, Parul University, Vadodara, Gujarat, India; [4] Department of Electronics and Telecommunication Engineering, G H Raisoni College of Engineering and Management, Pune, Maharashtra, India

11.1 Introduction

In recent years, there has been a significant shift away from traditional energy systems towards more sustainable and decentralized approaches. The primary challenge is to meet growing energy demands in ways that promote both stability and sustainability. Researchers and engineers have increasingly advocated for a transition from conventional systems reliant on fossil fuels to those powered by renewable energy sources (RES). This evolution has led to the development of MGs, which are small-scale power networks that integrate various renewable energy resources (RERs), energy storage systems (ESSs), and loads (Dev et al., 2024). These systems, typically generate between 10 and 100 MW, operate autonomously to serve localized regions. The adoption of MGs has captured the attention of the research community due to the significant advantages they offer, such as enhanced reliability, improved power quality, greater security, and higher operational efficiency (Han et al., 2015). Moreover, MGs reduce the environmental impact by utilizing clean energy sources like solar, wind, and hydro power. With their ability to function independently or in conjunction with the main grid, MGs represent a key solution

https://doi.org/10.1016/B978-0-443-36476-1.00006-7

in modernizing energy infrastructure to better accommodate the rising demand for electricity in a more environmentally responsible manner.

In the MG systems, LFC plays a crucial role in ensuring stable and reliable power supply by maintaining the balance between power generation and demand, thereby stabilizing system frequency. The MGs are prone to frequency deviations due to the intermittent and unpredictable nature of renewable sources. In traditional power systems, large synchronous generators with inherent inertia help to mitigate frequency fluctuations. However, the MGs often operate with minimal or no inertia, especially in islanded mode, making them more sensitive to load disturbances and renewable energy fluctuations. The LFC in MGs must compensate for these issues while also ensuring synchronization with the main grid when in grid-connected mode. The key objectives of the MG LFC include (Bevrani et al., 2016):

- Frequency stabilization: Ensuring that the system frequency remains within acceptable limits despite changes in load or generation.
- Power sharing: Balancing power among different distributed energy resources (DERs) while maintaining quality and reliability.
- Seamless integration: Coordinating the transition between grid-connected and islanded modes without frequency disturbances.

The schematic diagram of a AC and DC MG systems are shown in Figs. 11.1 and 11.2 respectively with LFC block diagram in Fig. 11.3.

A typical MG system is composed of various DERs such as a solar photovoltaic (PV) array, wind turbine generator (WTG), diesel engine generator (DEG), fuel cells (FC), and energy storage units like flywheels (FESS) and batteries (BESS). These DERs are linked to a central bus,

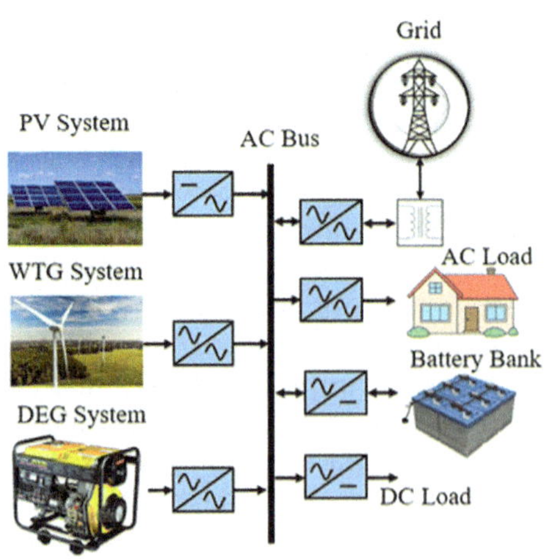

FIGURE 11.1 AC Micorgrid.

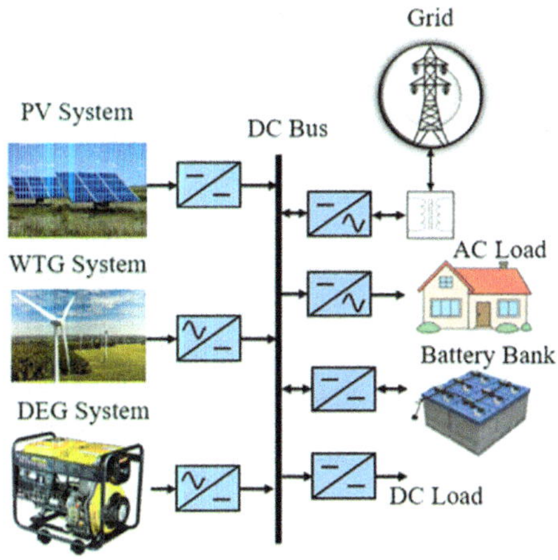

FIGURE 11.2 DC Micorgrid.

serving as the core of energy distribution in the MG. Power electronic devices are critical for synchronizing AC sources and enabling seamless energy transfer between DC sources like PV panels and fuel cells, and the AC bus. The BESS, for instance, are integrated with the AC bus using power electronic converters, which handle the bidirectional conversion of AC to DC for charging the battery, and DC back to AC during discharging. This converter system allows for efficient, stable energy exchange within the MGs, ensuring reliability, especially when renewable energy sources fluctuate. Through this combination of DERs and power electronics, the MGs can dynamically manage energy flows, balancing generation and demand while maintaining system stability. This versatile architecture enhances both energy efficiency and resilience in modern power systems.

Frequency is a vital aspect of power quality because it reflects the balance between power supply and demand in an electrical grid (Hemmati et al., 2018). In alternating current (AC) systems, frequency must be kept stable-typically at 50 or 60 Hz-because deviations from this range can have serious consequences. When frequency increases or decreases beyond acceptable limits, it can indicate that generation does not match consumption, leading to instability in the grid. This imbalance can damage sensitive equipment, disrupt the operation of industrial processes, and even trigger system-wide outages. Furthermore, frequency deviations can affect the performance of protection devices and compromise the reliability of power delivery. With the increasing integration of intermittent renewable energy sources such as wind and solar, maintaining frequency stability has become more challenging, as their unpredictable output can lead to fluctuations. Therefore, proper

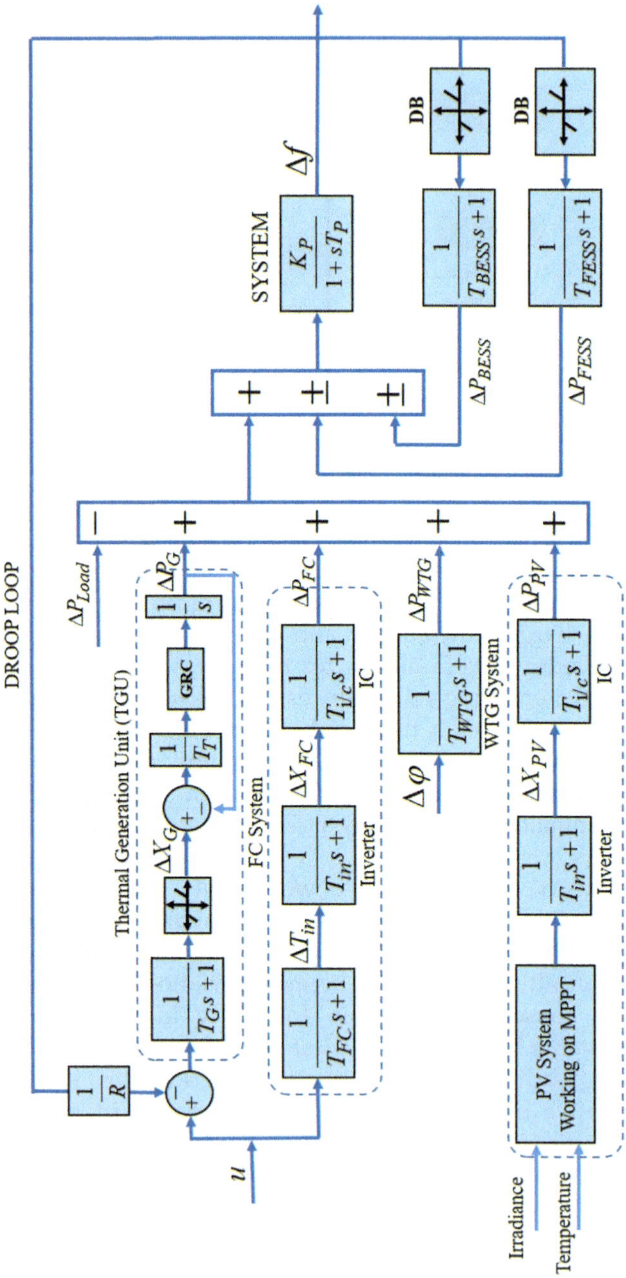

FIGURE 11.3 MG LFC model (Armin et al., 2018).

frequency regulation is essential not only for ensuring power quality but also for the smooth integration of renewable energy, safeguarding both the grid and connected equipment.

When it comes to maintaining stability in MG systems, components like solar PV and wind turbine generator systems are typically excluded from the control loop due to their intermittent and unpredictable nature. Instead, more stable sources like fuel cells and diesel engine generators are primarily involved in the control loop. Traditional droop control strategies, which rely on fixed gain control, often fail to effectively manage frequency stability, especially under load disturbances. To enhance the stability of MG systems, auxiliary control mechanisms are required in addition to droop control, ensuring robust performance across different load conditions. In modeling MG systems, using first-order approximations for components like PV, FC, WTG, DEG, and energy storage systems such as battery energy storage and flywheel energy storage is a practical approach. This is particularly feasible when only minor load fluctuations are anticipated during normal operation, allowing for a simplified yet effective control design.

The variation in load demand (expressed in per unit) is managed by the combined power output from distributed energy resources like PV systems, WTG systems, FC systems, and DEG systems, in addition to the energy transfers between the battery energy storage system and flywheel energy storage system, as outlined below (Zaheeruddin, 2022):

$$\Delta P_{\text{load}} = \Delta P_{PV} + \Delta P_{\text{WTG}} + \Delta P_{FC} + \Delta P_{G} \pm \Delta P_{\text{BESS}} \pm \Delta P_{\text{FESS}}$$

11.2 Review of existing technologies

Control in MG systems is crucial for ensuring reliable, efficient, and stable operation. Given the complexity of integrating diverse energy sources such as solar, wind, and conventional generators, effective control strategies are necessary to manage power flow, maintain voltage and frequency stability, and ensure seamless transitions between grid-connected and islanded modes. Moreover, control systems play a pivotal role in optimizing energy use, reducing operational costs, and minimizing the impact of disturbances or fluctuations in energy supply. Without robust control mechanisms, MGs would struggle to achieve their potential in enhancing energy security, supporting renewable integration, and providing sustainable power solutions.

Classical control techniques are often used in MG operation to ensure stable and reliable performance. These control methods rely on mathematical models and feedback loops to regulate various aspects of the

MGs. Proportional integral derivative (PID) or proportional integral (PI) controllers, are often used to regulate the voltage and frequency of the MG by adjusting the output of generators and power electronic converters (Murugesan et al., 2023; Chen et al., 2022). Using a PID controller is straightforward and doesn't require advanced technical skills for controller design. However, its limitations include poor robustness and restricted bandwidth, which can limit its applicability. Fractional order PID controllers have garnered growing interest in recent times, given their superior control performance compared to traditional integer-order PID controllers. They can offer advantages in terms of improved transient response, reduced steady-state error, and better handling of nonlinear and time-varying systems (Lam et al., 2020; Prasad and Parimi, 2022). Unlike classical integer-order controls, fractional-order control allows for more flexibility and precision in capturing the complex dynamics. Optimal and adaptive control in MGs are essential to address the dynamic nature of these systems, optimize energy resources, enhance reliability, and promote sustainable energy practices. They enable MGs to operate efficiently while adapting to changing conditions, contributing to a more resilient and sustainable energy future. The model predictive control (MPC) is an optimization based control approach consisting of a feedback mechanism to counter uncertainties, disturbances, and physical constraints such as battery storage capacity. Apart from this, MPC can incorporate generation and demand forecast, and is based on future behavior of the system which is crucial for systems dependent on demand and renewable energy generation (Sarkar et al., 1216). Robust controls are essential for MGs to ensure reliable, resilient, and efficient operation in the face of uncertainties, variability, and various operational challenges. They help maintain power quality, stability, and the ability to adapt to changing conditions, making MGs a viable and attractive option for reliable energy supply in diverse settings. Various robust control approaches to handle uncertainties and disturbances while maintaining frequency and voltage stability in an MG system such as (Cucuzzella et al., 2015), - synthesis control (Shah et al., 2021), sliding mode control (SMC) (Taghizadegan et al., 2024; Das et al., 2012), backstepping control (Kanwal et al., 2024), active disturbance rejection control (ADRC) (Li et al., 2008) are present in literature. In the last few decades, soft computing approaches have become valuable in controller tuning because they offer flexibility, robustness, adaptability, and the ability to handle complex and nonlinear systems. Soft computing techniques have been combined with traditional control methods to create hybrid control systems that leverage the strengths of both approaches. Unlike traditional hard computing methods that rely on precise mathematical models and binary logic, soft computing methods embrace uncertainty, imprecision, and approximation. Among different soft computing approaches, genetic algorithm (GA) (Al-Saedi et al., 2013), artificial neural network (ANN) (Arya and Kumar, 2017), fuzzy logic (Dhanasekaran et al., 2020), particle swarm optimization (PSO) (Shayeghi et al., 2021), bacteria foraging (BF) (Orimi and Marzbali, 2023), ant colony (Abdelaziz and Ali, 2015), mayfly (Mekhamer et al., 2024), grasshopper optimization (Bhagat et al., 2023), cuckoo search optimization algorithm (Khadanga et al., 2018), teaching learning (Armin et al., 2018), coati optimization (Matijašević et al., 2022), artificial hummingbird optimization (Xu et al., 2018), whale optimization (Babalola et al., 2017) including many others are well reported in literature. In recent years, there has been a growing focus on utilizing the ML algorithms for the analysis of distribution networks owing to its ability of handling large amount of data with much simpler and time effective models (He et al., 2017). The ML algorithms have been used in wide range of applications such as electric load and price forecasting (Kurt et al., 2018), fault and failure prevention (Alshareef et al., 2014), securing cyberspace threats (Ahmed et al., 2023), theft detection (Ahmed et al., 2021), and islanding detection (Bal et al., 2023) among many others.

11.2.1 Notable achievements in frequency control

Notable achievements in LFC for MGs have emerged across multiple dimensions, driven by advancements in control strategies, technological integration, and optimization techniques as shown in Fig. 11.4.

One significant milestone is the development of advanced control strategies such as MPC and SMC, which enhance the robustness and adaptability of frequency control in MGs, particularly under unpredictable renewable energy inputs. Additionally, Vehicle-to-Grid (V2G) technology has enabled electric vehicles (EVs) to act as mobile energy storage units, supporting grid frequency by feeding excess power back to the grid when demand rises (Pérez and Conde, 2020; Kikusato et al., 2022). Furthermore, DR mechanisms have introduced the ability to dynamically adjust loads in response to frequency fluctuations, improving grid stability (Anwar and Pan, 2015). Optimized control techniques ensure better system performance under varying operational conditions. The integration of BESS and FESS has also played a pivotal role in achieving faster

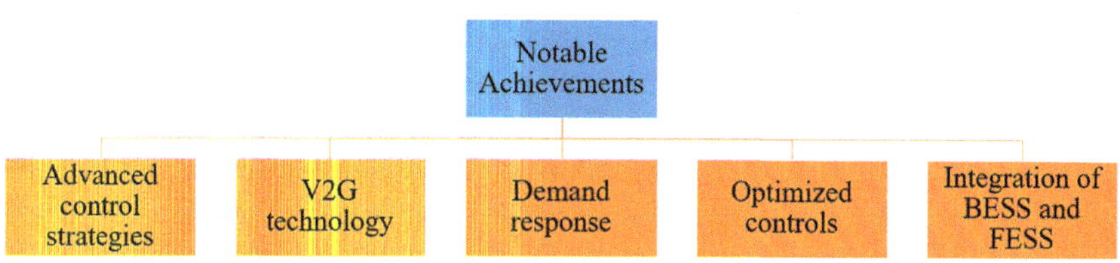

FIGURE 11.4 Notable achievements.

response times to frequency deviations, buffering the inherent variability in renewable energy generation and improving overall system resilience (Çam and Kocaarslan, 2005; Siti et al., 2019). These advancements collectively contribute to enhancing the reliability and efficiency of LFC in modern MGs, making them more capable of handling the complexities of distributed energy resources and renewable integration.

11.2.2 Challenges and limitations in current approaches

Current control approaches for LFC in MGs face several limitations, particularly in handling the complexities of modern decentralized energy systems. One key limitation is the inability to fully manage the intermittency of renewable energy sources such as solar and wind, leading to frequent power imbalances and challenges in maintaining stable frequency. Traditional control methods like PID controllers often struggle with the nonlinearity and rapid fluctuations of renewable power inputs, causing slower response times and less efficient regulation. Additionally, many control strategies rely on fixed gain settings that are not adaptive to varying load conditions, reducing their effectiveness in dynamic environments. Although advanced strategies like SMC and MPC offer improvements, they often come with increased computational complexity, making real-time implementation difficult in larger systems. Moreover, these approaches may still lack robustness in the face of sudden disturbances or unmodeled system dynamics. Table 11.1 summarizes the advantages and limitations of different control approaches reported so far in field of frequency regulation in MGs.

The MGs system face several other technical challenges that hinder their efficient operation. Challenges other than frequency regulation in MG is shown in Fig. 11.5.

Cybersecurity is a significant concern, as MGs rely on interconnected digital technologies that can be vulnerable to cyberattacks, leading to potential disruptions and data breaches protection against such threats is crucial for maintaining system integrity (Bharathi and Sasikumar, 2021). Power quality is another challenge, particularly due to the integration of various renewable energy sources, which can introduce harmonics and voltage fluctuations into the system. This can adversely affect sensitive electrical equipment and reduce overall system reliability (Rehman, 2025). As far as stability issues are concerned, MGs must maintain frequency and voltage levels despite the variable nature of renewable generation and load demand as discussed earlier. Load balancing in MG systems involves the efficient distribution of power between generation sources and loads to maintain system stability, minimize energy losses, and optimize resource utilization (Valta et al., 2018). Finally, the requirement for bidirectional flow

TABLE 11.1 Control approaches used for frequency regulation in MG.

References	Control approach	Advantages	Limitations
(Murugesan et al., 2023; Najafirad and Mahdian Dehkordi, 2023; Shi et al., 2024)	PI, PID	Simple in design and wide application	Poor transient behavior; sensitivity to parameter tuning
(Lam et al., 2020; Prasad and Parimi, 2022)	Fractional order PID	Enhanced flexibility, better control performance	Complex design and implementation; higher computation demand
(Rouhani et al., 2024)	Optimal control	Adaptability, effective handling of system constraints	High computational complexity; dependency on accurate system modeling
(Zaheeruddin, 2022; Cucuzzella et al., 2015; Shah et al., 2021)	And - synthesis	Robustness against uncertainty	Complexity
(Sarkar et al., 1216; Lakshmi Satya Nagasri and Marimuthu, 2023)	Model predictive control	Better constraint handling ability	Tuning complexity; dependency on accurate system modeling
(Li et al., 2008)	Active disturbance rejection control	Robustness to disturbances and uncertainties; model independence; fast response and adaptability	Parameter tuning challenges; computational burden for complex systems; noise sensitivity

Continued

TABLE 11.1 Control approaches used for frequency regulation in MG.—cont'd

References	Control approach	Advantages	Limitations
(Armin et al., 2018; Das et al., 2012)	Sliding mode control	Robustness to uncertainties and disturbances; fast response and high accuracy; simple design for nonlinear systems	Chattering phenomenon; complex tuning and implementation
(Choudhury and Sahoo, 2024; Mbungu et al., 2022)	Event triggered control	Reduced resource utilization; improved efficiency in distributed systems	Complex design; risk of zeno behavior; delayed responses to fast dynamics

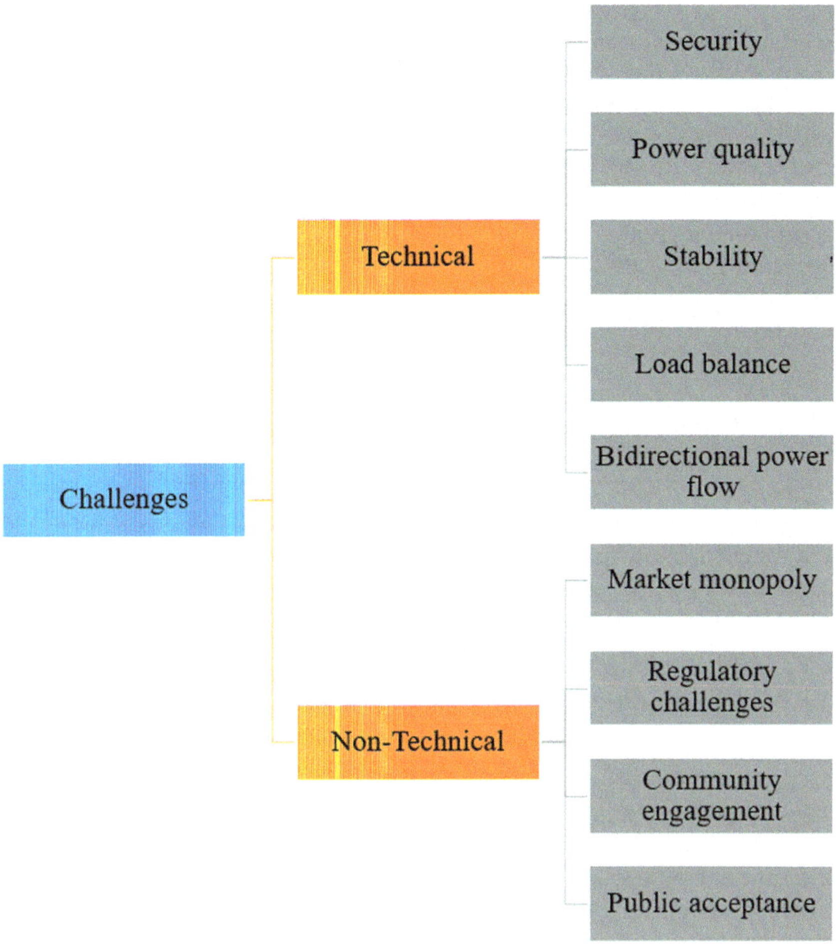

FIGURE 11.5 Persistent challenges.

introduces complexity in control strategies, necessitating advanced technologies to manage energy exchange between the grid and local resources efficiently (Tahir et al., 2024). Addressing these challenges is vital for the operation of MG systems in the future.

Non-technical challenges in the deployment of MGs are significant and multifaceted. Market monopoly can pose a substantial hurdle, as a few dominant players may control energy markets, limiting competition. This lack of competition can hinder the development and integration of MG technologies as new entrants struggle to gain market share (Muttaqee et al., 2023). Additionally, regulatory challenges often arise due to outdated policies that do not accommodate the unique characteristics of

MGs. These systems may not fit neatly into existing regulatory frameworks, leading to ambiguity and inefficiencies in operation and investment (Babaei et al., 2024). Community engagement is also crucial for the successful implementation of MGs. Effective communication and involvement of local stakeholders can foster a sense of ownership and support for MG initiatives. However, without adequate outreach and transparency, public trust can be eroded (Zhang et al., 2023). Public acceptance is another critical factor, as individuals and communities must be willing to embrace new technologies and changes in their energy landscape (Fani et al., 2022). Overcoming skepticism about the benefits and reliability of MGs is essential for their widespread adoption.

11.3 Emerging technologies and innovations in microgrid LFC

This section explores the latest advancements and innovations in MG LFC, focusing on three areas: advancements in MG components, integration with smart grid and IoT technologies, and the transformative role of AI in LFC.

11.3.1 Technological advances in microgrid components

Significant progress has been made in the components that constitute MGs, such as energy storage systems, inverters, and power electronics. These advancements are enhancing the efficiency and reliability of LFC. Energy storage is critical for LFC in MGs, especially when integrating intermittent renewable energy sources like wind and solar. Recent improvements in energy storage technologies, such as lithium-ion and solid-state batteries, have enhanced the response time and energy density of these systems. For LFC, energy storage plays a vital role by providing fast-response reserves that stabilize frequency during supply-demand imbalances. Hybrid energy storage systems, which combine technologies like batteries and super capacitors, offer even more flexibility (Nosratabadi et al., 2017). While super capacitors address short-term frequency deviations, batteries manage longer-term energy discharge, allowing for a more holistic LFC approach (Krishna Rao et al., 2024). Inverters, especially grid-forming inverters, are essential for managing the power output from renewable energy sources in MGs. These inverters have evolved to become more intelligent, capable of autonomously adjusting their output to match real-time frequency variations (Feng et al., 2021). Technological improvements in power electronics have resulted in more efficient converters that are critical for managing energy flow in hybrid AC/DC MGs. These converters allow seamless energy exchange between AC and DC subsystems, enabling MGs to efficiently manage energy from diverse sources, such as DC-based solar panels and AC-based traditional generators. The virtual power plants (VPPs) are gaining traction as a promising innovation for LFC in MGs. The VPPs aggregate various DERs into a single, manageable entity, allowing for coordinated control of load and generation. By integrating multiple energy sources-like solar, wind, and energy storage-VPPs optimize the energy balance within an MG, thereby stabilizing frequency (Ohanu et al., 2024). This aggregated approach also enhances the ability to predict and manage fluctuations, improving LFC reliability across interconnected MGs. Table 11.2 summarizes the key advancements.

TABLE 11.2 Summary of key advancements.

Technology	Advantages	Impact on frequency regulation
Energy storage systems	Enables energy exchange and load leveling; stores excess energy for future use; provides quick response to power fluctuations	Enhances system inertia and damping; mitigates frequency deviations during sudden load or generation changes
Inverters	Supports bidirectional power flow; improves power quality	Balances supply-demand in islanded or grid-connected modes
Power electronics converters	Ensures efficient energy transfer; provides fast response to dynamic system changes; enables integration of renewables	Improves frequency stability through precise control of power flow; reduces transients during disturbances
Super capacitors	Delivers high power density and rapid charge-discharge cycles; requires low maintenance; provides short-term energy support	Offers fast frequency response during sudden load changes; stabilizes frequency in low-inertia MGs
Virtual power plant (VPP)	Aggregates distributed energy resources for optimized operation; enhances grid reliability; facilitates real-time control and flexibility	Coordinates multiple DERs for dynamic frequency support; improves system resilience during disturbances

11.3.2 Integration with smart grid and IoT technologies

The integration of MGs with smart grids and the IoT is revolutionizing LFC by enabling real-time monitoring, decentralized control, and improved communication. The widespread deployment of IoT devices within MGs infrastructure has transformed the monitoring of MGs and also manages stability. The IoT sensors embedded in energy generation units, storage systems, and consumer loads collect real-time data on energy usage, load demand, and frequency deviations (Antonopoulos et al., 2020). This data is processed to make informed LFC decisions. With edge computing, which processes data locally rather than through centralized servers ensures faster decisions on control. This speed is crucial for LFC, where real-time adjustments are required to maintain frequency stability (Khan et al., 2024). For example, if frequency dips below a specific threshold, edge devices can automatically trigger load shedding or initiate energy storage discharge to restore balance. As MGs become increasingly integrated with the larger smart grid, they benefit from enhanced communication and coordination with the main grid. Smart grids enable two-way communication, allowing MGs to not only draw power from the main grid but also provide ancillary services like frequency regulation (Dankir et al., 2024). During periods of high demand or generation fluctuation, MGs can collaborate with the main grid to balance supply and demand, improving overall grid stability. This integration also facilitates MG participation in demand response programs, where loads can be reduced or shifted during peak times to help stabilize frequency across the entire energy network.

11.3.3 Role of AI in future LFC

AI is transforming LFC by enabling predictive control. AI-driven algorithms can forecast energy generation and consumption based on historical data, weather patterns, and other variables (Ronay et al., 2017). For example, AI can predict the output of a solar panel based on upcoming weather conditions or anticipate load demand fluctuations based on time-of-day patterns. With this foresight, MGs can proactively adjust generation and consumption, minimizing the risk of frequency deviations. AI is also enabling the development of fully autonomous MGs, where LFC and other operations can be managed without human intervention (Oladapo et al., 2024). In such systems, AI algorithms continuously monitor grid conditions, control loads, regulate generation, and stabilize frequency, ensuring reliable operation even in islanded or remote locations. By automating decision-making, AI minimizes the risk of human error and provides a more consistent and reliable approach to LFC. AI-driven control algorithms, such as MPC and neural networks, bring advanced capabilities to LFC in complex MG environments. MPC, for instance, optimizes control inputs by predicting future grid conditions, allowing MGs to adjust load and generation in real-time (Ahmethodzic and Music, 2021). Neural networks excel at managing non-linear and uncertain conditions, making them especially useful in MGs with high reliance on intermittent renewable energy sources (Hu et al., 2021). The ML models excel in managing the dynamic conditions of MGs, where fluctuations in renewable energy generation or load demand occur frequently. These models learn from past data and adapt their control strategies to optimize LFC in real-time (Gupta et al., 2022). As ML models accumulate more data, they improve in predicting and responding to grid changes, offering a flexible and robust approach to frequency regulation.

11.4 Future trends in microgrid frequency control

The MGs, capable of operating independently or connected to the main grid, play an increasingly important role in enhancing energy reliability, resilience, and sustainability. A critical aspect of MG operation is LFC, which ensures a balance between energy demand and supply to maintain stable grid frequency. As the energy landscape evolves, advancements in control strategies, energy storage technologies, regulatory and policy developments will shape the future of MG frequency control. The following discussion examines these key future trends, highlighting innovations and their potential impacts. The future research trends are shown in Fig. 11.6.

11.4.1 Advancements in control strategies

As MGs become more common and increasingly reliant on renewable energy sources, control strategies must evolve to address the challenges associated with decentralized energy systems and unpredictable energy generation (Ullah et al., 2022). Future advancements in control strategies will focus on improving the responsiveness, efficiency, and intelligence of MG frequency control mechanisms. One of the most promising control strategies is MPC, which uses real-time data to predict future grid conditions and optimize control actions over a predetermined time frame. The MPC's ability to forecast potential frequency deviations based on changing load and generation conditions allows for proactive adjustments in power output and load shedding. This predictive capability is particularly valuable in MGs with a high share of renewable energy, where generation is often variable (Lakshmi Satya Nagasri and

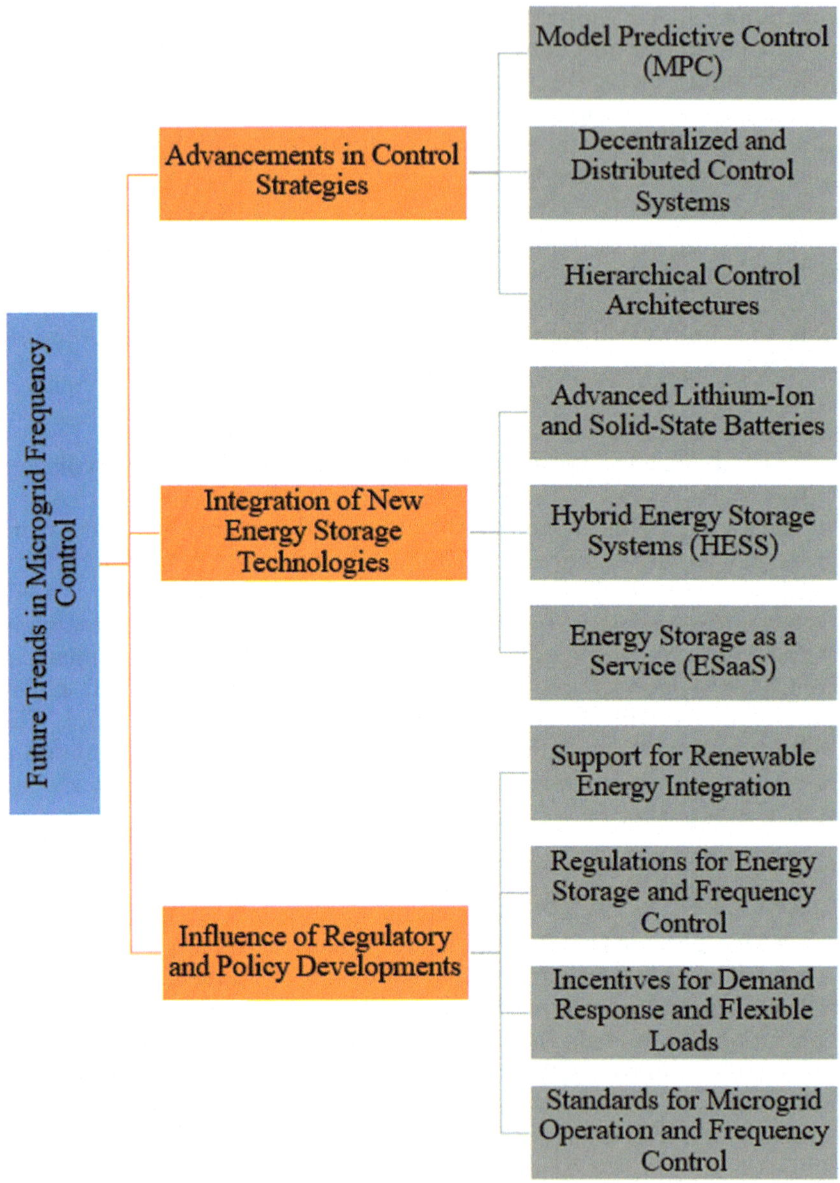

FIGURE 11.6 Future research trends.

Marimuthu, 2023). By anticipating these fluctuations, MPC enhances frequency control in increasingly complex MG environments.

Traditionally, power grids have relied on centralized control for managing frequency stability. However, as MGs proliferate and become more interconnected, decentralized and distributed control systems are emerging as a viable solution. In decentralized systems, individual MG components such as inverters, energy storage systems, and renewable generation units take on more responsibility for managing frequency control. These systems allow for faster, localized responses to frequency deviations, improving the adaptability and resilience of the MG (Rashidi et al., 2021). The distributed approach also reduces the risk of single points of failure. In the event of a fault or disruption, other components in the MG can continue to maintain frequency stability, ensuring reliable operation (Andishgar et al., 2017). This makes decentralized control systems particularly advantageous for islanded MGs, which operate independently of the main grid.

Another key trend is the development of hierarchical control systems, which distribute control responsibilities across different levels, typically primary, secondary, and tertiary layers. Primary control addresses immediate, real-time responses to frequency deviations, while secondary control manages slower adjustments, such as rebalancing load and generation over longer periods (Manshahia et al., 2022). Tertiary control optimizes the overall performance of the MG, often coordinating with the main grid when needed (Kumar et al., 2022). Hierarchical systems allow for more significant and efficient frequency control by addressing issues at appropriate timescales. Future advancements in hierarchical systems will likely incorporate AI and ML to enhance decision-making and automation, further improving the resilience and efficiency of MG frequency control.

11.4.2 Integration of new energy storage technologies

Energy storage technologies are critical to maintaining frequency stability in MGs, especially as the reliance on intermittent renewable energy sources like solar and wind grows. The integration of advanced storage systems will play a key role in shaping the future of MG frequency control. Lithium-ion batteries are currently the dominant energy storage technology in MGs. However, ongoing research and development are improving battery chemistry, leading to higher energy densities, faster response times, and longer life cycles. These improvements enable MGs to better store excess renewable energy and quickly release it when needed to stabilize frequency (Sakthivelnathan et al., 2024). Solid-state batteries, an emerging technology, offer even greater potential. With solid electrolytes instead of liquid ones, solid-state batteries are safer and more durable, making them ideal for MGs operating in challenging environments. Their faster response times are particularly beneficial for frequency control, as they can quickly provide energy to correct supply-demand imbalances. Hybrid energy storage systems (HESS) combine multiple types of storage technologies, such as batteries, super capacitors, and flywheels, to optimize performance across various timescales. Super capacitors, for example, deliver rapid bursts of power to address short-term frequency fluctuations, while batteries provide long term energy storage. By utilizing the strengths of each technology, HESS allows MGs to achieve more effective and reliable frequency regulation. HESS is especially valuable for MGs with high levels of renewable energy integration, as it enables efficient energy storage during periods of excess generation and quick energy discharge when generation decreases (Ton and Smith, 2012; Lu et al., 2023).

The concept of energy storage as a service (ESaaS) is gaining traction as a new business model that enables MG operators to lease energy storage capacity rather than owning it outright. ESaaS lowers the financial barrier to adopting large-scale energy storage systems, making it more accessible for smaller MGs or community-based projects (Tan et al., 2022). With ESaaS, third-party providers manage and maintain the storage systems while offering frequency control services on a subscription basis. This model provides MGs with greater flexibility in scaling their energy storage needs without incurring the significant upfront capital costs of purchasing and maintaining storage infrastructure.

11.4.3 Influence of regulatory and policy developments

Regulatory frameworks and policy decisions are playing an increasingly important role in the future of MG frequency control. Governments and energy regulators are shaping the deployment of MGs, the integration of renewable energy, and the adoption of advanced frequency control technologies through targeted policies and regulations (Mohapatra et al., 2024). Many governments are implementing policies that promote the adoption of renewable energy, including projects tied to MGs. Incentives such as feed-in tariffs, tax breaks, and renewable energy mandates are encouraging the deployment of solar, wind, and other clean energy sources. However, the variability of these sources presents challenges for frequency control. As a result, regulatory efforts are increasingly focused on supporting energy storage systems and advanced control technologies that can mitigate the impact of renewable energy fluctuations on grid stability. Governments are funding pilot programs and demonstration projects to test innovative frequency control solutions, helping to establish best practices and guide future regulatory decisions.

Policymakers are also reforming regulatory frameworks to support the deployment of energy storage systems and allow MGs to offer frequency control services. For example, in the United States, the Federal Energy

Regulatory Commission's (FERC) ordered 841 mandates that energy storage systems have equal access to wholesale electricity markets (Mohapatra et al., 2024). This opens up new opportunities for storage providers to offer frequency regulation services, making energy storage a more viable option for MGs. Similar initiatives are being implemented globally, accelerating the adoption of storage technologies and improving the ability of MGs to maintain frequency stability.

The demand response programs are becoming increasingly important for frequency control. These programs incentivize consumers to reduce or shift their energy consumption during peak periods, helping to balance supply and demand. In MGs, demand response can be used to manage frequency by automatically reducing non-essential loads when deviations occur (Fan et al., 2019). Regulators are promoting demand response participation by implementing policies that reward MGs for offering flexibility services to the main grid. As demand response becomes more integrated into regulatory frameworks, MGs will be better positioned to contribute to overall grid stability while generating new revenue streams. As MGs proliferate, regulatory agencies are developing standards to ensure their safe and efficient operation. Technical standards, such as IEEE 1547 in the U.S., define how MGs and DERs should interact with the main grid. These standards are evolving to include guidelines for voltage and frequency regulation, ensuring that MGs can integrate seamlessly with the broader energy system (Mohapatra et al., 2024).

11.5 Research directions for microgrid LFC

Addressing the technical challenges of MG operation requires a multifaceted and cross-disciplinary approach to ensure reliability, security, and scalability. Cybersecurity is a critical concern, as MGs rely on real-time data exchange between DERs, making them vulnerable to cyberattacks such as false data injection (FDI), which can disrupt system responses (Eklund et al., 2024). Advanced encryption techniques, real-time monitoring, and the ML based anomaly detection systems are essential for mitigating these risks. Islanding detection reviewed in (Sechilariu et al., 2018) poses another challenge, as undetected disconnections from the main grid can cause power quality issues and operational hazards; robust communication protocols and control systems are needed to address this. Power imbalances from intermittent renewable energy sources can lead to frequency and voltage deviations, requiring advanced LFC strategies and the integration of energy storage systems, such as BESS, for stabilization. Scalability also becomes increasingly complex as MGs expand, necessitating optimized control algorithms capable of adapting to larger, more distributed setups. Cross-disciplinary collaborations are pivotal in overcoming these challenges, integrating insights from power systems, control theory, AI, cybersecurity, energy economics, and engineering fields such as electronics, mechanical, and chemical engineering (Wolsink, 2020). For instance, advanced communication networks and sensors developed by electronics engineers can improve real-time monitoring and coordination, while chemical engineers can enhance energy storage technologies through higher-capacity, faster-charging batteries. Mechanical engineers can optimize renewable energy system designs for consistent power generation, and production engineers can streamline the manufacturing of MG components to enhance cost efficiency and scalability. Additionally, customized LFC strategies tailored for sectors like agriculture and dairy farming can leverage renewable resources such as biogas, solar, or wind energy to address their specific power needs, further expanding the applicability of MG technologies. These cross-disciplinary efforts, combined with economic and regulatory collaborations, are crucial for driving innovation and addressing the technical, economic, and operational challenges of modernizing MGs.

11.6 Societal impacts and human factors

Beyond technical challenges in MGs, it is equally important to consider the societal impacts and human factors that influence the success of MGs in real world applications. Addressing the human side of this technology can help foster acceptance, encourage widespread adoption, and ensure long term sustainability. Below are three key subtopics that probe into these considerations: community engagement, human-centric design principles, and social and economic implications.

11.6.1 Considerations for community engagement

Community engagement is critical to the successful implementation and operation of MGs, especially in areas where public acceptance and participation can make or break projects. MGs can vary greatly in size, from small community based systems to large industrial networks. The size and type of MG (whether grid-connected or islanded) directly influence the level of engagement needed. For instance, small MGs in rural or off grid communities often require higher involvement from local stakeholders to ensure that the system addresses their specific needs and integrates local renewable resources effectively. On the other hand, larger grid-connected MGs, such as those in urban settings, may involve more complex governance structures but less direct engagement with individual users (Mahfuz et al., 2022). One of the key challenges in community engagement is building trust and ensuring that the community understands the benefits and limitations of MG technologies. People must be informed about how the MG will function, the types of renewable energy involved (solar, wind, etc.), and the potential for integration with energy storage systems such as batteries. Educating the community on the concepts of V2G and G2V interactions, where EVs can act as mobile energy storage, can enhance participation and acceptance (Wang and Sun, 2023). Additionally, technical challenges like communication delays, scalability, and the intermittency of renewable energy need to be framed in a way that non-technical individuals can grasp. This is particularly important in rural areas where the success of islanded MGs may depend on community buy in and operational support. Several methods, such as public meetings, workshops, and participatory design approaches, can be employed to engage the community (Shahzad et al., 2023). However, each method comes with its own challenges. For instance, public meetings might not always represent all stakeholders, while workshops can be resource intensive. In terms of LFC, simplifying the technical aspects and ensuring that community members understand how LFC impacts their daily lives, such as by stabilizing power in case of fluctuations, is essential.

11.6.2 Human-centric design principles

The successful adoption of MGs also hinges on applying human-centric design principles that focus on usability, accessibility, and adaptability to local contexts. Designing systems with end users in mind ensures that the technology integrates seamlessly into people's daily lives and becomes an enabler rather than an obstacle (Farghali et al., 2023). One area where human centric design is particularly important is in material selection for energy storage and infrastructure. For instance, the choice of battery technology can significantly impact user experience. Lithium-ion batteries, while popular due to their energy density and long cycle life, come with concerns about safety and environmental impact. Emerging technologies, such as solid-state batteries and flow batteries, are promising alternatives that may be better suited for community use in certain scenarios (Khare and Chaturvedi, 2023). These technologies must be reliable, cost-effective, and safe for widespread adoption, especially in residential or sensitive environments. EVs can also play a pivotal role in human centric design within MGs, particularly through V2G and G2V systems. The interaction between EVs and the grid must be intuitive for users, with clear communication and control systems that allow individuals to understand when their vehicles are charging from the grid or discharging power back into it. Smart communication systems, which facilitate this interaction, need to be user friendly and secure, ensuring that personal data is protected while optimizing energy exchange. Renewable energy sources such as solar and wind are highly variable, and this variability poses challenges for LFC. Human-centric designs that include predictive models based on weather patterns or usage data can help balance the energy supply. These designs need to communicate information effectively to users, perhaps through smart home interfaces or mobile apps, so that they are aware of energy availability and can adjust their usage accordingly. Moreover, MG systems should be designed with modularity and scalability in mind. As community needs grow or shrink, the system should be adaptable without requiring significant overhauls. For instance, adding more solar panels, integrating new battery storage systems, or upgrading the communication network should be straightforward and cost effective.

11.6.3 Social and economic implications of microgrid LFC

The implementation of MG LFC also has far reaching social and economic implications. MGs can empower local communities by offering reliable and sustainable energy solutions, but they also bring about new economic models and opportunities for collaboration across various fields. One of the social implications of MG LFC is the potential for democratizing energy. MGs allow communities to generate and manage their own power, reducing dependency on centralized power grids. This could lead to increased energy equity, particularly in underserved or remote areas (Wouters, 2015). However, it also raises questions about governance, as communities will need frameworks for decision making, maintenance, and conflict resolution related to the MG's operation. The economic implications of MGs are both promising and challenging. On the one hand, MGs can lead to cost savings through reduced reliance on central power grids, enhanced energy efficiency, and the potential for lower electricity costs, particularly in remote areas. By utilizing renewable energy sources, MGs can decrease fuel costs and price volatility associated with fossil fuels. Additionally, energy storage solutions such as batteries enable better load management, reducing peak demand charges. However, the initial capital investment required for infrastructure, control systems, and energy storage can be high, posing a barrier to widespread adoption. Ongoing maintenance costs and the need for skilled personnel to manage MGs also add to the economic complexity. Despite these challenges MGs offer long-term economic benefits, including resilience to grid (Martín and de la Hoz, 2024; Bandura and Bonin, 2024).

11.7 Ethical and regulatory considerations

In the rapidly advancing field of MG technologies, ethical and regulatory considerations are critical to ensuring fair, safe, and sustainable deployment. Ethical issues in MG deployment primarily revolves around equitable resource distribution and environmental responsibility. MGs often serve off-grid or underserved communities, such as rural or underdeveloped areas, and hold the potential to alleviate energy poverty. Yet, disparities in access to MG technologies can exacerbate existing inequalities, leaving marginalized populations without reliable and affordable energy. Ensuring inclusivity and equitable access to MG benefits is a critical ethical obligation. Transparency and informed consent are also essential for ethical MG operation. Consumers should be made aware of how MGs function, their pricing mechanisms, and potential risks to foster trust and engagement within the communities they serve. Furthermore, MGs are frequently promoted as environmentally friendly due to their ability to integrate renewable energy sources. However, the environmental impact of their components, such as the extraction and disposal of rare earth metals used in batteries, must be addressed. Ethical deployment involves considering the entire lifecycle of MG components, minimizing environmental harm, and adopting sustainable practices. Readers are encouraged to refer (United Nations Climate Change, 2024) where in regulatory framework for MGs is discussed considering the case study if Singapore.

Regulations play a pivotal role in ensuring the safe, efficient, and fair deployment of MG technologies. Regulatory frameworks must govern various aspects of MG operation, including technical standards, pricing structures, cybersecurity, and consumer protection (2024). Clear guidelines are essential to prevent issues like energy monopolies, cyber threats, and inefficiencies in energy distribution. As energy systems evolve with advancements in smart technologies and the integration of DERs like solar panels and wind turbines, regulatory frameworks must adapt to accommodate these changes. Cybersecurity regulations are particularly vital, as MGs increasingly rely on digital infrastructure and the IoT for monitoring and control, making them potential targets for cyberattacks. Safeguarding these systems is essential to prevent disruptions to energy supply and ensure grid stability. While MG technologies have the potential to reduce carbon emissions, improve energy access, and enhance resilience to power outages, innovation must be pursued with societal well-being in mind. A purely technology-driven approach risks overlooking critical social considerations, such as job displacement in traditional energy sectors or the impact of new infrastructure on local communities. Policymakers must support retraining programs for workers transitioning from fossil fuel-based industries to renewable energy sectors, ensuring no one is left behind in the shift to a cleaner energy system (Warneryd et al., 2020). Moreover, inclusive strategies are necessary to prevent MG technologies from exacerbating socio-economic disparities. Wealthier communities may have greater financial capacity to adopt MG systems, potentially leaving lower-income communities vulnerable to energy instability. Policymakers and developers must implement frameworks that ensure equal access to MG benefits, promoting energy equity across all socio-economic groups. By addressing these ethical and regulatory challenges, the deployment of MG technologies can be guided toward creating a resilient, inclusive, and sustainable energy future.

The COP29 UN climate conference, held in Baku, Azerbaijan, marked a significant milestone in global climate action. A key achievement was the decision to triple climate finance to developing countries, from the previous goal of USD 100 billion annually, to USD 300 billion annually by 2035, and setting firm deadlines for achieving net-zero emissions (Onu et al., 2023). These measures will support developing nations in accessing finance and technical assistance for carbon market participation. The conference also advanced transparency through the Enhanced Transparency Framework, with several countries submitting their first biennial transparency reports. Additionally, new funding commitments, such as the UK's £3 million pledge to support forest conservation under the Reducing Emissions from Deforestation and Forest Degradation (REDD+) initiative, reinforced efforts to combat deforestation (Pan et al., 2023).

11.8 International collaborations and knowledge sharing

The development of MG technologies benefits significantly from global cooperation, as no single country possesses all the answers to the complex challenges of MG implementation. International collaboration allows countries to pool resources, share technical knowledge, and establish unified standards that facilitate the seamless development and integration of MGs worldwide. Joint research initiatives and shared experiences enhance efforts to address key technical challenges, such as improving energy storage solutions, enhancing grid resilience, and advancing cybersecurity measures for grid systems. Collaborative research is especially vital for tackling common issues like integrating renewable energy into MGs, optimizing storage systems, and developing cost-effective renewable energy technologies. By combining expertise and resources, researchers can devise innovative solutions for renewable energy storage and grid management, paving the way for more efficient and resilient systems. Standardization through global partnerships ensures compatibility across nations, promoting the scalable deployment of MG components. Moreover, sharing best practices and lessons learned is critical as countries and organizations gain experience with MG technologies. Each region faces unique technical, regulatory, and social challenges, and exchanging diverse insights allows for tailored solutions that address local contexts. For example, regions that have successfully integrated high percentages of renewable energy into their MGs or implemented grid stability measures with limited storage can serve as models for others. Global cooperation also plays a crucial role in addressing the energy needs of developing nations, where financial constraints and technical expertise often hinder MG adoption. International collaborations can bridge this gap by providing funding, technical assistance, and capacity-building initiatives. By fostering a cooperative environment, the global community can accelerate the transition to resilient, efficient, and sustainable MG systems while ensuring equitable access to modern energy solutions.

11.9 Conclusion and final thoughts on future of MG systems

This chapter synthesized key developments and future directions in LFC for MGs, highlighting the limitations of classical controllers like PI and PID in addressing the complexities and uncertainties of modern MG systems. The discussion underscored the pressing need for robust control strategies to manage the intermittent nature of renewable energy sources and the dynamic behavior of MGs. Emerging technologies such as AI, ML, and IoT were identified as transformative tools, offering predictive capabilities and adaptive control mechanisms to enhance LFC performance. The chapter also examined a range of technical and non-technical challenges, including system stability, power quality, cybersecurity, market dynamics, regulatory barriers, and community acceptance. It emphasized the importance of policy development and governance frameworks to support the integration and scalability of decentralized energy resources. Ultimately, the chapter envisioned a future where innovative, technology-driven control solutions play a central role in advancing MG LFC, paving the way for resilient, efficient, and sustainable energy systems.

The future of MGs holds immense potential as a cornerstone of sustainable energy systems, offering greater energy security, resilience, and flexibility. With advances in renewable energy integration, energy storage technologies like batteries and flywheels, and the growth of EVs, MGs can play a key role in decentralized power generation. The focus on smarter, more efficient LFC mechanisms, aided by robust control strategies and soft computing optimization techniques, will help address the variability of renewable resources and enhance system stability. Additionally, as MGs evolve, challenges such as cyber security, power quality, and regulatory barriers must be addressed through interdisciplinary research and collaboration across technical, economic, and social domains. Emerging technologies like V2G systems, improved demand response strategies, and enhanced energy storage solutions will continue to propel MGs forward as a vital component of modern energy infrastructure. Overall, MGs are positioned to contribute significantly to the transition toward greener, more resilient, and efficient energy networks globally.

References

Abdelaziz, A.Y., Ali, E.S., 2015. Cuckoo Search algorithm based load frequency controller design for nonlinear interconnected power system. Int. J. Electr. Power Energy Syst. 73, 632–643. https://doi.org/10.1016/j.ijepes.2015.05.050.

Ahmed, I., Ahmad, I., Ahmed, S., Adil, H.M.M., 2021. Robust nonlinear control of battery electric vehicle charger in grid to vehicle applications. J. Energy Storage 42. https://doi.org/10.1016/j.est.2021.103039.

Ahmed, I., Adnan, Md, Ali, M., Kaddoum, G., 2023. Super twisting sliding mode controller for grid-to-vehicle and vehicle-to-grid battery electric vehicle charger. J. Energy Storage 70.

Ahmethodzic, L., Music, M., 2021. Comprehensive review of trends in microgrid control. Renew. Energy Focus 38, 84–96. https://doi.org/10.1016/j.ref.2021.07.003.

Al-Saedi, W., Lachowicz, S.W., Habibi, D., Bass, O., 2013. Voltage and frequency regulation based DG unit in an autonomous microgrid operation using Particle Swarm Optimization. Int. J. Electr. Power Energy Syst. 53 (1), 742–751. https://doi.org/10.1016/j.ijepes.2013.06.002.

Alshareef, S., Talwar, S., Morsi, W.G., 2014. A new approach based on wavelet design and machine learning for islanding detection of distributed generation. IEEE Trans. Smart Grid 5 (4), 1575–1583. https://doi.org/10.1109/TSG.2013.2296598.

Andishgar, M.H., Gholipour, E., Hooshmand, R.a., 2017. An overview of control approaches of inverter-based microgrids in islanding mode of operation. Renew. Sustain. Energy Rev. 80, 1043–1060. https://doi.org/10.1016/j.rser.2017.05.267.

Antonopoulos, I., Robu, V., Couraud, B., Kirli, D., Norbu, S., Kiprakis, A., Flynn, D., Elizondo-Gonzalez, S., Wattam, S., 2020. Artificial intelligence and machine learning approaches to energy demand-side response: a systematic review. Renew. Sustain. Energy Rev. 130. https://doi.org/10.1016/j.rser.2020.109899.

Anwar, M.N., Pan, S., 2015. A new PID load frequency controller design method in frequency domain through direct synthesis approach. Int. J. Electr. Power Energy Syst. 67, 560–569. https://doi.org/10.1016/j.ijepes.2014.12.024.

Armin, M., Roy, P.N., Sarkar, S.K., Das, S.K., 2018. LMI-based robust PID controller design for voltage control of islanded microgrid. Asian J. Control 20 (5), 2014–2025. https://doi.org/10.1002/asjc.1710.

Arya, Y., Kumar, N., 2017. BFOA-scaled fractional order fuzzy PID controller applied to AGC of multi-area multi-source electric power generating systems. Swarm Evol. Comput. 32, 202–218. https://doi.org/10.1016/j.swevo.2016.08.002.

Babaei, M.R., Ghasemi-Marzbali, A., Abbasalizadeh, S., 2024. Control of a combined battery/supercapacitor storage system for DC microgrid application. J. Energy Storage 96. https://doi.org/10.1016/j.est.2024.112675.

Babalola, A.A., Belkacemi, R., Zarrabian, S., Craven, R., 2017. Adaptive immune system reinforcement learning-based algorithm for real-time cascading failures prevention. Eng. Appl. Artif. Intell. 57, 118–133. https://doi.org/10.1016/j.engappai.2016.09.003.

Bal, T., Ray, S., Sinha, N., Ramesh, D., Knypiński, Ł., 2023. Integrating demand response for enhanced load frequency control in micro-grids with heating, ventilation and air-conditioning systems. Energies 16 (15), 5767. https://doi.org/10.3390/en16155767.

Bandura, R., Bonin, C., 2024. Is the Global Workforce Ready for the Energy Transition? A Report by Center for Strategic & International Studies.

Bevrani, H., Feizi, M.R., Ataee, S., 2016. Robust frequency control in an islanded microgrid: H∞ and μ-synthesis approaches. IEEE Trans. Smart Grid 7 (2), 706–717. https://doi.org/10.1109/TSG.2015.2446984.

Bhagat, S.K., Saikia, L.C., Babu, N.R., 2023. Application of artificial hummingbird algorithm in a renewable energy source integrated multi-area power system considering Fuzzy based tilt integral derivative controller. E-Prime - Adv. Electr. Eng. Electr. Energy 4. https://doi.org/10.1016/j.prime.2023.100153.

Bharathi, K., Sasikumar, M., 2021. Power flow control based on bidirectional converter for hybrid power generation system using microcontroller. Microprocess. Microsyst. 82. https://doi.org/10.1016/j.micpro.2021.103950.

Çam, E., Kocaarslan, İ., 2005. Load frequency control in two area power systems using fuzzy logic controller. Energy Convers. Manag. 46 (2), 233–243. https://doi.org/10.1016/j.enconman.2004.02.022.

Chen, Z., Liu, Z., Wang, L., 2022. A modified model predictive control method for frequency regulation of microgrids under status feedback attacks and time-delay attacks. Int. J. Electr. Power Energy Syst. 137. https://doi.org/10.1016/j.ijepes.2021.107713.

Choudhury, S., Sahoo, G.K., 2024. A critical analysis of different power quality improvement techniques in microgrid. E-Prime - Adv. Electr. Eng. Electr. Energy 8. https://doi.org/10.1016/j.prime.2024.100520.

Cucuzzella, M., Incremona, G.P., Ferrara, A., 2015. Design of robust higher order sliding mode control for microgrids. IEEE J. Emerg. Selected Topics Circ. Syst. 5 (3), 393–401. https://doi.org/10.1109/JETCAS.2015.2450411.

Dankir, S., Puig, V., Lasri, R., Maatoui, Y., Chekenbah, H., 2024. Empowering microgrid energy management with artificial intelligence and model predictive control. IFAC-PapersOnLine 58 (13), 436–441. https://doi.org/10.1016/j.ifacol.2024.07.521.

Das, D.C., Roy, A.K., Sinha, N., 2012. GA based frequency controller for solar thermal-diesel-wind hybrid energy generation/energy storage system. Int. J. Electr. Power Energy Syst. 43 (1), 262–279. https://doi.org/10.1016/j.ijepes.2012.05.025.

Dev, A., Mondal, B., Verma, V.K., Kumar, V., 2024. Teaching learning optimization-based sliding mode control for frequency regulation in microgrid. Electr. Eng. 106 (6), 7009–7021. https://doi.org/10.1007/s00202-024-02422-8.

Dhanasekaran, B., Siddhan, S., Kaliannan, J., 2020. Ant colony optimization technique tuned controller for frequency regulation of single area nuclear power generating system. Microproc. Microsyst. 73. https://doi.org/10.1016/j.micpro.2019.102953.

Eklund, M., Khalilpour, K., Voinov, A., Hossain, M.J., 2024. Community microgrids: a decision framework integrating social capital with business models for improved resiliency. Appl. Energy 367, 123458. https://doi.org/10.1016/j.apenergy.2024.123458.

Fan, Y.V., Lee, C.T., Lim, J.S., Klemeš, J.J., Le, P.T.K., 2019. Cross-disciplinary approaches towards smart, resilient and sustainable circular economy. Republ. J. Cleaner Prod. 232, 1482–1491. https://doi.org/10.1016/j.jclepro.2019.05.266.

Fani, B., Shahgholian, G., Haes Alhelou, H., Siano, P., 2022. Inverter-based islanded microgrid: a review on technologies and control. E-Prime - Adv. Electr. Eng. Electr. Energy 2. https://doi.org/10.1016/j.prime.2022.100068.

Farghali, M., Osman, A.I., Chen, Z., Abdelhaleem, A., Ihara, I., Mohamed, I.M.A., Yap, P.S., Rooney, D.W., 2023. Social, environmental, and economic consequences of integrating renewable energies in the electricity sector: a review. Environ. Chem. Lett. 21 (3), 1381–1418. https://doi.org/10.1007/s10311-023-01587-1.

Feng, C., Wang, Y., Chen, Q., Ding, Y., Strbac, G., Kang, C., 2021. Smart grid encounters edge computing: opportunities and applications. Adv. Appl. Energy 1. https://doi.org/10.1016/j.adapen.2020.100006.

Gupta, N., Sagar Bhaskar, M., Padmanaban, S., Almakhles, D., 2022. A Systematic Approach for Solar and Hydro Resource Assessment for DC Microgrid Applications DC Microgrids: Advances, Challenges, and Applications. Wiley, pp. 159–182. https://doi.org/10.1002/9781119777618.ch6.

Han, Y., Young, P.M., Jain, A., Zimmerle, D., 2015. Robust control for microgrid frequency deviation reduction with attached storage system. IEEE Trans. Smart Grid 6 (2), 557–565. https://doi.org/10.1109/TSG.2014.2320984.

He, Y., Mendis, G.J., Wei, J., 2017. Real-time detection of false data injection attacks in smart grid: a deep learning-based intelligent mechanism. IEEE Trans. Smart Grid 8 (5), 2505–2516. https://doi.org/10.1109/TSG.2017.2703842.

Hemmati, R., Azizi, N., Shafie-khah, M., Catalão, J.P.S., 2018. Decentralized frequency-voltage control and stability enhancement of standalone wind turbine-load-battery. Int. J. Electr. Power Energy Syst. 102, 1–10. https://doi.org/10.1016/j.ijepes.2018.04.021.

Hu, J., Shan, Y., Guerrero, J.M., Ioinovici, A., Chan, K.W., Rodriguez, J., 2021. Model predictive control of microgrids – an overview. Renew. Sustain. Energy Rev. 136. https://doi.org/10.1016/j.rser.2020.110422.

Kanwal, S., Rauf, M.Q., Khan, B., Mokryani, G., 2024. Artificial neural network assisted robust droop control of autonomous microgrid. IET Renew. Power Gener. 18 (7), 1346–1369. https://doi.org/10.1049/rpg2.12739.

Khadanga, R.K., Padhy, S., Panda, S., Kumar, A., 2018. Design and analysis of multi-stage PID controller for frequency control in an islanded micro-grid using a novel hybrid whale optimization-pattern search algorithm. Int. J. Numer. Model. Electron. Network. Dev. Field. 31 (5). https://doi.org/10.1002/jnm.2349.

Khan, M.R., Haider, Z.M., Malik, F.H., Almasoudi, F.M., Alatawi, K.S.S., Bhutta, M.S., 2024. A comprehensive review of microgrid energy management strategies considering electric vehicles, energy storage systems, and AI techniques. Processes 12 (2). https://doi.org/10.3390/pr12020270.

Khare, V., Chaturvedi, P., 2023. Design, control, reliability, economic and energy management of microgrid: a review. E-Prime - Adv. Electr. Eng. Electr. Energy 5. https://doi.org/10.1016/j.prime.2023.100239.

Kikusato, H., Ustun, T.S., Suzuki, M., Sugahara, S., Hashimoto, J., Otani, K., Ikeda, N., Komuro, I., Yokoi, H., Takahashi, K., 2022. Flywheel energy storage system based microgrid controller design and PHIL testing. Energy Rep. 8, 470–475. https://doi.org/10.1016/j.egyr.2022.05.221.

Krishna Rao, C., Sahoo, S.K., Yanine, F.F., 2024. IoT enabled Intelligent Energy Management System employing advanced forecasting algorithms and load optimization strategies to enhance renewable energy generation. Unconvent. Resour. 4. https://doi.org/10.1016/j.uncres.2024.100101.

Kumar, V., Naresh, R., Sharma, V., 2022. Stochastic profit-based unit commitment problem considering renewable energy sources with battery storage systems and plug-in hybrid electric vehicles. Int. J. Energy Res. 46 (12), 16445–16460. https://doi.org/10.1002/er.8308.

Kurt, M.N., Ogundijo, O., Li, C., Wang, X., 2018. Online cyber-attack detection in smart grid: a reinforcement learning approach. IEEE Trans. Smart Grid 10 (5), 5174–5185. https://doi.org/10.1109/TSG.2018.2878570.

Lakshmi Satya Nagasri, D., Marimuthu, R., 2023. Review on advanced control techniques for microgrids. Energy Rep. 10, 3054–3072. https://doi.org/10.1016/j.egyr.2023.09.162.

Lam, Q.L., Bratcu, A.I., Riu, D., Boudinet, C., Labonne, A., Thomas, M., 2020. Primary frequency H∞ control in stand-alone microgrids with storage units: a robustness analysis confirmed by real-time experiments. Int. J. Electr. Power Energy Syst. 115. https://doi.org/10.1016/j.ijepes.2019.105507.

Li, X., Song, Y.J., Han, S.B., 2008. Frequency control in micro-grid power system combined with electrolyzer system and fuzzy PI controller. J. Power Sources 180 (1), 468–475. https://doi.org/10.1016/j.jpowsour.2008.01.092.

Lu, J., Hu, J., Yu, J., Cao, J., 2023. A dynamic demand response control strategy for isolated microgrid with primary frequency regulation. Elec. Power Syst. Res. 224. https://doi.org/10.1016/j.epsr.2023.109691.

Mahfuz, M.H., Ortmeyer, T., Hall, E., 2022. IEEE computer society United States development of a microgrid controller interface using human-centered design approach. In: IEEE Power and Energy Society General Meeting. https://doi.org/10.1109/PESGM48719.2022.9917191.

Manshahia, M.S., Kharchenko, V., Munapo, E., Joshua Thomas, J., Vasant, P., 2022. State-of-the-Art Optimization and Metaheuristic Algorithms Handbook of Intelligent Computing and Optimization for Sustainable Development. Wiley, pp. 509–536. https://doi.org/10.1002/9781119792642.ch25.

Martín, H., de la Hoz, J., 2024. Embedding regulatory frameworks in microgrids management. Power Systems. 3613, 347–369. https://doi.org/10.1007/978-981-97-6623-9_12.

Matijašević, T., Antić, T., Capuder, T., 2022. A systematic review of machine learning applications in the operation of smart distribution systems. Energy Rep. 8, 12379–12407. https://doi.org/10.1016/j.egyr.2022.09.068.

Mbungu, N.T., Madiba, T., Bansal, R.C., Bettayeb, M., Naidoo, R.M., Siti, M.W., Adefarati, T., 2022. Economic optimal load management control of microgrid system using energy storage system. J. Energy Storage 46. https://doi.org/10.1016/j.est.2021.103843.

Mekhamer, A.S., Hasanien, H.M., Alharbi, M., Tostado-Véliz, M., Jurado, F., Zaky, D.A., 2024. Coati optimization algorithm-based optimal frequency control of power systems including storage devices and electric vehicles. J. Energy Storage 93. https://doi.org/10.1016/j.est.2024.112367.

Mohapatra, S.S., Maharana, M.K., Pradhan, A., 2024. Review study on recent advancements in islanding detection and diagnosis in microgrids using signal processing and machine learning techniques. Elec. Power Compon. Syst. https://doi.org/10.1080/15325008.2024.2350061.

Murugesan, D., Jagatheesan, K., Shah, P., Sekhar, R., 2023. Fractional order controller for microgrid power system using cohort intelligence optimization. Res. Contr. Opt. 11.

Muttaqee, M., Furqan, M., Boudet, H., 2023. Community response to microgrid development: case studies from the U.S. Energy Pol. 181. https://doi.org/10.1016/j.enpol.2023.113690.

Najafirad, M.J., Mahdian Dehkordi, N., 2023. Distributed event-triggered control of DC microgrids with output saturation constraint. Int. J. Electr. Power Energy Syst. 148. https://doi.org/10.1016/j.ijepes.2022.108936.

Nosratabadi, S.M., Hooshmand, R.A., Gholipour, E., 2017. A comprehensive review on microgrid and virtual power plant concepts employed for distributed energy resources scheduling in power systems. Renew. Sustain. Energy Rev. 67, 341–363. https://doi.org/10.1016/j.rser.2016.09.025.

Ohanu, C.P., Rufai, S.A., Oluchi, U.C., 2024. A comprehensive review of recent developments in smart grid through renewable energy resources integration. Heliyon 10 (3). https://doi.org/10.1016/j.heliyon.2024.e25705.

Oladapo, B.I., Olawumi, M.A., Omigbodun, F.T., 2024. Machine learning for optimising renewable energy and grid efficiency. Atmosphere 15 (10). https://doi.org/10.3390/atmos15101250.

Onu, U.G., Zambroni de Souza, A.C., Bonatto, B.D., 2023. Drivers of microgrid projects in developed and developing economies. Util. Policy 80. https://doi.org/10.1016/j.jup.2022.101487.

Orimi, S.G., Marzbali, A.G., 2023. Load frequency control of multi-area multi-source system with nonlinear structures using modified Grasshopper optimization algorithm. Appl. Soft Comput. 137.

Pan, X., Shao, T., Zheng, X., Zhang, Y., Ma, X., Zhang, Q., 2023. Energy and sustainable development nexus: a review. Energy Strategy Rev. 47. https://doi.org/10.1016/j.esr.2023.101078.

Pérez, G., Conde, A., 2020. Frequency regulation of a weak microgrid through a distribution management system. Elec. Power Syst. Res. 184, 106320. https://doi.org/10.1016/j.epsr.2020.106320.

Prasad, P.S., Parimi, A.M., 2022. Robust μ-synthesis controller design and analysis for load frequency control in an AC microgrid system. IFAC-PapersOnLine 55 (1), 547–554. https://doi.org/10.1016/j.ifacol.2022.04.090.

Rashidi, R., Hatami, A., Abedini, M., 2021. Multi-microgrid energy management through tertiary-level control: structure and case study. Sustain. Energy Technol. Assessments 47. https://doi.org/10.1016/j.seta.2021.101395.

Rehman, A.U., 2025. Economic and strategic challenges in microgrid integration: insights from operational dynamics and renewable energy potential. Green Technol. Sust. 3 (1). https://doi.org/10.1016/j.grets.2024.100130.

Ronay, K., Bica, D., Munteanu, C., 2017. Micro-grid development using artificial neural network for renewable energy forecast and system control. Procedia Eng. 181, 818–823. https://doi.org/10.1016/j.proeng.2017.02.472.

Rouhani, S.H., Su, C.L., Mobayen, S., Razmjooy, N., Elsisi, M., 2024. Cyber resilience in renewable microgrids: a review of standards, challenges, and solutions. Energy 309. https://doi.org/10.1016/j.energy.2024.133081.

Sakthivelnathan, N., Arefi, A., Lund, C., Mehrizi-Sani, A., Muyeen, S.M., 2024. Energy storage as a service to achieve a required reliability level for renewable-rich standalone microgrids. J. Energy Storage 77. https://doi.org/10.1016/j.est.2023.109691.

Sarkar, M.K., Dev, A., Asthana, P. and Narzary, D. Chattering free robust adaptive integral higher order sliding mode control for load frequency problems in multi-area power systems. IET Contr. Theory Appl.. 12, 1216.

Sechilariu, M., Locment, F., Darene, N., 2018. Social acceptance of microgrids dedicated to electric vehicle charging stations. In: 7th International IEEE Conference on Renewable Energy Research and Applications, ICRERA 2018. Institute of Electrical and Electronics Engineers Inc., France, pp. 1374–1379. https://doi.org/10.1109/ICRERA.2018.8566787.

Shah, A.A., Han, X., Armghan, H., Aftab, A.A., 2021. A nonlinear integral backstepping controller to regulate the voltage and frequency of an islanded microgrid inverter. Electronics 10 (6), 660. https://doi.org/10.3390/electronics10060660.

Shahzad, S., Abbasi, M.A., Ali, H., Iqbal, M., Munir, R., Kilic, H., 2023. Possibilities, challenges, and future opportunities of microgrids: a review. Sustainability 15 (8), 6366. https://doi.org/10.3390/su15086366.

Shayeghi, H., Rahnama, A., Alhelou, H.H., 2021. Frequency control of fully-renewable interconnected microgrid using fuzzy cascade controller with demand response program considering. Energy Rep. 7, 6077–6094. https://doi.org/10.1016/j.egyr.2021.09.027.

Shi, J., Peng, C., Zhang, J., Xie, X., 2024. Model-free frequency regulation in islanded microgrids: an event-triggered adaptive dynamic programming approach. Int. J. Electr. Power Energy Syst. 155. https://doi.org/10.1016/j.ijepes.2023.109635.

Siti, M.W., Tungadio, D.H., Sun, Y., Mbungu, N.T., Tiako, R., 2019. Optimal frequency deviations control in microgrid interconnected systems. IET Renew. Power Gener. 13 (13), 2376–2382. https://doi.org/10.1049/iet-rpg.2018.5801.

Taghizadegan, N., Babaei, F., Safari, A., 2024. A linear active disturbance rejection control technique for frequency control of networked microgrids. Energy Syst. 15 (2), 807–826. https://doi.org/10.1007/s12667-023-00563-6.

Tahir, F., Dwyer, S., Kelly, S., 2024. Embracing complexity: microgrids and community engagement in Australia. Energy Res. Social Sci. 118. https://doi.org/10.1016/j.erss.2024.103811.

Tan, S., Xie, P., Guerrero, J.M., Vasquez, J.C., 2022. False data injection cyber-attacks detection for multiple DC microgrid clusters. Appl. Energy 310. https://doi.org/10.1016/j.apenergy.2021.118425.

Ton, D.T., Smith, M.A., 2012. The U.S. Department of energy's microgrid initiative. Electr. J. 25 (8), 84–94. https://doi.org/10.1016/j.tej.2012.09.013.

Ullah, Z., Wang, S., Wu, G., Xiao, M., Lai, J., Elkadeem, M.R., 2022. Advanced energy management strategy for microgrid using real-time monitoring interface. J. Energy Storage 52. https://doi.org/10.1016/j.est.2022.104814.

United Nations. Climate Change. (2024).

Valta, J., Mäkinen, S., Kotilainen, K., Järventausta, P., Mendes, G., 2018. In: 2018 IEEE PES Innovative Smart Grid Technologies Conference Europe (ISGT-Europe) Comparison of Regulatory Challenges Faced by Different Microgrid Ownership Models.

Wang, C., Sun, X., 2023. The promise of solid-state batteries for safe and reliable energy storage. Engineering 21, 32–35. https://doi.org/10.1016/j.eng.2022.10.008.

Warneryd, M., Håkansson, M., Karltorp, K., 2020. Unpacking the complexity of community microgrids: a review of institutions' roles for development of microgrids. Renew. Sustain. Energy Rev. 121. https://doi.org/10.1016/j.rser.2019.109690.

Wolsink, M., 2020. Distributed energy systems as common goods: socio-political acceptance of renewables in intelligent microgrids. Renew. Sustain. Energy Rev. 127. https://doi.org/10.1016/j.rser.2020.109841.

Wouters, C., 2015. Towards a regulatory framework for microgrids - the Singapore experience. Sustain. Cities Soc. 15, 22–32. https://doi.org/10.1016/j.scs.2014.10.007.

Xu, F.Y., Cun, X., Yan, M., Yuan, H., Wang, Y., Lai, L.L., 2018. Power market load forecasting on neural network with beneficial correlated regularization. IEEE Trans. Ind. Inf. 14 (11), 5050–5059. https://doi.org/10.1109/TII.2017.2789297.

Zaheeruddin, K.S., 2022. Intelligent fractional-order-based centralized frequency controller for microgrid. IETE J. Res. 68 (4), 2848–2862. https://doi.org/10.1080/03772063.2020.1730249.

Zhang, J., Gu, M., Chen, X., 2023. Supercapacitors for renewable energy applications: a review. Micro Nano Eng. 21. https://doi.org/10.1016/j.mne.2023.100229.

Index

'*Note:* Page numbers followed by "f" indicate figures and "t" indicate tables.'

D

Printed and bound by CPI Group (UK) Ltd, Croydon, CR0 4YY
14/09/2026
02216409-0005